Applied Machine Learning

David Forsyth

Applied Machine Learning

 **Springer**

David Forsyth
Computer Science Department
University of Illinois Urbana Champaign
Urbana, IL, USA

ISBN 978-3-030-18116-1   ISBN 978-3-030-18114-7   (eBook)
https://doi.org/10.1007/978-3-030-18114-7

This Springer imprint is published by the registered company Springer Nature Switzerland AG.
The registered company address is: Gewerbestrasse 11, 6330 Cham, Switzerland

Preface

Machine learning methods are now an important tool for scientists, researchers, engineers, and students in a wide range of areas. Many years ago, one could publish papers introducing (say) classifiers to one of the many fields that hadn't heard of them. Now, you need to know what a classifier is to get started in most fields. This book is written for people who want to adopt and use the main tools of machine learning but aren't necessarily going to want to be machine learning researchers—as of writing, this seems like almost everyone. There is no new fact about machine learning here, but the selection of topics is my own. I think it's different from what one sees in other books.

The book is revised and corrected from class notes written for a course I've taught on numerous occasions to very broad audiences. Most students were at the final year undergraduate/first year graduate student level in a US university. About half of each class consisted of students who weren't computer science students but still needed a background in learning methods. The course stressed applying a wide range of methods to real datasets, and the book does so, too.

The key principle in choosing what to write about was to cover the ideas in machine learning that I thought everyone who was going to use learning tools should have seen, whatever their chosen specialty or career. Although it's never a good thing to be ignorant of anything, an author must choose. Most people will find a broad shallow grasp of this field more useful than a deep and narrow grasp, so this book is broad, and coverage of many areas is shallow. I think that's fine, because my purpose is to ensure that all have seen enough to know that, say, firing up a classification package will make many problems go away. So I've covered enough to get you started and to get you to realize that it's worth knowing more.

The notes I wrote have been useful to more experienced students as well. In my experience, many learned some or all of this material without realizing how useful it was and then forgot it. If this happened to you, I hope the book is a stimulus to your memory. You really should have a grasp of all of this material. You might need to know more, but you certainly shouldn't know less.

This Book

I wrote this book to be taught, or read, by starting at the beginning and proceeding to the end. In a 15-week semester, I cover a lot and usually set 12 assignments, always programming assignments. Different instructors or readers may have different needs, and so I sketch some pointers to what can be omitted below.

What You Need to Know Before You Start

This book assumes you have a moderate background in probability and statistics before you start. I wrote a companion book, *Probability and Statistics for Computer Science*, which covers this background. There is a little overlap, because not everyone will read both books cover to cover (a mistake—you should!). But I've kept the overlap small (about 40 pp.) and confined to material that is better repeated anyway. Here's what you should know from that book (or some other, if you insist):

- Various descriptive statistics (mean, standard deviation, variance) and visualization methods for 1D datasets
- Scatter plots, correlation, and prediction for 2D datasets
- A little discrete probability
- A very little continuous probability (rough grasp of probability density functions and how to interpret them)
- Random variables and expectations
- A little about samples and populations
- Maximum likelihood
- Simple Bayesian inference
- A selection of facts about an assortment of useful probability distributions, or where to look them up

General Background: Your linear algebra should be reasonably fluent at a practical level. Fairly soon, we will see matrices, vectors, orthonormal matrices, eigenvalues, eigenvectors, and the singular value decomposition. All of these ideas will be used without too much comment.

Programming Background: You should be able to pick up a practical grasp of a programming environment without too much fuss. I use either R or MATLAB for this sort of thing, depending on how reliable I think a particular package is. You should, too. In some places, it's a good idea to use Python.

Survival Skills: Most simple questions about programming can be answered by searching. I usually use a web search to supply details of syntax, particular packages, etc. which I forget easily. You should, too. When someone asks me, say, "how do I write a loop in R?" in office hours, I very often answer by searching for R loop (or whatever) and then pointing out that I'm not actually needed. The questioner is often embarrassed at this point. You could save everyone trouble and embarrassment by cutting out the middleman in this transaction.

Datasets and Broken Links

I think using real datasets is an important feature of this book. But real life is messy, and datasets I refer to here may have been moved by the time you read this. Generally, a little poking around using a web search engine will find datasets or items that have been moved. I will also try to provide pointers to missing or moved datasets on a webpage which will also depend from my home page. It's easy to find me on the Internet by searching for my name, and ignoring the soap opera star (that's really not me).

Citations

Generally, I have followed the natural style of a textbook and tried not to cite papers in the text. I'm not up to providing a complete bibliography of modern machine learning and didn't want to provide an incomplete one. However, I have mentioned papers in some places and have done so when it seemed very important for users to be aware of the paper or when I was using a dataset whose compilers asked for a citation. I will try to correct errors or omissions in citation on a webpage, which will depend from my home page.

What Has Been Omitted

A list of everything omitted would be impractically too long. There are three topics I most regret omitting: kernel methods, reinforcement learning, and neural sequence models like LSTM. I omitted each because I thought that, while each is an important part of the toolbox of a practitioner, there are other topics with more claim on space. I may well write additional chapters on these topics after I recover from finishing this book. When they're in a reasonable shape, I'll put them on a webpage that depends from my home page.

There is very little learning theory here. While learning theory is very important (and I put in a brief chapter to sketch what's there and give readers a flavor), it doesn't directly change what practitioners *do*. And there's quite a lot of machinery with weak theoretical underpinnings that is extremely useful.

Urbana, IL, USA   David Forsyth

Acknowledgments

I acknowledge a wide range of intellectual debts, starting at kindergarten. Important figures in the very long list of my creditors include Gerald Alanthwaite, Mike Brady, Tom Fair, Margaret Fleck, Jitendra Malik, Joe Mundy, Jean Ponce, Mike Rodd, Charlie Rothwell, and Andrew Zisserman.

I have benefited from looking at a variety of sources, though this work really is my own. I particularly enjoyed the following books:

- *The Nature of Statistical Learning Theory*, V. Vapnik; Springer, 1999
- *Machine Learning: A Probabilistic Perspective*, K. P. Murphy; MIT Press, 2012
- *Pattern Recognition and Machine Learning*, C. M. Bishop; Springer, 2011
- *The Elements of Statistical Learning: Data Mining, Inference, and Prediction*, Second Edition, T. Hastie, R. Tibshirani, and J. Friedman; Springer, 2016
- *An Introduction to Statistical Learning: With Applications in R*, G. James, D. Witten, T. Hastie, and R. Tibshirani; Springer, 2013
- *Deep Learning*, I. Goodfellow, Y. Bengio, and A. Courville; MIT Press, 2016
- *Probabilistic Graphical Models: Principles and Techniques*, D. Koller and N. Friedman; MIT Press, 2009
- *Artificial Intelligence: A Modern Approach,* Third Edition, S. J. Russell and P. Norvig; Pearson, 2015
- *Data Analysis and Graphics Using R: An Example-Based Approach*, J. Maindonald and W. J. Braun; Cambridge University Press, 2e, 2003

A wonderful feature of modern scientific life is the willingness of people to share data on the Internet. I have roamed the Internet widely looking for datasets and have tried to credit the makers and sharers of data accurately and fully when I use the dataset. If, by some oversight, I have left you out, please tell me and I will try and fix this. I have been particularly enthusiastic about using data from the following repositories:

- *The UC Irvine Machine Learning Repository*, at http://archive.ics.uci.edu/ml/
- *Dr. John Rasp's Statistics Website*, at http://www2.stetson.edu/~jrasp/
- *OzDasl: The Australasian Data and Story Library*, at http://www.statsci.org/data/
- *The Center for Genome Dynamics, at the Jackson Laboratory*, at http://cgd.jax.org/ (which contains staggering amounts of information about mice) and the datasets listed and described in Sects. 17.2 and 18.1

I looked at Wikipedia regularly when preparing this manuscript, and I've pointed readers to neat stories there when they're relevant. I don't think one could learn the material in this book by reading Wikipedia, but it's been tremendously helpful in restoring ideas that I have mislaid, mangled, or simply forgotten.

When I did the first version of this course, Alyosha Efros let me look at notes for a learning course he taught, and these affected my choices of topic. Ben Recht gave me advice on several choices of topic. I co-taught this class with Trevor Walker for one semester, and his comments were extremely valuable. Eric Huber has made numerous suggestions in his role as course lead for an offering that included an online component. TA's for various versions of this and related classes have also helped improve the notes. Thanks to: Jyoti Aneja, Lavisha Aggarwal, Xiaoyang Bai, Christopher Benson, Shruti Bhargava, Anand Bhattad, Daniel Calzada, Binglin Chen, Taiyu Dong, Tanmay Gangwani, Sili Hui, Ayush Jain, Krishna Kothapalli, Maghav Kumar, Ji Li, Qixuan Li, Jiajun Lu, Shreya Rajpal, Jason Rock, Daeyun Shin, Mariya Vasileva, and Anirud Yadav. Typo's were spotted by (at least!): Johnny Chang, Yan Geng Niv Hadas, Vivian Hu, Eric Huber, Michael McCarrin, Thai Duy Cuong Nguyen, Jian Peng, and Victor Sui.

Several people commented very helpfully on the deep network part of the book, including Mani Golparvar Fard, Tanmay Gupta, Arun Mallya, Amin Sadeghi, Sepehr Sameni, and Alex Schwing.

I have benefited hugely from reviews organized by the publisher. Reviewers made many extremely helpful suggestions, which I have tried to adopt, by cutting chapters, moving chapters around, and general re-engineering of topics. Reviewers were anonymous to me at the time of review, but their names were later revealed so I can thank them by name. Thanks to:
Xiaoming Huo, Georgia Institute of Technology
Georgios Lazarou, University of South Alabama
Ilias Tagkopoulos, University of California, Davis
Matthew Turk, University of California, Santa Barbara
George Tzanetakis, University of Victoria
Qin Wang, University of Alabama
Guanghui Wang, University of Kansas
Jie Yang, University of Illinois at Chicago
Lisa Zhang, University of Toronto, Mississauga

A long list of people have tried to help me make this book better, and I'm very grateful for their efforts. But what remains is my fault, not theirs. Sorry.

Contents

V   Graphical Models   303

13 Hidden Markov Models   305

14 Learning Sequence Models Discriminatively   333

About the Author

David Forsyth grew up in Cape Town. He received his B.Sc. (Elec. Eng.) and M.Sc. (Elec. Eng.) from the University of the Witwatersrand, Johannesburg, in 1984 and in 1986, respectively and D.Phil. from Balliol College, Oxford, in 1989. He spent 3 years on the Faculty at the University of Iowa and 10 years on the Faculty at the University of California at Berkeley and then moved to the University of Illinois. He served as program co-chair for IEEE Computer Vision and Pattern Recognition in 2000, 2011, 2018, and 2021; general co-chair for CVPR 2006 and ICCV 2019, and program co-chair for the European Conference on Computer Vision 2008 and is a regular member of the program committee of all major international conferences on computer vision. He has also served six terms on the SIGGRAPH program committee. In 2006, he received an IEEE Technical Achievement Award and in 2009 and 2014 he was named an IEEE Fellow and an ACM Fellow, respectively; he served as editor in chief of IEEE TPAMI from 2014 to 2017; he is lead coauthor of *Computer Vision: A Modern Approach*, a textbook of computer vision that ran to two editions and four languages; and is sole author of *Probability and Statistics for Computer Science*, which provides the background for this book. Among a variety of odd hobbies, he is a compulsive diver, certified up to normoxic trimix level.

PART ONE

Classification

CHAPTER 1

Learning to Classify

A **classifier** is a procedure that accepts a set of features and produces a class label for them. Classifiers are immensely useful, and find wide application, because many problems are naturally classification problems. For example, if you wish to determine whether to place an advert on a webpage or not, you would use a classifier (i.e., look at the page, and say yes or no according to some rule). As another example, if you have a program that you found for free on the web, you would use a classifier to decide whether it was safe to run it (i.e., look at the program, and say yes or no according to some rule). As yet another example, credit card companies must decide whether a transaction is good or fraudulent.

All these examples are two-class classifiers, but in many cases it is natural to have more classes. You can think of sorting laundry as applying a multiclass classifier. You can think of doctors as complex multiclass classifiers: a doctor accepts a set of features (your complaints, answers to questions, and so on) and then produces a response which we can describe as a class. The grading procedure for any class is a multiclass classifier: it accepts a set of features—performance in tests, homeworks, and so on—and produces a class label (the letter grade).

A classifier is usually trained by obtaining a set of labelled training examples and then searching for a classifier that optimizes some cost function which is evaluated on the training data. What makes training classifiers interesting is that performance on training data doesn't really matter. What matters is performance on run-time data, which may be extremely hard to evaluate because one often does not know the correct answer for that data. For example, we wish to classify credit card transactions as safe or fraudulent. We could obtain a set of transactions with true labels, and train with those. But what we care about is new transactions, where it would be very difficult to know whether the classifier's answers are right. To be able to do anything at all, the set of labelled examples must be representative of future examples in some strong way. We will always assume that the labelled examples are an IID sample from the set of all possible examples, though we never use the assumption explicitly.

> **Remember This:** *A classifier is a procedure that accepts a set of features and produces a label. Classifiers are trained on labelled examples, but the goal is to get a classifier that performs well on data which is not seen at the time of training. Training a classifier requires labelled data that is representative of future data.*

© Springer Nature Switzerland AG 2019
D. Forsyth, *Applied Machine Learning*,
https://doi.org/10.1007/978-3-030-18114-7_1

1.1 Classification: The Big Ideas

We will write the training dataset $(\mathbf{x}_i, y_i)$. For the i'th example, $\mathbf{x}_i$ represents the values taken by a collection of features. In the simplest case, $\mathbf{x}_i$ would be a vector of real numbers. In some cases, $\mathbf{x}_i$ could contain categorical data or even unknown values. Although $\mathbf{x}_i$ isn't guaranteed to be a vector, it's usually referred to as a **feature vector**. The y_i are labels giving the type of the object that generated the example. We must use these labelled examples to come up with a classifier.

1.1.1 The Error Rate and Other Summaries of Performance

We can summarize the performance of any particular classifier using the **error** or **total error rate** (the percentage of classification attempts that gave the wrong answer) and the **accuracy** (the percentage of classification attempts that gave the right answer). For most practical cases, even the best choice of classifier will make mistakes. For example, an alien tries to classify humans into male and female, using only height as a feature. Whatever the alien's classifier does with that feature, it will make mistakes. This is because the classifier must choose, for each value of height, whether to label the humans with that height male or female. But for the vast majority of heights, there are some males and some females with that height, and so the alien's classifier must make some mistakes.

As the example suggests, a particular feature vector $\mathbf{x}$ may appear with different labels (so the alien will see six foot males and six foot females, quite possibly in the training dataset and certainly in future data). Labels appear with some probability conditioned on the observations, $P(y|\mathbf{x})$. If there are parts of the feature space where $P(\mathbf{x})$ is relatively large (so we expect to see observations of that form) *and* where $P(y|\mathbf{x})$ has relatively large values for more than one label, even the best possible classifier will have a high error rate. If we knew $P(y|\mathbf{x})$ (which is seldom the case), we could identify the classifier with the smallest error rate and compute its error rate. The minimum expected error rate obtained with the best possible classifier applied to a particular problem is known as the **Bayes risk** for that problem. In most cases, it is hard to know what the Bayes risk is, because to compute it requires knowing $P(y|\mathbf{x})$, which isn't usually known.

The error rate of a classifier is not that meaningful on its own, because we don't usually know the Bayes risk for a problem. It is more helpful to compare a particular classifier with some natural alternatives, sometimes called **baselines**. The choice of baseline for a particular problem is almost always a matter of application logic. The simplest general baseline is a know-nothing strategy. Imagine classifying the data without using the feature vector at all—how well does this strategy do? If each of the C classes occurs with the same frequency, then it's enough to label the data by choosing a label uniformly and at random, and the error rate for this strategy is $1 - 1/C$. If one class is more common than the others, the lowest error rate is obtained by labelling everything with that class. This comparison is often known as **comparing to chance**.

It is very common to deal with data where there are only two labels. You should keep in mind this means the highest possible error rate is 50%—if you have a classifier with a higher error rate, you can improve it by switching the outputs. If one class is much more common than the other, training becomes more complicated because the best strategy—labelling everything with the common class—becomes hard to beat.

> **Remember This:** *Classifier performance is summarized by either the total error rate or the accuracy. You will very seldom know what the best possible performance for a classifier on a problem is. You should always compare performance to baselines. Chance is one baseline that can be surprisingly strong.*

1.1.2 More Detailed Evaluation

The error rate is a fairly crude summary of the classifier's behavior. For a two-class classifier and a 0-1 loss function, one can report the **false positive rate** (the percentage of negative test data that was classified positive) and the **false negative rate** (the percentage of positive test data that was classified negative). Note that it is important to provide both, because a classifier with a low false positive rate tends to have a high false negative rate, and vice versa. As a result, you should be suspicious of reports that give one number but not the other. Alternative numbers that are reported sometimes include the **sensitivity** (the percentage of true positives that are classified positive) and the **specificity** (the percentage of true negatives that are classified negative).

Predict

		0	1	2	3	4	Class error
True	0	151	7	2	3	1	7.9%
	1	32	5	9	9	0	91%
	2	10	9	7	9	1	81%
	3	6	13	9	5	2	86%
	4	2	3	2	6	0	100%

TABLE 1.1: The class-confusion matrix for a multiclass classifier. This is a table of cells, where the i, j'th cell contains the count of cases where the true label was i and the predicted label was j (some people show the fraction of cases rather than the count). Further details about the dataset and this example appear in Worked Example 2.1

The false positive and false negative rates of a two-class classifier can be generalized to evaluate a multiclass classifier, yielding the **class-confusion matrix**. This is a table of cells, where the i, j'th cell contains the count of cases where the true label was i and the predicted label was j (some people show the fraction of cases rather than the count). Table 1.1 gives an example. This is a class-confusion matrix from a classifier built on a dataset where one tries to predict the degree of heart disease from a collection of physiological and physical measurements. There are five classes $(0, \ldots, 4)$. The i, j'th cell of the table shows the number of data points of true class i that were classified to have class j. As I find it hard to recall whether rows or columns represent true or predicted classes, I have marked this on the table. For each row, there is a **class error rate**, which is the percentage of data points of that class that were misclassified. The first thing to look at in a table like this is the diagonal; if the largest values appear there, then the classifier is working well. This clearly isn't what is happening for Table 1.1. Instead, you can see that the method is very good at telling whether a data point is in class 0 or not (the class error rate is rather small), but cannot distinguish between the other classes. This is a strong hint that the data can't be used to draw the distinctions that we want. It might be a lot better to work with a different set of classes.

Remember This: *When more detailed evaluation of a classifier is required, look at the false positive rate and the false negative rate. Always look at both, because doing well at one number tends to result in doing poorly on the other. The class-confusion matrix summarizes errors for multiclass classification.*

1.1.3 Overfitting and Cross-Validation

Choosing and evaluating a classifier takes some care. The goal is to get a classifier that works well on future data *for which we might never know the true label*, using a training set of labelled examples. This isn't necessarily easy. For example, think about the (silly) classifier that takes any data point and, if it is the same as a point in the training set, emits the class of that point; otherwise, it chooses randomly between the classes.

The **training error** of a classifier is the error rate on examples used to train the classifier. In contrast, the **test error** is error on examples not used to train the classifier. Classifiers that have small training error might not have small test error, because the classification procedure is chosen to do well on the training data. This effect is sometimes called **overfitting**. Other names include **selection bias**, because the training data has been selected and so isn't exactly like the test data, and **generalizing badly**, because the classifier must generalize from the training data to the test data. The effect occurs because the classifier has been chosen to perform well *on the training dataset*. An efficient training procedure is quite likely to find special properties of the training dataset that aren't representative of the test dataset, because the training dataset is not the same as the test dataset. The training dataset is typically a sample of all the data one might like to have classified, and so is quite likely a lot smaller than the test dataset. Because it is a sample, it may have quirks that don't appear in the test dataset. One consequence of overfitting is that classifiers should always be evaluated on data that was not used in training.

Now assume that we want to estimate the error rate of the classifier on test data. We cannot estimate the error rate of the classifier using data that was used to train the classifier, because the classifier has been trained to do well on that data, which will mean our error rate estimate will be too low. An alternative is to separate out some training data to form a **validation set** (confusingly, this is sometimes called a test set), then train the classifier on the rest of the data, and evaluate on the validation set. The error estimate on the validation set is the value of a random variable, because the validation set is a sample of all possible data you might classify. But this error estimate is **unbiased**, meaning that the expected value of the error estimate is the true value of the error (details in Sect. 3.1).

However, separating out some training data presents the difficulty that the classifier will not be the best possible, because we left out some training data when we trained it. This issue can become a significant nuisance when we are trying to tell which of a set of classifiers to use—did the classifier perform poorly on validation data because it is not suited to the problem representation or because it was trained on too little data?

We can resolve this problem with **cross-validation**, which involves repeatedly: splitting data into training and validation sets uniformly and at random, training a classifier on the training set, evaluating it on the validation set, and then averaging the error over all splits. Each different split is usually called a **fold**. This procedure yields an estimate of the likely future performance of a classifier, at the expense of substantial computation. A common form of this algorithm uses a single data item to form a validation set. This is known as **leave-one-out cross-validation**.

> **Remember This:** *Classifiers usually perform better on training data than on test data, because the classifier was chosen to do well on the training data. This effect is known as overfitting. To get an accurate estimate of future performance, classifiers should always be evaluated on data that was not used in training.*

1.2 Classifying with Nearest Neighbors

Assume we have a labelled dataset consisting of N pairs $(\mathbf{x}_i, y_i)$. Here $\mathbf{x}_i$ is the i'th feature vector, and y_i is the i'th class label. We wish to predict the label y for any new example $\mathbf{x}$; this is often known as a query example or query. Here is a really effective strategy: Find the labelled example $\mathbf{x}_c$ that is closest to $\mathbf{x}$, and report the class of that example.

How well can we expect this strategy to work? A precise analysis would take us way out of our way, but simple reasoning is informative. Assume there are two classes, 1 and -1 (the reasoning will work for more, but the description is slightly more involved). We expect that, if $\mathbf{u}$ and $\mathbf{v}$ are sufficiently close, then $p(y|\mathbf{u})$ is similar to $p(y|\mathbf{v})$. This means that if a labelled example $\mathbf{x}_i$ is close to $\mathbf{x}$, then $p(y|\mathbf{x})$ is similar to $p(y|\mathbf{x}_i)$. Furthermore, we expect that queries are "like" the labelled dataset, in the sense that points that are common (resp. rare) in the labelled data will appear often (resp. seldom) in the queries.

Now imagine the query comes from a location where $p(y = 1|\mathbf{x})$ is large. The closest labelled example $\mathbf{x}_c$ should be nearby (because queries are "like" the labelled data) and should be labelled with 1 (because nearby examples have similar label probabilities). So the method should produce the right answer with high probability.

Alternatively, imagine the query comes from a location where $p(y = 1|\mathbf{x})$ is about the same as $p(y = -1|\mathbf{x})$. The closest labelled example $\mathbf{x}_c$ should be nearby (because queries are "like" the labelled data). But think about a set of examples that are about as close. The labels in this set should vary significantly (because $p(y = 1|\mathbf{x})$ is about the same as $p(y = -1|\mathbf{x})$. This means that, if the query is labelled 1 (resp. -1), a small change in the query will cause it to be labelled -1 (resp. 1). In these regions the classifier will tend to make mistakes more often, as it should. Using a great deal more of this kind of reasoning, nearest neighbors can be shown to produce an error that is no worse than twice the best error rate, if the method has enough examples. There is no prospect of seeing enough examples in practice for this result to apply.

One important generalization is to find the k nearest neighbors, then choose a label from those. A (k, l) nearest neighbor classifier finds the k example points closest to the point being considered, and classifies this point with the class that has the highest number of votes, as long as this class has more than l votes (otherwise, the point is classified as unknown). In practice, one seldom uses more than three nearest neighbors.

Remember This: *Classifying with nearest neighbors can be straightforward and accurate. With enough training data, there are theoretical guarantees that the error rate is no worse than twice the best possible error. These usually don't apply to practice.*

1.2.1 Practical Considerations for Nearest Neighbors

One practical difficulty in using nearest neighbor classifiers is you need a lot of labelled examples for the method to work. For some problems, this means you can't use the method. A second practical difficulty is you need to use a sensible choice of distance. For features that are obviously of the same type, such as lengths, the usual metric may be good enough. But what if one feature is a length, one is a color, and one is an angle? It is almost always a good idea to scale each feature independently so that the variance of each feature is the same, or at least consistent; this prevents features with very large scales dominating those with very small scales. Another possibility is to transform the features so that the covariance matrix is the identity (this is sometimes known as **whitening**; the method follows from the ideas of Chap. 4). This can be hard to do if the dimension is so large that the covariance matrix is hard to estimate.

A third practical difficulty is you need to be able to find the nearest neighbors for your query point. This is surprisingly difficult to do faster than simply checking the distance to each training example separately. If your intuition tells you to use a tree and the difficulty will go away, your intuition isn't right. It turns out that nearest neighbor in high dimensions is one of those problems that is a lot harder than it seems, because high dimensional spaces are quite hard to reason about informally. There's a long history of methods that appear to be efficient but, once carefully investigated, turn out to be bad.

Fortunately, it is usually enough to use an **approximate nearest neighbor**. This is an example that is, with high probability, almost as close to the query point as the nearest neighbor is. Obtaining an approximate nearest neighbor is very much easier than obtaining a nearest neighbor. We can't go into the details here, but there are several distinct methods for finding approximate nearest neighbors. Each involves a series of tuning constants and so on, and, on different datasets, different methods and different choices of tuning constant produce the best results. If you want to use a nearest neighbor classifier on a lot of run-time data, it is usually worth a careful search over methods and tuning constants to find an algorithm that yields a very fast response to a query. It is known how to do this search, and there is excellent software available (FLANN, http://www.cs.ubc.ca/~mariusm/index.php/FLANN/FLANN, by Marius Muja and David G. Lowe).

It is straightforward to use cross-validation to estimate the error rate of a nearest neighbor classifier. Split the labelled training data into two pieces, a (typically large) training set and a (typically small) validation set. Now take each element of the validation set and label it with the label of the closest element of the training set. Compute the fraction of these attempts that produce an error (the true label and predicted labels differ). Now repeat this for a different split, and average the errors over splits. With care, the code you'll write is shorter than this description.

Worked Example 1.1   *Classifying Using Nearest Neighbors*

Build a nearest neighbor classifier to classify the MNIST digit data. This dataset is very widely used to check simple methods. It was originally constructed by Yann Lecun, Corinna Cortes, and Christopher J.C. Burges. It has been extensively studied. You can find this dataset in several places. The original dataset is at http://yann.lecun.com/exdb/mnist/. The version I used was used for a Kaggle competition (so I didn't have to decompress Lecun's original format). I found it at http://www.kaggle.com/c/digit-recognizer.

Solution: I used R for this problem. As you'd expect, R has nearest neighbor code that seems quite good (I haven't had any real problems with it, at least). There isn't really all that much to say about the code. I used the R FNN package. I trained on 1000 of the 42,000 examples in the Kaggle version, and I tested on the next 200 examples. For this (rather small) case, I found the following class-confusion matrix:

Predict

True	0	1	2	3	4	5	6	7	8	9
0	12	0	0	0	0	0	0	0	0	0
1	0	20	4	1	0	1	0	2	2	1
2	0	0	20	1	0	0	0	0	0	0
3	0	0	0	12	0	0	0	0	4	0
4	0	0	0	0	18	0	0	0	1	1
5	0	0	0	0	0	19	0	0	1	0
6	1	0	0	0	0	0	18	0	0	0
7	0	0	1	0	0	0	0	19	0	2
8	0	0	1	0	0	0	0	0	16	0
9	0	0	0	2	3	1	0	1	1	14

There are no class error rates here, because I couldn't recall the magic line of R to get them. However, you can see the classifier works rather well for this case. MNIST is comprehensively explored in the exercises.

Remember This: *Nearest neighbor has good properties. With enough training data and a low enough dimension, the error rate is guaranteed to be no more than twice the best error rate. The method is wonderfully flexible about the labels the classifier predicts. Nothing changes when you go from a two-class classifier to a multiclass classifier.*
There are important difficulties. You need a large training dataset. If you don't have a reliable measure of how far apart two things are, you shouldn't be doing nearest neighbors. And you need to be able to query a large dataset of examples to find the nearest neighbor of a point.

1.3 Naive Bayes

One straightforward source of a classifier is a probability model. For the moment, assume we know $p(y|\mathbf{x})$ for our data. Assume also that all errors in classification are equally important. Then the following rule produces smallest possible expected classification error rate:

> For a test example $\mathbf{x}$, report the class y that has the highest value of $(p(y|\mathbf{x}))$. If the largest value is achieved by more than one class, choose randomly from that set of classes.

Usually, we do not have $p(y|\mathbf{x})$. If we have $p(\mathbf{x}|y)$ (often called either a **likelihood** or **class conditional probability**) and $p(y)$ (often called a **prior**), then we can use Bayes' rule to form

$$p(y|\mathbf{x}) = \frac{p(\mathbf{x}|y)p(y)}{p(\mathbf{x})}$$

(the **posterior**). This isn't much help in this form, but write $x^{(j)}$ for the j'th component of $\mathbf{x}$. Now *assume* that features are conditionally independent conditioned on the class of the data item. Our assumption is

$$p(\mathbf{x}|y) = \prod_j p(x^{(j)}|y).$$

It is very seldom the case that this assumption is true, but it turns out to be fruitful to pretend that it is. This assumption means that

$$
\begin{aligned}
p(y|\mathbf{x}) &= \frac{p(\mathbf{x}|y)p(y)}{p(\mathbf{x})} \\
&= \frac{\left(\prod_j p(x^{(j)}|y)\right)p(y)}{p(\mathbf{x})} \\
&\propto \left(\prod_j p(x^{(j)}|y)\right)p(y).
\end{aligned}
$$

Now to make a decision, we need to choose the class that has the largest value of $p(y|\mathbf{x})$. In turn, this means we need only to know the posterior values up to scale at $\mathbf{x}$, so we don't need to estimate $p(\mathbf{x})$. In the case of where all errors have the same cost, this yields the rule

choose y such that $\left[\left(\prod_j p(x^{(j)}|y)\right)p(y)\right]$ is largest.

This rule suffers from a practical problem. You can't actually multiply a large number of probabilities and expect to get an answer that a floating point system thinks is different from zero. Instead, you should add the log probabilities. Notice that the logarithm function has one nice property: it is monotonic, meaning that $a > b$ is equivalent to $\log a > \log b$. This means the following, more practical, rule is equivalent:

choose y such that $\left[\left(\sum_j \log p(x^{(j)}|y)\right) + \log p(y)\right]$ is largest.

To use this rule, we need models for $p(y)$ and for $p(x^{(j)}|y)$ for each j. The usual way to find a model of $p(y)$ is to count the number of training examples in each class, then divide by the number of classes.

It turns out that simple parametric models work really well for $p(x^{(j)}|y)$. For example, one could use a normal distribution for each $x^{(j)}$ in turn, for each possible value of y, using the training data. The parameters of this normal distribution are chosen using maximum likelihood. The logic of the measurements might suggest other distributions, too. If one of the $x^{(j)}$'s was a count, we might fit a Poisson distribution (again, using maximum likelihood). If it was a 0-1 variable, we might fit a Bernoulli distribution. If it was a discrete variable, then we might use a multinomial model. Even if the $x^{(j)}$ is continuous, we can use a multinomial model by quantizing to some fixed set of values; this can be quite effective.

A naive Bayes classifier that has poorly fitting models for each feature could classify data very well. This (reliably confusing) property occurs because classification doesn't require a good model of $p(\mathbf{x}|y)$, or even of $p(y|\mathbf{x})$. All that needs to happen is that, at any $\mathbf{x}$, the score for the right class is higher than the score for all other classes. Figure 1.1 shows an example where a normal model of the class-conditional histograms is poor, but the normal model will result in a good naive Bayes classifier. This works because a data item from (say) class one will reliably have a larger probability under the normal model for class one than it will for class two.

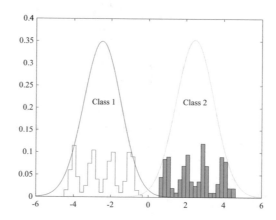

Figure 1.1: Naive Bayes can work well even if the class-conditional models are poor. The figure shows class-conditional histograms of a feature x for two different classes. The histograms have been normalized so that the counts sum to one, so you can think of them as probability distributions. It should be fairly obvious that a normal model (superimposed) doesn't describe these histograms well. However, the normal model will result in a good naive Bayes classifier

Worked Example 1.2   *Classifying Breast Tissue Samples*

The "breast tissue" dataset at https://archive.ics.uci.edu/ml/datasets/Breast+Tissue contains measurements of a variety of properties of six different classes of breast tissue. Build and evaluate a naive Bayes classifier to distinguish between the classes automatically from the measurements.

Solution: I used R for this example, because I could then use packages easily. The main difficulty here is finding appropriate packages, understanding their documentation, and checking they're right (unless you want to write the source yourself, which really isn't all that hard). I used the R package `caret` to do train–test splits, cross-validation, etc. on the naive Bayes classifier in the R package `klaR`. I separated out a test set randomly (approx 20% of the cases for each class, chosen at random), then trained with cross-validation on the remainder. I used a normal model for each feature. The class-confusion matrix on the test set was

		Predict					
		adi	car	con	fad	gla	mas
	adi	2	0	0	0	0	0
	car	0	3	0	0	0	1
True	con	2	0	2	0	0	0
	fad	0	0	0	0	1	0
	gla	0	0	0	0	2	1
	mas	0	1	0	3	0	1

which is fairly good. The accuracy is 52%. In the training data, the classes are nearly balanced and there are six classes, meaning that chance is about 17%. These numbers, and the class-confusion matrix, will vary with test–train split. I have not averaged over splits, which would give a somewhat more accurate estimate of accuracy.

1.3.1 Cross-Validation to Choose a Model

Naive Bayes presents us with a new problem. We can choose from several different types of model for $p(x^{(j)}|y)$ (e.g., normal models vs. Poisson models), and we need to know which one produces the best classifier. We also need to know how well that classifier will work. It is natural to use cross-validation to estimate how well each type of model works. You can't just look at every type of model for every variable, because that would yield too many models. Instead, choose M types of model that seem plausible (for example, by looking at histograms of feature components conditioned on class and using your judgement). Now compute a cross-validated error for each of M types of model, and choose the type of model with lowest cross-validated error. Computing the cross-validated error involves repeatedly splitting the training set into two pieces, fitting the model on one and computing the error on the other, then averaging the errors. Notice this means the model you fit to each fold will have slightly different parameter values, because each fold has slightly different training data.

However, once we have chosen the type of model, we have two problems. First, we do not know the correct values for the parameters of the best type of model. For each fold in the cross-validation, we estimated slightly different parameters because we trained on slightly different data, and we don't know which estimate is right. Second, we do not have a good estimate of how well the best model works. This is because we chose the type of model with the smallest error estimate, which is likely smaller than the true error estimate for that type of model.

This problem is easily dealt with if you have a reasonably sized dataset. Split the labelled dataset into two pieces. One (call it the training set) is used for training and for choosing a model type, the other (call it the test set) is used only for evaluating the final model. Now for each type of model, compute the cross-validated error on the training set.

Now use the cross-validated error to choose the type of model. Very often this just means you choose the type that produces the lowest cross-validated error, but there might be cases where two types produce about the same error and one is a lot faster to evaluate, etc. Take the entire training set, and use this to estimate the parameters for that type of model. This estimate should be (a little) better than any of the estimates produced in the cross-validation, because it uses (slightly) more data. Finally, evaluate the resulting model on the test set.

This procedure is rather harder to describe than to do (there's a pretty natural set of nested loops here). There are some strong advantages. First, the estimate of how well a particular model type works is unbiased, because we evaluated on data not used on training. Second, once you have chosen a type of model, the parameter estimate you make is the best you can because you used all the training set to obtain it. Finally, your estimate of how well that particular model works is unbiased, too, because you obtained it using data that wasn't used to train or to select a model.

Procedure: 1.1 *Cross-Validation to Choose a Model*

Divide your dataset $\mathcal{D}$ into two parts at random: one is the training set ($\mathcal{R}$); and one the test set ($\mathcal{E}$).
For each model in the collection that interests you:

- repeatedly
    - split $\mathcal{R}$ into two parts at random, $\mathcal{R}_t$ and $\mathcal{V}$;
    - fit the model using $\mathcal{R}_t$ to train;
    - compute an estimate of the error using that model with $\mathcal{V}$.
- Now report the average of these errors as the error of the model.

Now use these average errors (and possibly other criteria like speed, etc.) to choose a model. Compute an estimate of the error rate using your chosen model on $\mathcal{E}$.

1.3.2 Missing Data

Missing data occurs when some values in the training data are unknown. This can happen in a variety of ways. Someone didn't record the value; someone recorded it incorrectly, and you know the value is wrong but you don't know what the right one is; the dataset was damaged in storage or transmission; instruments failed; and so on. This is quite typical of data where the feature values are obtained by measuring effects in the real world. It's much less common where the feature values are computed from signals—for example, when one tries to classify digital images, or sound recordings.

Missing data can be a serious nuisance in classification problems, because many methods cannot handle incomplete feature vectors. For example, nearest neighbor has no real way of proceeding if some components of the feature vector are unknown. If there are relatively few incomplete feature vectors, one could just drop them from the dataset and proceed, but this should strike you as inefficient.

Naive Bayes is rather good at handling data where there are many incomplete feature vectors in quite a simple way. For example, assume for some i, we wish to fit $p(x_i|y)$ with a normal distribution. We need to estimate the mean and standard deviation of that normal distribution (which we do with maximum likelihood, as one should). If not every example has a known value of x_i, this really doesn't matter; we simply omit the unknown number from the estimate. Write $x_{i,j}$ for the value of x_i for the j'th example. To estimate the mean, we form

$$\frac{\sum_{j \in \text{cases with known values}} x_{i,j}}{\text{number of cases with known values}}$$

and so on.

Dealing with missing data during classification is easy, too. We need to look for the y that produces the largest value of $\sum_i \log p(x_i|y)$. We can't evaluate $p(x_i|y)$ if the value of that feature is missing—but it is missing for each class. We can just leave that term out of the sum, and proceed. This procedure is fine if data is missing as a result of "noise" (meaning that the missing terms are independent of class). If the missing terms depend on the class, there is much more we could do—for example, we might build a model of the class-conditional density of missing terms.

Notice that if some values of a discrete feature x_i don't appear for some class, you could end up with a model of $p(x_i|y)$ that had zeros for some values. This almost inevitably leads to serious trouble, because it means your model states you cannot ever observe that value for a data item of that class. This isn't a safe property: it is hardly ever the case that not observing something means you cannot observe it. A simple, but useful, fix is to add one to all small counts. More sophisticated methods are available, but well beyond our scope.

> **Remember This:** *Naive Bayes classifiers are straightforward to build, and very effective. Dealing with missing data is easy. Experience has shown they are particularly effective at high dimensional data. A straightforward variant of cross-validation helps select which particular model to use.*

1.4 You Should

1.4.1 Remember These Terms

1.4.2 Remember These Facts

1.4.3 Remember These Procedures

1.4.4 Be Able to

- build a nearest neighbors classifier using your preferred software package, and produce a cross-validated estimate of its error rate or its accuracy;
- build a naive Bayes classifier using your preferred software package, and produce a cross-validated estimate of its error rate or its accuracy.

Programming Exercises

1.1. The UC Irvine machine learning data repository hosts a famous collection of data on whether a patient has diabetes (the Pima Indians dataset), originally owned by the National Institute of Diabetes and Digestive and Kidney Diseases and donated by Vincent Sigillito. This can be found at http://archive.ics.uci.edu/ml/datasets/Pima+Indians+Diabetes. This data has a set of attributes of patients, and a categorical variable telling whether the patient is diabetic or not. This is an exercise oriented to users of R, because you can use some packages to help.

   (a) Build a simple naive Bayes classifier to classify this dataset. You should hold out 20% of the data for evaluation, and use the other 80% for training. You should use a normal distribution to model each of the class-conditional distributions. You should write this classifier by yourself.

   (b) Now use the `caret` and `klaR` packages to build a naive Bayes classifier for this data. The `caret` package does cross-validation (look at `train`) and can be used to hold out data. The `klaR` package can estimate class-conditional densities using a density estimation procedure that I will describe much later in the course. Use the cross-validation mechanisms in `caret` to estimate the accuracy of your classifier.

1.2. The UC Irvine machine learning data repository hosts a collection of data on student performance in Portugal, donated by Paulo Cortez, University of Minho, in Portugal. You can find this data at https://archive.ics.uci.edu/ml/datasets/Student+Performance. It is described in P. Cortez and A. Silva. "Using Data Mining to Predict Secondary School Student Performance," In A. Brito and J. Teixeira Eds., *Proceedings of 5th FUture BUsiness TEChnology Conference (FUBUTEC 2008)* pp. 5–12, Porto, Portugal, April, 2008, There are two datasets (for grades in mathematics and for grades in Portuguese). There are 30 attributes each for 649 students, and 3 values that can be predicted (G1, G2, and G3). Of these, ignore G1 and G2.

   (a) Use the mathematics dataset. Take the G3 attribute, and quantize this into two classes, $G3 > 12$ and $G3 \leq 12$. Build and evaluate a naive Bayes classifier that predicts G3 from all attributes except G1 and G2. You should build this classifier from scratch (i.e., DON'T use the packages described in the code snippets). For binary attributes, you should use a binomial model. For the attributes described as "numeric," which take a small set of values, you should use a multinomial model. For the attributes described as "nominal," which take a small set of values, you should again use a multinomial model. Ignore the "absence" attribute. Estimate accuracy by cross-validation. You should use at least 10 folds, excluding 15% of the data at random to serve as test data, and average the accuracy over those folds. Report the mean and standard deviation of the accuracy over the folds.

(b) Now revise your classifier of the previous part so that, for the attributes described as "numeric," which take a small set of values, you use a multinomial model. For the attributes described as "nominal," which take a small set of values, you should still use a multinomial model. Ignore the "absence" attribute. Estimate accuracy by cross-validation. You should use at least 10 folds, excluding 15% of the data at random to serve as test data, and average the accuracy over those folds. Report the mean and standard deviation of the accuracy over the folds.

(c) Which classifier do you believe is more accurate and why?

1.3. The UC Irvine machine learning data repository hosts a collection of data on heart disease. The data was collected and supplied by Andras Janosi, M.D., of the Hungarian Institute of Cardiology, Budapest; William Steinbrunn, M.D., of the University Hospital, Zurich, Switzerland; Matthias Pfisterer, M.D., of the University Hospital, Basel, Switzerland; and Robert Detrano, M.D., Ph.D., of the V.A. Medical Center, Long Beach and Cleveland Clinic Foundation. You can find this data at https://archive.ics.uci.edu/ml/datasets/Heart+Disease.

Use the processed Cleveland dataset, where there are a total of 303 instances with 14 attributes each. The irrelevant attributes described in the text have been removed in these. The 14'th attribute is the disease diagnosis. There are records with missing attributes, and you should drop these.

(a) Take the disease attribute, and quantize this into two classes, num = 0 and num > 0. Build and evaluate a naive Bayes classifier that predicts the class from all other attributes Estimate accuracy by cross-validation. You should use at least 10 folds, excluding 15% of the data at random to serve as test data, and average the accuracy over those folds. Report the mean and standard deviation of the accuracy over the folds.

(b) Now revise your classifier to predict each of the possible values of the disease attribute (0–4 as I recall). Estimate accuracy by cross-validation. You should use at least 10 folds, excluding 15% of the data at random to serve as test data, and average the accuracy over those folds. Report the mean and standard deviation of the accuracy over the folds.

1.4. The UC Irvine machine learning data repository hosts a collection of data on breast cancer diagnostics, donated by Olvi Mangasarian, Nick Street, and William H. Wolberg. You can find this data at http://archive.ics.uci.edu/ml/datasets/Breast+Cancer+Wisconsin+(Diagnostic). For each record, there is an id number, 10 continuous variables, and a class (benign or malignant). There are 569 examples. Separate this dataset randomly into 100 validation, 100 test, and 369 training examples.

Write a program to train a support vector machine on this data using stochastic gradient descent. You should not use a package to train the classifier (you don't really need one), but your own code. You should ignore the id number, and use the continuous variables as a feature vector. You should scale these variables so each has unit variance. You should search for an appropriate value of the regularization constant, trying at least the values $\lambda = [1e-3, 1e-2, 1e-1, 1]$. Use the validation set for this search.

You should use at least 50 epochs of at least 100 steps each. In each epoch, you should separate out 50 training examples at random for evaluation. You should compute the accuracy of the current classifier on the set held out for the epoch every 10 steps. You should produce:

(a) A plot of the accuracy every 10 steps, for each value of the regularization constant.

(b) Your estimate of the best value of the regularization constant, together with a brief description of why you believe that is a good value.

(c) Your estimate of the accuracy of the best classifier on held-out data

CHAPTER 2

SVMs and Random Forests

2.1 The Support Vector Machine

Assume we have a labelled dataset consisting of N pairs $(\mathbf{x}_i, y_i)$. Here $\mathbf{x}_i$ is the i'th feature vector, and y_i is the i'th class label. We will assume that there are two classes, and that y_i is either 1 or -1. We wish to predict the sign of y for any point $\mathbf{x}$. We will use a linear classifier, so that for a new data item $\mathbf{x}$, we will predict

$$\mathsf{sign}\left(\mathbf{a}^T \mathbf{x} + b\right)$$

and the particular classifier we use is given by our choice of $\mathbf{a}$ and b.

You should think of $\mathbf{a}$ and b as representing a hyperplane, given by the points where $\mathbf{a}^T\mathbf{x} + b = 0$. Notice that the magnitude of $\mathbf{a}^T\mathbf{x} + b$ grows as the point $\mathbf{x}$ moves further away from the hyperplane. This hyperplane separates the positive data from the negative data, and is an example of a **decision boundary**. When a point crosses the decision boundary, the label predicted for that point changes. All classifiers have decision boundaries. Searching for the decision boundary that yields the best behavior is a fruitful strategy for building classifiers.

Example: 2.1 *A Linear Model with a Single Feature*

Assume we use a linear model with one feature. For an example with feature value x, this predicts $\mathsf{sign}\left(ax + b\right)$. Equivalently, the model tests x against the threshold $-b/a$.

Example: 2.2 *A Linear Model with Two Features*

Assume we use a linear model with two features. For an example with feature vector $\mathbf{x}$, the model predicts $\mathsf{sign}\left(\mathbf{a}^T\mathbf{x} + b\right)$. The sign changes along the line $\mathbf{a}^T\mathbf{x} + b = 0$. You should check that this is, indeed, a line. On one side of this line, the model makes positive predictions; on the other, negative. Which side is which can be swapped by multiplying $\mathbf{a}$ and b by -1.

This family of classifiers may look bad to you, and it is easy to come up with examples that it misclassifies badly. In fact, the family is extremely strong. First, it is easy to estimate the best choice of rule for very large datasets. Second, linear

© Springer Nature Switzerland AG 2019
D. Forsyth, *Applied Machine Learning*,
https://doi.org/10.1007/978-3-030-18114-7_2

classifiers have a long history of working very well in practice on real data. Third, linear classifiers are fast to evaluate.

In practice, examples that are classified badly by the linear rule usually are classified badly because there are too few features. Remember the case of the alien who classified humans into male and female by looking at their heights; if that alien had looked at their chromosomes as well as height, the error rate would have been smaller. In practical examples, experience shows that the error rate of a poorly performing linear classifier can usually be improved by adding features to the vector $\mathbf{x}$.

We will choose $\mathbf{a}$ and b by choosing values that minimize a cost function. The cost function must achieve two goals. First, the cost function needs a term that ensures each training example should be on the right side of the decision boundary (or, at least, not be too far on the wrong side). Second, the cost function needs a term that should penalize errors on query examples. The appropriate cost function has the form:

$$\text{Training error cost} + \lambda \text{ penalty term}$$

where λ is an unknown weight that balances these two goals. We will eventually set the value of λ by a search process.

2.1.1 The Hinge Loss

Write

$$\gamma_i = \mathbf{a}^T \mathbf{x}_i + b$$

for the value that the linear function takes on example i. Write $C(\gamma_i, y_i)$ for a function that compares γ_i with y_i. The training error cost will be of the form

$$(1/N) \sum_{i=1}^{N} C(\gamma_i, y_i).$$

A good choice of C should have some important properties.

- If γ_i and y_i have different signs, then C should be large, because the classifier will make the wrong prediction for this training example. Furthermore, if γ_i and y_i have different signs *and* γ_i has large magnitude, then the classifier will very likely make the wrong prediction for test examples that are close to $\mathbf{x}_i$. This is because the magnitude of $(\mathbf{a}^T\mathbf{x} + b)$ grows as $\mathbf{x}$ gets further from the decision boundary. So C should get larger as the magnitude of γ_i gets larger in this case.
- If γ_i and y_i have the same signs, but γ_i has small magnitude, then the classifier will classify $\mathbf{x}_i$ correctly, but might not classify points that are nearby correctly. This is because a small magnitude of γ_i means that $\mathbf{x}_i$ is close to the decision boundary, so there will be points nearby that are on the other side of the decision boundary. We want to discourage this, so C should not be zero in this case.
- Finally, if γ_i and y_i have the same signs and γ_i has large magnitude, then C can be zero because $\mathbf{x}_i$ is on the right side of the decision boundary and so are all the points near to $\mathbf{x}_i$.

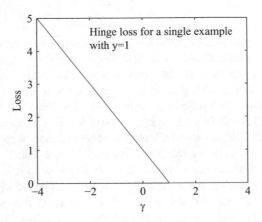

Figure 2.1: The hinge loss, plotted for the case $y_i = 1$. The horizontal variable is the $\gamma_i = \mathbf{a}^T \mathbf{x}_i + b$ of the text. Notice that giving a strong negative response to this positive example causes a loss that grows linearly as the magnitude of the response grows. Notice also that giving an insufficiently positive response also causes a loss. Giving a strongly positive response is free

The **hinge loss** (Fig. 2.1), which takes the form

$$C(y_i, \gamma_i) = \max(0, 1 - y_i \gamma_i),$$

has these properties.

- If γ_i and y_i have different signs, then C will be large. Furthermore, the cost grows linearly as $\mathbf{x}_i$ moves further away from the boundary on the wrong side.
- If γ_i and y_i have the same sign, but $y_i \gamma_i < 1$ (which means that $\mathbf{x}_i$ is close to the decision boundary), there is some cost, which gets larger as $\mathbf{x}_i$ gets closer to the boundary.
- If $y_i \gamma_i > 1$ (so the classifier predicts the sign correctly *and* $\mathbf{x}_i$ is far from the boundary) there is no cost.

A classifier trained to minimize this loss is encouraged to (a) make strong positive (or negative) predictions for positive (or negative) examples and (b) for examples it gets wrong, make predictions with the smallest magnitude that it can. A linear classifier trained with the hinge loss is known as a **support vector machine** or **SVM**.

Remember This: *An SVM is a linear classifier trained with the hinge loss. The hinge loss is a cost function that evaluates errors made by two-class classifiers. If an example is classified with the right sign and a large magnitude, the loss is zero; if the magnitude is small, the loss is larger; and if the example has the wrong sign, the loss is larger still. When the loss is zero, it grows linearly in the magnitude of the prediction.*

2.1.2 Regularization

The penalty term is needed, because the hinge loss has one odd property. Assume that the pair $\mathbf{a}$, b correctly classifies all training examples, so that $y_i(\mathbf{a}^T x_i + b) > 0$. Then we can always ensure that the hinge loss for the dataset is zero, by scaling $\mathbf{a}$ and b, because you can choose a scale so that $y_j(\mathbf{a}^T x_j + b) > 1$ for *every* example index j. This scale hasn't changed the result of the classification rule on the training data. Now if $\mathbf{a}$ and b result in a hinge loss of zero, then so do $2\mathbf{a}$ and $2b$. This should worry you, because it means we can't choose the classifier parameters uniquely.

Now think about future examples. We don't know what their feature values will be, and we don't know their labels. But we do know that the hinge loss for an example with feature vector $\mathbf{x}$ and unknown label y will be $\max(0, 1 - y\left[\mathbf{a}^T\mathbf{x} + b\right])$. Now imagine the hinge loss for this example *isn't* zero. If the example is classified correctly, then it is close to the decision boundary. We expect that there are fewer of these examples than examples that are far from the decision boundary and on the wrong side, so we concentrate on examples that are misclassified. For misclassified examples, if $\|\mathbf{a}\|$ is small, then at least the hinge loss will be small. By this argument, we would like to achieve a small value of the hinge loss on the training examples using an $\mathbf{a}$ that has small length, because that $\mathbf{a}$ will yield smaller hinge loss on errors in test.

We can do so by adding a penalty term to the hinge loss to favor solutions where $\|\mathbf{a}\|$ is small. To obtain an $\mathbf{a}$ of small length, it is enough to ensure that $(1/2)\mathbf{a}^T\mathbf{a}$ is small (the factor of $1/2$ makes the gradient cleaner). This penalty term will ensure that there is a unique choice of classifier parameters in the case the hinge loss is zero. Experience (and some theory we can't go into here) shows that having a small $\|\mathbf{a}\|$ helps even if there is no pair that classifies all training examples correctly. Doing so improves the error on future examples. Adding a penalty term to improve the solution of a learning problem is sometimes referred to as **regularization**. The penalty term is often referred to as a **regularizer**, because it tends to discourage solutions that are large (and so have possible high loss on future test data) but are not strongly supported by the training data. The parameter λ is often referred to as the **regularization parameter**.

Using the hinge loss to form the training cost, and regularizing with a penalty term $(1/2)\mathbf{a}^T\mathbf{a}$ means our cost function is

$$S(\mathbf{a}, b; \lambda) = \left[(1/N) \sum_{i=1}^{N} \max(0, 1 - y_i\left(\mathbf{a}^T\mathbf{x}_i + b\right))\right] + \lambda\left(\frac{\mathbf{a}^T\mathbf{a}}{2}\right).$$

There are now two problems to solve. First, assume we know λ; we will need to find $\mathbf{a}$ and b that minimize $S(\mathbf{a}, b; \lambda)$. Second, we have no theory that tells us how to choose λ, so we will need to search for a good value.

Remember This: *A regularization term is a penalty that discourages a classifier from making large errors on future data. Because the future data is unknown, the best available strategy is to force future examples that are misclassified to have small hinge loss. This is achieved by discouraging classifiers with large values of $\|a\|$.*

2.1.3 Finding a Classifier with Stochastic Gradient Descent

The usual recipes for finding a minimum are ineffective for our cost function. First, write $\mathbf{u} = [\mathbf{a}, b]$ for the vector obtained by stacking the vector $\mathbf{a}$ together with b. We have a function $g(\mathbf{u})$, and we wish to obtain a value of $\mathbf{u}$ that achieves the minimum for that function. Sometimes we can solve a problem like this by constructing the gradient and finding a value of $\mathbf{u}$ that makes the gradient zero, but not this time (try it; the max creates problems). We must use a numerical method.

Typical numerical methods take a point $\mathbf{u}^{(n)}$, update it to $\mathbf{u}^{(n+1)}$, then check to see whether the result is a minimum. This process is started from a start point. The choice of start point may or may not matter for general problems, but for our problem a random start point is fine. The update is usually obtained by computing a direction $\mathbf{p}^{(n)}$ such that for small values of η, $g(\mathbf{u}^{(n)} + \eta \mathbf{p}^{(n)})$ is smaller than $g(\mathbf{u}^{(n)})$. Such a direction is known as a **descent direction**. We must then determine how far to go along the descent direction, a process known as **line search**.

Obtaining a Descent Direction: One method to choose a descent direction is **gradient descent**, which uses the negative gradient of the function. Recall our notation that

$$\mathbf{u} = \begin{pmatrix} u_1 \\ u_2 \\ \dots \\ u_d \end{pmatrix}$$

and that

$$\nabla g = \begin{pmatrix} \frac{\partial g}{\partial u_1} \\ \frac{\partial g}{\partial u_2} \\ \dots \\ \frac{\partial g}{\partial u_d} \end{pmatrix}.$$

We can write a Taylor series expansion for the function $g(\mathbf{u}^{(n)} + \eta \mathbf{p}^{(n)})$. We have that

$$g(\mathbf{u}^{(n)} + \eta \mathbf{p}^{(n)}) = g(\mathbf{u}^{(n)}) + \eta \left[(\nabla g)^T \mathbf{p}^{(n)} \right] + O(\eta^2)$$

This means that we can expect that if

$$\mathbf{p}^{(n)} = -\nabla g(\mathbf{u}^{(n)}),$$

we expect that, at least for small values of h, $g(\mathbf{u}^{(n)} + \eta \mathbf{p}^{(n)})$ will be less than $g(\mathbf{u}^{(n)})$. This works (as long as g is differentiable, and quite often when it isn't) because g must go down for at least small steps in this direction.

But recall that our cost function is a sum of a penalty term and one error cost per example. This means the cost function looks like

$$g(\mathbf{u}) = \left[(1/N) \sum_{i=1}^{N} g_i(\mathbf{u}) \right] + g_0(\mathbf{u}),$$

as a function of $\mathbf{u}$. Gradient descent would require us to form

$$-\nabla g(\mathbf{u}) = -\left(\left[(1/N)\sum_{i=1}^{N}\nabla g_i(\mathbf{u})\right] + \nabla g_0(\mathbf{u})\right)$$

and then take a small step in this direction. But if N is large, this is unattractive, as we might have to sum a lot of terms. This happens a lot in building classifiers, where you might quite reasonably expect to deal with millions (billions; perhaps trillions) of examples. Touching each example at each step really is impractical.

Stochastic gradient descent is an algorithm that replaces the exact gradient with an approximation that has a random error, but is simple and quick to compute. The term

$$\left(\frac{1}{N}\right)\sum_{i=1}^{N}\nabla g_i(\mathbf{u}).$$

is a population mean, and we know (or should know!) how to deal with those. We can estimate this term by drawing a random sample (a **batch**) of N_b (the **batch size**) examples, with replacement, from the population of N examples, then computing the mean for that sample. We approximate the population mean by

$$\left(\frac{1}{N_b}\right)\sum_{j\in\text{batch}}\nabla g_j(\mathbf{u}).$$

The batch size is usually determined using considerations of computer architecture (how many examples fit neatly into cache?) or of database design (how many examples are recovered in one disk cycle?). One common choice is $N_b = 1$, which is the same as choosing one example uniformly and at random. We form

$$\mathbf{p}_{N_b}^{(n)} = -\left(\left[(1/N_b)\sum_{j\in\text{batch}}\nabla g_i(\mathbf{u})\right] + \nabla g_0(\mathbf{u})\right)$$

and then take a small step along $\mathbf{p}_{N_b}^{(n)}$. Our new point becomes

$$\mathbf{u}^{(n+1)} = \mathbf{u}^{(n)} + \eta\mathbf{p}_{N_b}^{(n)},$$

where η is called the **steplength** (or sometimes **stepsize** or **learning rate**, even though it isn't the size or the length of the step we take, or a rate!).

Because the expected value of the sample mean is the population mean, if we take many small steps along $\mathbf{p}_{N_b}$, they should average out to a step backwards along the gradient. This approach is known as stochastic gradient descent because we're not going along the gradient, but along a random vector which is the gradient only in expectation. It isn't obvious that stochastic gradient descent is a good idea. Although each step is easy to take, we may need to take more steps. The question is then whether we gain in the increased speed of the step what we lose by having to take more steps. Not much is known theoretically, but in practice the approach is hugely successful for training classifiers.

Choosing a Steplength: Choosing a steplength η takes some work. We can't search for the step that gives us the best value of g, because we don't want to evaluate the function g (doing so involves looking at each of the g_i terms). Instead, we use an η that is large at the start—so that the method can explore large changes in the values of the classifier parameters—and small steps later—so that it settles down. The choice of how η gets smaller is often known as a **steplength schedule** or **learning schedule**.

Here are useful examples of steplength schedules. Often, you can tell how many steps are required to have seen the whole dataset; this is called an **epoch**. A common steplength schedule sets the steplength in the e'th epoch to be

$$\eta^{(e)} = \frac{m}{e+n},$$

where m and n are constants chosen by experiment with small subsets of the dataset. When there are a lot of examples, an epoch is a long time to fix the steplength, and this approach can reduce the steplength too slowly. Instead, you can divide training into what I shall call **seasons** (blocks of a fixed number of iterations, smaller than epochs), and make the steplength a function of the season number.

There is no good test for whether stochastic gradient descent has converged to the right answer, because natural tests involve evaluating the gradient and the function, and doing so is expensive. More usual is to plot the error as a function of iteration on the validation set, and interrupt or stop training when the error has reached an acceptable level. The error (resp. accuracy) should vary randomly (because the steps are taken in directions that only approximate the gradient) but should decrease (resp. increase) overall as training proceeds (because the steps do approximate the gradient). Figures 2.2 and 2.3 show examples of these curves, which are sometimes known as **learning curves**.

Remember This: *Stochastic gradient descent is the dominant training paradigm for classifiers. Stochastic gradient descent uses a sample of the training data to estimate the gradient. Doing so results in a fast but noisy gradient estimate. This is particularly valuable when training sets are very large (as they should be). Steplengths are chosen according to a steplength schedule, and there is no test for convergence other than evaluating classifier behavior on validation data.*

2.1.4 Searching for λ

We do not know a good value for λ. We will obtain a value by choosing a set of different values, fitting an SVM using each value, and taking the λ value that will yield the best SVM. Experience has shown that the performance of a method is not profoundly sensitive to the value of λ, so that we can look at values spaced quite far apart. It is usual to take some small number (say, $1e-4$), then multiply by powers of 10 (or 3, if you're feeling fussy and have a fast computer). So, for example, we might look at $\lambda \in \{1e-4, 1e-3, 1e-2, 1e-1\}$. We know how to fit an SVM to a particular value of λ (Sect. 2.1.3). The problem is to choose the value that yields the best SVM, and to use that to get the best classifier.

We have seen a version of this problem before (Sect. 1.3.1). There, we chose from several different types of model to obtain the best naive Bayes classifier. The recipe from that section is easily adapted to the current problem. We regard each different λ value as representing a different model. We split the data into two pieces: one is a training set, used for fitting and choosing models; the other is a test set, used for evaluating the final chosen model.

Now for each value of λ, compute the cross-validated error of an SVM using that λ on the training set. Do this by repeatedly splitting the training set into two pieces (training and validation); fitting the SVM with that λ to the training piece using stochastic gradient descent; evaluating the error on the validation piece; and averaging these errors. Now use the cross-validated error to choose the best λ value. Very often this just means you choose the value that produces the lowest cross-validated error, but there might be cases where two values produce about the same error and one is preferred for some other reason. Notice that you can compute the standard deviation of the cross-validated error as well as the mean, so you can tell whether differences between cross-validated errors are significant.

Now take the entire training set, and use this to fit an SVM for the chosen λ value. This should be (a little) better than any of the SVMs obtained in the cross-validation, because it uses (slightly) more data. Finally, evaluate the resulting SVM on the test set.

This procedure is rather harder to describe than to do (there's a pretty natural set of nested loops here). There are some strong advantages. First, the estimate of how well a particular SVM type works is unbiased, because we evaluated on data not used on training. Second, once you have chosen the cross-validation parameter, the SVM you fit is the best you can fit because you used all the training set to obtain it. Finally, your estimate of how well that particular SVM works is unbiased, too, because you obtained it using data that wasn't used to train or to select a model.

Procedure: 2.1 *Choosing a λ*

Divide your dataset $\mathcal{D}$ into two parts at random: one is the training set ($\mathcal{R}$) and one the test set ($\mathcal{E}$). Choose a set of λ values that you will evaluate (typically, differing by factors of 10).
For each λ in the collection that interests you:

- repeatedly

  - split $\mathcal{R}$ into two parts at random, $\mathcal{R}_t$ and $\mathcal{V}$;
  - fit the model using $\mathcal{R}_t$ to train;
  - compute an estimate of the error using that model with $\mathcal{V}$.

- Now report the average of these errors as the error of the model.

> Now use these average errors (and possibly other criteria like speed, etc.) to choose a value of λ. Compute an estimate of the error rate using your chosen model on $\mathcal{E}$.

2.1.5 Summary: Training with Stochastic Gradient Descent

I have summarized the SVM training procedure in a set of boxes, below. You should be aware that the recipe there admits many useful variations, though. One useful practical trick is to rescale the feature vector components so each has unit variance. This doesn't change anything conceptual as the best choice of decision boundary for rescaled data is easily derived from the best choice for unscaled, and vice versa. Rescaling very often makes stochastic gradient descent perform better because the method takes steps that are even in each component.

It is quite usual to use packages to fit SVMs, and good packages may use a variety of tricks which we can't go into to make training more efficient. Nonetheless, you should have a grasp of the overall process, because it follows a pattern that is useful for training other models (among other things, most deep networks are trained using this pattern).

Procedure: 2.2 *Training an SVM: Overall*

Start with a dataset containing N pairs $(\mathbf{x}_i, y_i)$. Each $\mathbf{x}_i$ is a d-dimensional feature vector, and each y_i is a label, either 1 or -1. Optionally, rescale the $\mathbf{x}_i$ so that each component has unit variance. Choose a set of possible values of the regularization weight λ. Separate the dataset into two sets: test and training. Reserve the test set. For each value of the regularization weight, use the training set to estimate the accuracy of an SVM with that λ value, using cross-validation as in Procedure 2.3 and stochastic gradient descent. Use this information to choose λ_0, the best value of λ (usually, the one that yields the highest accuracy). Now use the training set to fit the best SVM using λ_0 as the regularization constant. Finally, use the test set to compute the accuracy or error rate of that SVM, and report that

---

Procedure: 2.3 *Training an SVM: Estimating the Accuracy*

Repeatedly: split the training dataset into two components (training and validation), at random; use the training component to train an SVM; and compute the accuracy on the validation component. Now average the resulting accuracy values.

Procedure: 2.4 *Training an SVM: Stochastic Gradient Descent*

Obtain $\mathbf{u} = (\mathbf{a}, b)$ by stochastic gradient descent on the cost function

$$g(\mathbf{u}) = \left[(1/N) \sum_{i=1}^{N} g_i(\mathbf{u}) \right] + g_0(\mathbf{u})$$

where $g_0(\mathbf{u}) = \lambda(\mathbf{a}^T \mathbf{a})/2$ and $g_i(\mathbf{u}) = \max(0, 1 - y_i\left(\mathbf{a}^T \mathbf{x}_i + b\right))$.
Do so by first choosing a fixed number of items per batch N_b, the number of steps per season N_s, and the number of steps k to take before evaluating the model (this is usually a lot smaller than N_s). Choose a random start point. Now iterate:

- Update the stepsize. In the s'th season, the stepsize is typically $\eta^{(s)} = \frac{m}{s+n}$ for constants m and n chosen by small-scale experiments.
- Split the training dataset into a training part and a validation part. This split changes each season. Use the validation set to get an unbiased estimate of error during that season's training.
- Now, until the end of the season (i.e., until you have taken N_s steps):

  - Take k steps. Each step is taken by selecting a batch of N_b data items uniformly and at random from the training part for that season. Write $\mathcal{D}$ for this set. Now compute

  $$\mathbf{p}^{(n)} = -\frac{1}{N_b} \left(\sum_{i \in \mathcal{D}} \nabla g_i(\mathbf{u}^{(n)}) \right) - \lambda \mathbf{u}^{(n)},$$

  and update the model by computing

  $$\mathbf{u}^{(n+1)} = \mathbf{u}^{(n)} + \eta \mathbf{p}^{(n)}$$

  - Evaluate the current model $\mathbf{u}^{(n)}$ by computing the accuracy on the validation part for that season. Plot the accuracy as a function of step number.

There are two ways to stop. You can choose a fixed number of seasons (or of epochs) and stop when that is done. Alternatively, you can watch the error plot and stop when the error reaches some level or meets some criterion.

2.1.6 Example: Adult Income with an SVM

Here is an example in some detail. I downloaded the dataset at http://archive. ics.uci.edu/ml/datasets/Adult. This dataset apparently contains 48,842 data items, but I worked with only the first 32,000. Each consists of a set of numerical and categorical features describing a person, together with whether their annual income is larger than or smaller than 50K\$. I ignored the categorical features to prepare these figures. This isn't wise if you want a good classifier, but it's fine for an example. I used these features to predict whether income is over or under 50K\$. I split the data into 5000 test examples and 27,000 training examples. It's important to do so at random. There are 6 numerical features.

Renormalizing the Data: I subtracted the mean (which doesn't usually make much difference) and rescaled each so that the variance was 1 (which is often very important). You should try this example both ways; if you don't renormalize the data, it doesn't work very well.

Setting up Stochastic Gradient Descent: We have estimates $\mathbf{a}^{(n)}$ and $b^{(n)}$ of the classifier parameters, and we want to improve the estimates. I used a batch size of $N_b = 1$. Pick the r'th example at random. The gradient is

$$\nabla \left(\max(0, 1 - y_r \left(\mathbf{a}^T \mathbf{x}_r + b \right)) + \frac{\lambda}{2} \mathbf{a}^T \mathbf{a} \right).$$

Assume that $y_k \left(\mathbf{a}^T \mathbf{x}_r + b \right) > 1$. In this case, the classifier predicts a score with the right sign, and a magnitude that is greater than one. Then the first term is zero, and the gradient of the second term is easy. Now if $y_k \left(\mathbf{a}^T \mathbf{x}_r + b \right) < 1$, we can ignore the max, and the first term is $1 - y_r \left(\mathbf{a}^T \mathbf{x}_r + b \right)$; the gradient is again easy. If $y_r \left(\mathbf{a}^T \mathbf{x}_r + b \right) = 1$, there are two distinct values we could choose for the gradient, because the max term isn't differentiable. It does not matter which value we choose because this situation hardly ever happens. We choose a steplength η, and update our estimates using this gradient. This yields:

$$\mathbf{a}^{(n+1)} = \mathbf{a}^{(n)} - \eta \begin{cases} \lambda \mathbf{a} & \text{if } y_k \left(\mathbf{a}^T \mathbf{x}_k + b \right) \geq 1 \\ \lambda \mathbf{a} - y_k \mathbf{x} & \text{otherwise} \end{cases}$$

and

$$b^{(n+1)} = b^{(n)} - \eta \begin{cases} 0 & \text{if } y_k \left(\mathbf{a}^T \mathbf{x}_k + b \right) \geq 1 \\ -y_k & \text{otherwise} \end{cases}.$$

Training: I used two different training regimes. In the first training regime, there were 100 seasons. In each season, I applied 426 steps. For each step, I selected one data item uniformly at random (sampling with replacement), then stepped down the gradient. This means the method sees a total of 42,600 data items. This means that there is a high probability it has touched each data item once (27,000 isn't enough, because we are sampling with replacement, so some items get seen more than once). I chose 5 different values for the regularization parameter and trained with a steplength of $1/(0.01 * s + 50)$, where s is the season. At the end of each season, I computed $\mathbf{a}^T \mathbf{a}$ and the accuracy (fraction of examples correctly classified) of the current classifier on the held-out test examples. Figure 2.2 shows the results. You should notice that the accuracy changes slightly each season; that for larger regularizer values $\mathbf{a}^T \mathbf{a}$ is smaller; and that the accuracy settles down to about 0.8 very quickly.

In the second training regime, there were 100 seasons. In each season, I applied 50 steps. For each step, I selected one data item uniformly at random (sampling

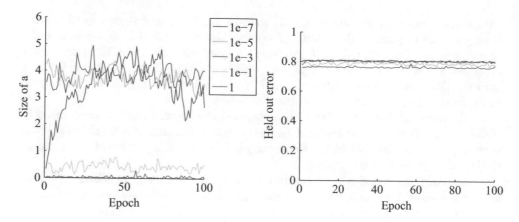

Figure 2.2: On the **left**, the magnitude of the weight vector **a** at the end of each season for the first training regime described in the text. On the **right**, the accuracy on held-out data at the end of each season. Notice how different choices of regularization parameter lead to different magnitudes of **a**; how the method isn't particularly sensitive to choice of regularization parameter (they change by factors of 100); how the accuracy settles down fairly quickly; and how overlarge values of the regularization parameter do lead to a loss of accuracy

with replacement), then stepped down the gradient. This means the method sees a total of 5000 data items, and about 3000 unique data items—it hasn't seen the whole training set. I chose 5 different values for the regularization parameter and trained with a steplength of $1/(0.01 * s + 50)$, where s is the season. At the end of each season, I computed $\mathbf{a}^T\mathbf{a}$ and the accuracy (fraction of examples correctly classified) of the current classifier on the held-out test examples. Figure 2.3 shows the results.

This is an easy classification example. Points worth noting are

- the accuracy makes large changes early, then settles down to make slight changes each season;
- quite large changes in regularization constant have small effects on the outcome, but there is a best choice;
- for larger values of the regularization constant, $\mathbf{a}^T\mathbf{a}$ is smaller;
- there isn't much difference between the two training regimes;
- normalizing the data is important;
- and the method doesn't need to see all the training data to produce a classifier that is about as good as it would be if the method *had* seen all training data.

All of these points are relatively typical of SVMs trained using stochastic gradient descent with very large datasets.

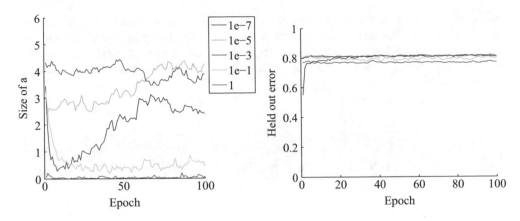

Figure 2.3: On the **left**, the magnitude of the weight vector **a** at the end of each season for the second training regime described in the text. On the **right**, the accuracy on held-out data at the end of each season. Notice how different choices of regularization parameter lead to different magnitudes of **a**; how the method isn't particularly sensitive to choice of regularization parameter (they change by factors of 100); how the accuracy settles down fairly quickly; and how overlarge values of the regularization parameter do lead to a loss of accuracy

---

Remember This: *Linear SVMs are a go-to classifier. When you have a binary classification problem, the first step should be to try a linear SVM. Training with stochastic gradient descent is straightforward and extremely effective. Finding an appropriate value of the regularization constant requires an easy search. There is an immense quantity of good software available.*

2.1.7   Multiclass Classification with SVMs

I have shown how one trains a linear SVM to make a binary prediction (i.e., predict one of two outcomes). But what if there are three, or more, labels? In principle, you could write a binary code for each label, then use a different SVM to predict each bit of the code. It turns out that this doesn't work terribly well, because an error by one of the SVMs is usually catastrophic.

There are two methods that are widely used. In the **all-vs-all** approach, we train a binary classifier for each pair of classes. To classify an example, we present it to each of these classifiers. Each classifier decides which of two classes the example belongs to, then records a vote for that class. The example gets the class label with the most votes. This approach is simple, but scales very badly with the number of classes (you have to build $O(N^2)$ different SVMs for N classes).

In the **one-vs-all** approach, we build a binary classifier for each class. This classifier must distinguish its class from all the other classes. We then take the class with the largest classifier score. One can think up quite good reasons this approach shouldn't work. For one thing, the classifier isn't told that you intend to use the score to tell similarity between classes. In practice, the approach works rather well and is quite widely used. This approach scales a bit better with the number of classes $(O(N))$.

Remember This: *It is straightforward to build a multiclass classifier out of binary classifiers. Any decent SVM package will do this for you.*

2.2   Classifying with Random Forests

I described a classifier as a rule that takes a feature, and produces a class. One way to build such a rule is with a sequence of simple tests, where each test is allowed to use the results of all previous tests. This class of rule can be drawn as a tree (Fig. 2.4), where each node represents a test, and the edges represent the possible outcomes of the test. To classify a test item with such a tree, you present it to the first node; the outcome of the test determines which node it goes to next; and so on, until the example arrives at a leaf. When it does arrive at a leaf, we label the test item with the most common label in the leaf. This object is known as a **decision tree**. Notice one attractive feature of this decision tree: it deals with multiple class labels quite easily, because you just label the test item with the most common label in the leaf that it arrives at when you pass it down the tree.

Figure 2.5 shows a simple 2D dataset with four classes, next to a decision tree that will correctly classify at least the training data. Actually classifying data with a tree like this is straightforward. We take the data item, and pass it down the tree. Notice it can't go both left and right, because of the way the tests work.

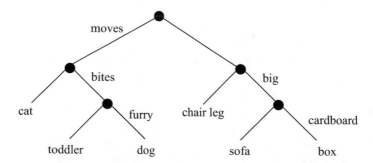

Figure 2.4: This—the household robot's guide to obstacles—is a typical decision tree. I have labelled only one of the outgoing branches, because the other is the negation. So if the obstacle moves, bites, but isn't furry, then it's a toddler. In general, an item is passed down the tree until it hits a leaf. It is then labelled with the leaf's label

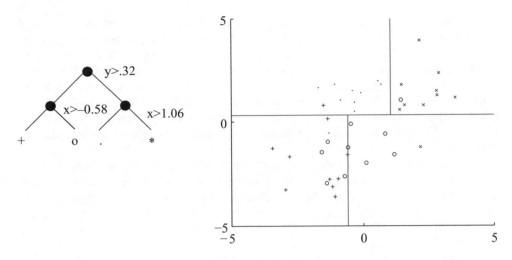

Figure 2.5: A straightforward decision tree, illustrated in two ways. On the **left**, I have given the rules at each split; on the **right**, I have shown the data points in two dimensions, and the structure that the tree produces in the feature space

This means each data item arrives at a single leaf. We take the most common label at the leaf, and give that to the test item. In turn, this means we can build a geometric structure on the feature space that corresponds to the decision tree. I have illustrated that structure in Fig. 2.5, where the first decision splits the feature space in half (which is why the term split is used so often), and then the next decisions split each of those halves into two.

The important question is how to get the tree from data. It turns out that the best approach for building a tree incorporates a great deal of randomness. As a result, we will get a different tree each time we train a tree on a dataset. None of the individual trees will be particularly good (they are often referred to as "weak learners"). The natural thing to do is to produce many such trees (a **decision forest**), and allow each to vote; the class that gets the most votes, wins. This strategy is extremely effective.

2.2.1 Building a Decision Tree

There are many algorithms for building decision trees. We will use an approach chosen for simplicity and effectiveness; be aware there are others. We will always use a binary tree, because it's easier to describe and because that's usual (it doesn't change anything important, though). Each node has a **decision function**, which takes data items and returns either 1 or -1.

We train the tree by thinking about its effect on the training data. We pass the whole pool of training data into the root. Any node splits its incoming data into two pools, left (all the data that the decision function labels 1) and right (ditto, -1). Finally, each leaf contains a pool of data, which it can't split because it is a leaf.

Training the tree uses a straightforward algorithm. First, we choose a class of decision functions to use at each node. It turns out that a very effective algorithm is to choose a single feature at random, then test whether its value is larger than, or smaller than a threshold. For this approach to work, one needs to be quite careful about the choice of threshold, which is what we describe in the next section. Some minor adjustments, described below, are required if the feature chosen isn't ordinal. Surprisingly, being clever about the choice of *feature* doesn't seem add a great deal of value. We won't spend more time on other kinds of decision function, though there are lots.

Now assume we use a decision function as described, and we know how to choose a threshold. We start with the root node, then recursively either split the pool of data at that node, passing the left pool left and the right pool right, or stop splitting and return. Splitting involves choosing a decision function from the class to give the "best" split for a leaf. The main questions are how to choose the best split (next section), and when to stop.

Stopping is relatively straightforward. Quite simple strategies for stopping are very good. It is hard to choose a decision function with very little data, so we must stop splitting when there is too little data at a node. We can tell this is the case by testing the amount of data against a threshold, chosen by experiment. If all the data at a node belongs to a single class, there is no point in splitting. Finally, constructing a tree that is too deep tends to result in generalization problems, so we usually allow no more than a fixed depth D of splits. Choosing the best splitting threshold is more complicated.

Figure 2.6 shows two possible splits of a pool of training data. One is quite obviously a lot better than the other. In the good case, the split separates the pool

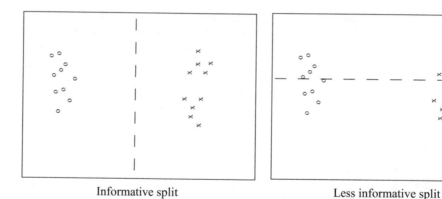

Informative split   Less informative split

Figure 2.6: Two possible splits of a pool of training data. Positive data is represented with an "x," negative data with a "o." Notice that if we split this pool with the informative line, all the points on the left are "o"s, and all the points on the right are "x"s. This is an excellent choice of split—once we have arrived in a leaf, everything has the same label. Compare this with the less informative split. We started with a node that was half "x" and half "o," and now have two nodes each of which is half "x" and half "o"—this isn't an improvement, because we do not know more about the label as a result of the split

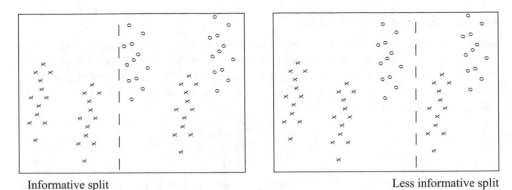

Informative split   Less informative split

Figure 2.7: Two possible splits of a pool of training data. Positive data is represented with an "x," negative data with a "o." Notice that if we split this pool with the informative line, all the points on the left are "x"s, and two-thirds of the points on the right are "o"s. This means that knowing which side of the split a point lies would give us a good basis for estimating the label. In the less informative case, about two-thirds of the points on the left are "x"s and about half on the right are "x"s—knowing which side of the split a point lies is much less useful in deciding what the label is

into positives and negatives. In the bad case, each side of the split has the same number of positives and negatives. We cannot usually produce splits as good as the good case here. What we are looking for is a split that will make the proper label more certain.

Figure 2.7 shows a more subtle case to illustrate this. The splits in this figure are obtained by testing the horizontal feature against a threshold. In one case, the left and the right pools contain about the same fraction of positive ("x") and negative ("o") examples. In the other, the left pool is all positive, and the right pool is mostly negative. This is the better choice of threshold. If we were to label any item on the left side positive and any item on the right side negative, the error rate would be fairly small. If you count, the best error rate for the informative split is 20% on the training data, and for the uninformative split it is 40% on the training data.

But we need some way to score the splits, so we can tell which threshold is best. Notice that, in the uninformative case, knowing that a data item is on the left (or the right) does not tell me much more about the data than I already knew. We have that $p(1|\text{left pool, uninformative}) = 2/3 \approx 3/5 = p(1|\text{parent pool})$ and $p(1|\text{right pool, uninformative}) = 1/2 \approx 3/5 = p(1|\text{parent pool})$. For the informative pool, knowing a data item is on the left classifies it completely, and knowing that it is on the right allows us to classify it an error rate of $1/3$. The informative split means that my uncertainty about what class the data item belongs to is significantly reduced if I know whether it goes left or right. To choose a good threshold, we need to keep track of how informative the split is.

2.2.2 Choosing a Split with Information Gain

Write $\mathcal{P}$ for the set of all data at the node. Write $\mathcal{P}_l$ for the left pool, and $\mathcal{P}_r$ for the right pool. The entropy of a pool $\mathcal{C}$ scores how many bits would be required to represent the class of an item in that pool, on average. Write $n(i; \mathcal{C})$ for the number of items of class i in the pool, and $N(\mathcal{C})$ for the number of items in the pool. Then the entropy $H(\mathcal{C})$ of the pool $\mathcal{C}$ is

$$-\sum_i \frac{n(i;\mathcal{C})}{N(\mathcal{C})} \log_2 \frac{n(i;\mathcal{C})}{N(\mathcal{C})}.$$

It is straightforward that $H(\mathcal{P})$ bits are required to classify an item in the parent pool $\mathcal{P}$. For an item in the left pool, we need $H(\mathcal{P}_l)$ bits; for an item in the right pool, we need $H(\mathcal{P}_r)$ bits. If we split the parent pool, we expect to encounter items in the left pool with probability

$$\frac{N(\mathcal{P}_l)}{N(\mathcal{P})}$$

and items in the right pool with probability

$$\frac{N(\mathcal{P}_r)}{N(\mathcal{P})}.$$

This means that, on average, we must supply

$$\frac{N(\mathcal{P}_l)}{N(\mathcal{P})} H(\mathcal{P}_l) + \frac{N(\mathcal{P}_r)}{N(\mathcal{P})} H(\mathcal{P}_r)$$

bits to classify data items if we split the parent pool. Now a good split is one that results in left and right pools that are informative. In turn, we should need fewer bits to classify once we have split than we need before the split. You can see the difference

$$I(\mathcal{P}_l, \mathcal{P}_r; \mathcal{P}) = H(\mathcal{P}) - \left(\frac{N(\mathcal{P}_l)}{N(\mathcal{P})} H(\mathcal{P}_l) + \frac{N(\mathcal{P}_r)}{N(\mathcal{P})} H(\mathcal{P}_r) \right)$$

as the **information gain** caused by the split. This is the average number of bits that you *don't* have to supply if you know which side of the split an example lies. Better splits have larger information gain.

Recall that our decision function is to choose a feature at random, then test its value against a threshold. Any data point where the value is larger goes to the left pool; where the value is smaller goes to the right. This may sound much too simple to work, but it is actually effective and popular. Assume that we are at a node, which we will label k. We have the pool of training examples that have reached that node. The i'th example has a feature vector $\mathbf{x}_i$, and each of these feature vectors is a d-dimensional vector.

We choose an integer j in the range $1, \ldots, d$ uniformly and at random. We will split on this feature, and we store j in the node. Recall we write $x_i^{(j)}$ for the value of the j'th component of the i'th feature vector. We will choose a threshold t_k, and split by testing the sign of $x_i^{(j)} - t_k$. Choosing the value of t_k is easy.

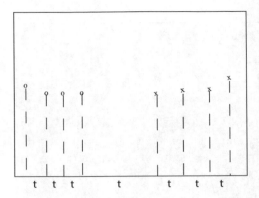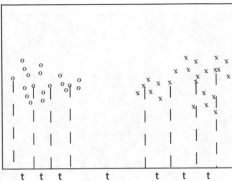

Figure 2.8: We search for a good splitting threshold by looking at values of the chosen component that yield different splits. On the **left**, I show a small dataset and its projection onto the chosen splitting component (the horizontal axis). For the 8 data points here, there are only 7 threshold values that produce interesting splits, and these are shown as "t"s on the axis. On the **right**, I show a larger dataset; in this case, I have projected only a subset of the data, which results in a small set of thresholds to search

Assume there are N_k examples in the pool. Then there are $N_k - 1$ possible values of t_k that lead to different splits. To see this, sort the N_k examples by $x^{(j)}$, then choose values of t_k halfway between example values (Fig. 2.8). For each of these values, we compute the information gain of the split. We then keep the threshold with the best information gain.

We can elaborate this procedure in a useful way, by choosing m features at random, finding the best split for each, then keeping the feature and threshold value that is best. It is important that m is a lot smaller than the total number of features—a usual root of thumb is that m is about the square root of the total number of features. It is usual to choose a single m, and choose that for all the splits.

Now assume we happen to have chosen to work with a feature that isn't ordinal, and so can't be tested against a threshold. A natural, and effective, strategy is as follows. We can split such a feature into two pools by flipping an unbiased coin for each value—if the coin comes up H, any data point with that value goes left, and if it comes up T, any data point with that value goes right. We chose this split at random, so it might not be any good. We can come up with a good split by repeating this procedure F times, computing the information gain for each split, then keeping the one that has the best information gain. We choose F in advance, and it usually depends on the number of values the categorical variable can take.

We now have a relatively straightforward blueprint for an algorithm, which I have put in a box. It's a blueprint, because there are a variety of ways in which it can be revised and changed.

Procedure: 2.5 *Building a Decision Tree: Overall*

We have a dataset containing N pairs $(\mathbf{x}_i, y_i)$. Each $\mathbf{x}_i$ is a d-dimensional feature vector, and each y_i is a label. Call this dataset a **pool**. Now recursively apply the following procedure:

- If the pool is too small, or if all items in the pool have the same label, or if the depth of the recursion has reached a limit, stop.
- Otherwise, search the features for a good split that divides the pool into two, then apply this procedure to each child.

We search for a good split by the following procedure:

- Choose a subset of the feature components at random. Typically, one uses a subset whose size is about the square root of the feature dimension.
- For each component of this subset, search for a good split. If the component is ordinal, do so using the procedure of Box 2.6, otherwise use the procedure of Box 2.7.

Procedure: 2.6 *Splitting an Ordinal Feature*

We search for a good split on a given ordinal feature by the following procedure:

- Select a set of possible values for the threshold.
- For each value split the dataset (every data item with a value of the component below the threshold goes left, others go right), and compute the information gain for the split.

Keep the threshold that has the largest information gain.
A good set of possible values for the threshold will contain values that separate the data "reasonably." If the pool of data is small, you can project the data onto the feature component (i.e., look at the values of that component alone), then choose the $N-1$ distinct values that lie between two data points. If it is big, you can randomly select a subset of the data, then project that subset on the feature component and choose from the values between data points.

Procedure: 2.7 *Splitting a Non-ordinal Feature*

Split the values this feature takes into sets by flipping an unbiased coin for each value—if the coin comes up H, any data point with that value goes left, and if it comes up T, any data point with that value goes right. Repeat this procedure F times, compute the information gain for each split, then keep the split that has the best information gain. We choose F in advance, and it usually depends on the number of values the categorical variable can take.

2.2.3 Forests

A single decision tree tends to yield poor classifications. One reason is because the tree is not chosen to give the best classification of its training data. We used a random selection of splitting variables at each node, so the tree can't be the "best possible." Obtaining the best possible tree presents significant technical difficulties. It turns out that the tree that gives the best possible results on the training data can perform rather poorly on test data. The training data is a small subset of possible examples, and so must differ from the test data. The best possible tree on the training data might have a large number of small leaves, built using carefully chosen splits. But the choices that are best for training data might not be best for test data.

Rather than build the best possible tree, we have built a tree efficiently, but with number of random choices. If we were to rebuild the tree, we would obtain a different result. This suggests the following extremely effective strategy: build many trees, and classify by merging their results.

2.2.4 Building and Evaluating a Decision Forest

There are two important strategies for building and evaluating decision forests. I am not aware of evidence strongly favoring one over the other, but different software packages use different strategies, and you should be aware of the options. In one strategy, we separate labelled data into a training and a test set. We then build multiple decision trees, training each using the whole training set. Finally, we evaluate the forest on the test set. In this approach, the forest has not seen some fraction of the available labelled data, because we used it to test. However, each tree has seen every training data item.

Procedure: 2.8 *Building a Decision Forest*

We have a dataset containing N pairs $(\mathbf{x}_i, y_i)$. Each $\mathbf{x}_i$ is a d-dimensional feature vector, and each y_i is a label. Separate the dataset into a test set and a training set. Train multiple distinct decision trees on the training set, recalling that the use of a random set of components to find a good split means you will obtain a distinct tree each time.

In the other strategy, sometimes called **bagging**, each time we train a tree we randomly subsample the labelled data with replacement, to yield a training set the same size as the original set of labelled data. Notice that there will be duplicates in this training set, which is like a bootstrap replicate. This training set is often called a **bag**. We keep a record of the examples that do not appear in the bag (the "out of bag" examples). Now to evaluate the forest, we evaluate each tree on its out of bag examples, and average these error terms. In this approach, the entire forest has seen all labelled data, and we also get an estimate of error, but no tree has seen all the training data.

Procedure: 2.9 *Building a Decision Forest Using Bagging*

We have a dataset containing N pairs $(\mathbf{x}_i, y_i)$. Each $\mathbf{x}_i$ is a d-dimensional feature vector, and each y_i is a label. Now build k bootstrap replicates of the training dataset. Train one decision tree on each replicate.

2.2.5 Classifying Data Items with a Decision Forest

Once we have a forest, we must classify test data items. There are two major strategies. The simplest is to classify the item with each tree in the forest, then take the class with the most votes. This is effective, but discounts some evidence that might be important. For example, imagine one of the trees in the forest has a leaf with many data items with the same class label; another tree has a leaf with exactly one data item in it. One might not want each leaf to have the same vote.

Procedure: 2.10 *Classification with a Decision Forest*

Given a test example $\mathbf{x}$, pass it down each tree of the forest. Now choose one of the following strategies.

- Each time the example arrives at a leaf, record one vote for the label that occurs most often at the leaf. Now choose the label with the most votes.
- Each time the example arrives at a leaf, record N_l votes for each of the labels that occur at the leaf, where N_l is the number of times the label appears in the training data at the leaf. Now choose the label with the most votes.

An alternative strategy that takes this observation into account is to pass the test data item down each tree. When it arrives at a leaf, we record one vote for each of the training data items in that leaf. The vote goes to the class of the training data item. Finally, we take the class with the most votes. This approach allows big, accurate leaves to dominate the voting process. Both strategies are in use, and

I am not aware of compelling evidence that one is always better than the other. This may be because the randomness in the training process makes big, accurate leaves uncommon in practice.

Worked Example 2.1 *Classifying Heart Disease Data*

Build a random forest classifier to classify the "heart" dataset from the UC Irvine machine learning repository. The dataset is at http://archive.ics.uci.edu/ ml/datasets/Heart+Disease. There are several versions. You should look at the processed Cleveland data, which is in the file "processed.cleveland.data.txt".

Solution: I used the R random forest package. This uses a bagging strategy. This package makes it quite simple to fit a random forest, as you can see. In this dataset, variable 14 (V14) takes the value 0, 1, 2, 3, or 4 depending on the severity of the narrowing of the arteries. Other variables are physiological and physical measurements pertaining to the patient (read the details on the website). I tried to predict all five levels of variable 14, using the random forest as a multivariate classifier. This works rather poorly, as the out-of-bag class-confusion matrix below shows. The total out-of-bag error rate was 45%.

<table>
<tr><td></td><td></td><td colspan="5" align="center">Predict</td><td></td></tr>
<tr><td></td><td></td><td>0</td><td>1</td><td>2</td><td>3</td><td>4</td><td>Class error</td></tr>
<tr><td rowspan="5">True</td><td>0</td><td>151</td><td>7</td><td>2</td><td>3</td><td>1</td><td>7.9%</td></tr>
<tr><td>1</td><td>32</td><td>5</td><td>9</td><td>9</td><td>0</td><td>91%</td></tr>
<tr><td>2</td><td>10</td><td>9</td><td>7</td><td>9</td><td>1</td><td>81%</td></tr>
<tr><td>3</td><td>6</td><td>13</td><td>9</td><td>5</td><td>2</td><td>86%</td></tr>
<tr><td>4</td><td>2</td><td>3</td><td>2</td><td>6</td><td>0</td><td>100%</td></tr>
</table>

This is the example of a class-confusion matrix from Table 1.1. Fairly clearly, one can predict narrowing or no narrowing from the features, but not the degree of narrowing (at least, not with a random forest). So it is natural to quantize variable 14 to two levels, 0 (meaning no narrowing) and 1 (meaning any narrowing, so the original value could have been 1, 2, or 3). I then built a random forest to predict this quantized variable from the other variables. The total out-of-bag error rate was 19%, and I obtained the following out-of-bag class-confusion matrix

<table>
<tr><td></td><td></td><td colspan="2" align="center">Predict</td><td></td></tr>
<tr><td></td><td></td><td>0</td><td>1</td><td>Class error</td></tr>
<tr><td rowspan="2">True</td><td>0</td><td>138</td><td>26</td><td>16%</td></tr>
<tr><td>1</td><td>31</td><td>108</td><td>22%</td></tr>
</table>

Notice that the false positive rate (16%, from 26/164) is rather better than the false negative rate (22%). You might wonder whether it is better to train on and predict $0, \ldots, 4$, then quantize the predicted value. If you do this, you will find you get a false positive rate of 7.9%, but a false negative rate that is much higher (36%, from 50/139). In this application, a false negative is likely more of a problem than a false positive, so the trade-off is unattractive.

> **Remember This:** *Random forests are straightforward to build and very effective. They can predict any kind of label. Good software implementations are easily available.*

2.3 You Should

2.3.1 Remember These Terms

2.3.2 Remember These Facts

2.3.3 Use These Procedures

2.3.4 Be Able to

- build an SVM using your preferred software package, and produce a cross-validated estimate of its error rate or its accuracy;
- write code to train an SVM using stochastic gradient descent, and produce a cross-validated estimate of its error rate or its accuracy;
- and build a decision forest using your preferred software package, and produce a cross-validated estimate of its error rate or its accuracy.

Programming Exercises

2.1. The UC Irvine machine learning data repository hosts a collection of data on breast cancer diagnostics, donated by Olvi Mangasarian, Nick Street, and William H. Wolberg. You can find this data at http://archive.ics.uci.edu/ml/datasets/Breast+Cancer+Wisconsin+(Diagnostic). For each record, there is an id number, 10 continuous variables, and a class (benign or malignant). There are 569 examples. Separate this dataset randomly into 100 validation, 100 test, and 369 training examples.

Write a program to train a support vector machine on this data using stochastic gradient descent. You should not use a package to train the classifier (you don't really need one), but your own code. You should ignore the id number, and use the continuous variables as a feature vector. You should scale these variables so each has unit variance. You should search for an appropriate value of the regularization constant, trying at least the values $\lambda = [1e-3, 1e-2, 1e-1, 1]$. Use the validation set for this search.

You should use at least 50 epochs of at least 100 steps each. In each epoch, you should separate out 50 training examples at random for evaluation. You should compute the accuracy of the current classifier on the set held out for the epoch every 10 steps. You should produce:

(a) A plot of the accuracy every 10 steps, for each value of the regularization constant.

(b) Your estimate of the best value of the regularization constant, together with a brief description of why you believe that is a good value.

(c) Your estimate of the accuracy of the best classifier on held-out data

2.2. The UC Irvine machine learning data repository hosts a collection of data on adult income, donated by Ronny Kohavi and Barry Becker. You can find this data at https://archive.ics.uci.edu/ml/datasets/Adult. For each record, there is a set of continuous attributes, and a class $\geq$50K or $<$50K. There are 48,842 examples. You should use only the continuous attributes (see the description on the web page) and drop examples where there are missing values of the continuous attributes. Separate the resulting dataset randomly into 10% validation, 10% test, and 80% training examples.

Write a program to train a support vector machine on this data using stochastic gradient descent. You should not use a package to train the classifier (you don't really need one), but your own code. You should ignore the id number, and use the continuous variables as a feature vector. You should scale these variables so that each has unit variance. You should search for an appropriate value of the regularization constant, trying at least the values $\lambda = [1e-3, 1e-2, 1e-1, 1]$. Use the validation set for this search

You should use at least 50 epochs of at least 300 steps each. In each epoch, you should separate out 50 training examples at random for evaluation. You should compute the accuracy of the current classifier on the set held out for the epoch every 30 steps. You should produce:

(a) A plot of the accuracy every 30 steps, for each value of the regularization constant.

(b) Your estimate of the best value of the regularization constant, together with a brief description of why you believe that is a good value.

(c) Your estimate of the accuracy of the best classifier on held-out data

2.3. The UC Irvine machine learning data repository hosts a collection of data on the whether p53 expression is active or inactive. You can find out what this means, and more information about the dataset, by reading: Danziger, S.A., Baronio, R., Ho, L., Hall, L., Salmon, K., Hatfield, G.W., Kaiser, P., and Lathrop, R.H. "Predicting Positive p53 Cancer Rescue Regions Using Most Informative Positive (MIP) Active Learning," *PLOS Computational Biology*, 5(9), 2009; Danziger, S.A., Zeng, J., Wang, Y., Brachmann, R.K. and Lathrop, R.H. "Choosing where to look next in a mutation sequence space: Active Learning of informative p53 cancer rescue mutants", *Bioinformatics*, 23(13), 104–114, 2007; and Danziger, S.A., Swamidass, S.J., Zeng, J., Dearth, L.R., Lu, Q., Chen, J.H., Cheng, J., Hoang, V.P., Saigo, H., Luo, R., Baldi, P., Brachmann, R.K. and Lathrop, R.H. "Functional census of mutation sequence spaces: the example of p53 cancer rescue mutants," *IEEE/ACM transactions on computational biology and bioinformatics*, 3, 114–125, 2006.

You can find this data at https://archive.ics.uci.edu/ml/datasets/p53+Mutants. There are a total of 16,772 instances, with 5409 attributes per instance. Attribute 5409 is the class attribute, which is either active or inactive. There are several versions of this dataset. You should use the version K8.data.

(a) Train an SVM to classify this data, using stochastic gradient descent. You will need to drop data items with missing values. You should estimate a regularization constant using cross-validation, trying at least three values. Your training method should touch at least 50% of the training set data. You should produce an estimate of the accuracy of this classifier on held-out data consisting of 10% of the dataset, chosen at random.

(b) Now train a naive Bayes classifier to classify this data. You should produce an estimate of the accuracy of this classifier on held-out data consisting of 10% of the dataset, chosen at random.

(c) Compare your classifiers. Which one is better? why?

2.4. The UC Irvine machine learning data repository hosts a collection of data on whether a mushroom is edible, donated by Jeff Schlimmer and to be found at http://archive.ics.uci.edu/ml/datasets/Mushroom. This data has a set of categorical attributes of the mushroom, together with two labels (poisonous or edible). Use the R random forest package (as in the example in the chapter) to build a random forest to classify a mushroom as edible or poisonous based on its attributes.

  (a) Produce a class-confusion matrix for this problem. If you eat a mushroom based on your classifier's prediction it is edible, what is the probability of being poisoned?

MNIST Exercises

The following exercises are elaborate, but rewarding. The MNIST dataset is a dataset of 60,000 training and 10,000 test examples of handwritten digits, originally constructed by Yann Lecun, Corinna Cortes, and Christopher J.C. Burges. It is very widely used to check simple methods. There are 10 classes in total ("0" to "9"). This dataset has been extensively studied, and there is a history of methods and feature constructions at https://en.wikipedia.org/wiki/MNIST_database and at http://yann.lecun.com/exdb/mnist/. You should notice that the best methods perform extremely well. The original dataset is at http://yann.lecun.com/exdb/mnist/. It is stored in an unusual format, described in detail on that website. Writing your own reader is pretty simple, but web search yields readers for standard packages. There is reader code in matlab available (at least) at http://ufldl.stanford.edu/wiki/index.php/Using_the_MNIST_Dataset. There is reader code for R available (at least) at https://stackoverflow.com/questions/21521571/how-to-read-mnist-database-in-r.

The dataset consists of 28×28 images. These were originally binary images, but appear to be grey level images as a result of some anti-aliasing. I will ignore mid grey pixels (there aren't many of them) and call dark pixels "ink pixels," and light pixels "paper pixels." The digit has been centered in the image by centering the center of gravity of the image pixels. Here are some options for re-centering the digits that I will refer to in the exercises.

- **Untouched:** do not re-center the digits, but use the images as is.
- **Bounding box:** construct a $b \times b$ bounding box so that the horizontal (resp. vertical) range of ink pixels is centered in the box.
- **Stretched bounding box:** construct an $b \times b$ bounding box so that the horizontal (resp. vertical) range of ink pixels runs the full horizontal (resp. vertical) range of the box. Obtaining this representation will involve rescaling image pixels: you find the horizontal and vertical ink range, cut that out of the original image, then resize the result to $b \times b$.

Once the image has been re-centered, you can compute features. For this exercise, we will use raw pixels as features.

  2.5. Investigate classifying MNIST using naive Bayes. Use the procedures of Sect. 1.3.1 to compare four cases on raw pixel image features. These cases are obtained by choosing either normal model or binomial model for every feature, and untouched images or stretched bounding box images.
  (a) Which is the best case?
  (b) How accurate is the best case? (remember, the answer to this is *not* obtained by taking the best accuracy from the previous subexercise— check Sect. 1.3.1 if you're vague on this point).
  2.6. Investigate classifying MNIST using nearest neighbors. You will use approximate nearest neighbors. Obtain the FLANN package for approximate nearest

neighbors from http://www.cs.ubc.ca/~mariusm/index.php/FLANN/FLANN. To use this package, you should consider first using a function that builds an index for the training dataset (`flann_build_index()`, or variants), then querying with your test points (`flann_find_nearest_neighbors_index()`, or variants). The alternative (`flann_find_nearest_neighbors()`, etc.) builds the index then throws it away, which can be inefficient if you don't use it correctly.

    (a) Compare untouched raw pixels with bounding box raw pixels and with stretched bounding box raw pixels. Which works better? Why? Is there a difference in query times?

    (b) Does rescaling each feature (i.e., each pixel value) so that it has unit variance improve either classifier from the previous subexercise?

2.7. Investigate classifying MNIST using an SVM. Compare the following cases: untouched raw pixels and stretched bounding box raw pixels. Which works best? Why?

2.8. Investigate classifying MNIST using a decision forest. Using the same parameters for your forest construction (i.e., same depth of tree; same number of trees; etc.), compare the following cases: untouched raw pixels and stretched bounding box raw pixels. Which works best? Why?

2.9. If you've done all four previous exercises, you're likely tired of MNIST, but very well informed. Compare your methods to the table of methods at http://yann.lecun.com/exdb/mnist/. What improvements could you make?

CHAPTER 3

A Little Learning Theory

The key thing to know about a classifier is how well it will work on future test data. There are two cases to look at: how error on held-out training data predicts test error, and how training error predicts test error. Error on held-out training data is a very good predictor of test error. It's worth knowing why this should be true, and Sect. 3.1 deals with that. Our training procedures assume that a classifier that achieves good training error is going to behave well on test—we need some reason to be confident that this is the case. It is possible to bound test error from training error. The bounds are all far too loose to have any practical significance, but their presence is reassuring.

A classifier with a small gap between training error and test error is said to generalize well. It is possible to bound test error from training error alone, but the bounds are loose. The problem is that the training error for a given classifier is a biased estimate of test error—the classifier was *chosen* to make the training error small. If the classifier is chosen from a small finite family, it is relatively straightforward to bound the test error using the training error. There are two steps. First, we bound the probability that there is a large gap between test error and observed error for a single classifier. Then, we bound the probability that any of the classifiers in our family has a large gap. Equivalently, we can assert that with high probability, the test error will be smaller than some value. Section 3.2 sketches (without proof) the reasoning.

This reasoning doesn't deal with the most important case, which is a classifier chosen from an infinite family. Doing so requires some technical cleverness. One should think of a classifier as a map from a finite dataset to a string of labels (one for each data item). We can now ask, for a given dataset, how many strings a family of classifiers can produce. The key to reasoning about the gap is not the number of classifiers in the family, but the number of strings it can produce on a fixed dataset. It turns out that many families produce far fewer strings than you would expect, and this leads to a bound on the held-out error. Section 3.3 sketches (without proof) the reasoning.

3.1 Held-Out Loss Predicts Test Loss

It is helpful to generalize to a **predictor**—a function that accepts features and reports something. If the report is a label, then the predictor is a classifier. We will see predictors that produce other reports later. There are many kinds of predictor—linear functions, trees, and so on. We now take the view that the kind of predictor you use is just a matter of convenience (what package you have available, what math you feel like doing, etc.). Once you know what kind of predictor you will use, you must choose the parameters of that predictor, which you do by minimizing the **training loss** (the cost function used to evaluate errors on training data).

© Springer Nature Switzerland AG 2019
D. Forsyth, *Applied Machine Learning*,
https://doi.org/10.1007/978-3-030-18114-7_3

We will work with losses that are on the average over the training data of a **pointwise loss**—a function ℓ that accepts three arguments: the true y-value that the predictor should have produced, a feature vector $\mathbf{x}$, and a prediction $F(\mathbf{x})$. This average is an estimate of the expected value of that pointwise loss over all data. Currently, we have seen only one pointwise loss, for the linear SVM

$$\ell_h(y, \mathbf{x}, F) = \max(0, 1 - yF(\mathbf{x})).$$

We will see a number of others, with more examples on page 285. The material of this section applies to all pointwise losses.

Here is the simplest setup. Assume we have used training data to construct some predictor F. We have a pointwise loss $\ell(y, \mathbf{x}, F)$. We have N pairs $(\mathbf{x}_i, y_i)$ of held-out data items. We assume that none of these were used in constructing the predictor, and we assume that these pairs are IID samples from the distribution of test data, which we write $P(X, Y)$. We now evaluate the held-out loss

$$\frac{1}{N} \sum_i \ell(y_i, \mathbf{x}_i, F).$$

Under almost all circumstances, this is a rather good estimate of the true expected loss on all possible test data, which is

$$\mathbb{E}_{P(X,Y)}[\ell].$$

In particular, quite simple methods yield bounds on how often the estimate will be very different from the true value. In turn, this means that the held-out loss is a good estimate of the test loss.

3.1.1 Sample Means and Expectations

Write L for the random variable whose value is obtained by drawing a sample $(\mathbf{x}_i, y_i)$ from $P(X, Y)$ and then evaluating $\ell(y_i, \mathbf{x}_i, F)$. We will study the relationship between the expected value, $\mathbb{E}[L]$, and the approximation obtained by drawing N IID samples from $P(L)$ and computing $\frac{1}{N} \sum_i L_i$. The value of the approximation is a random variable, because we would get a different value if we drew a different set of samples. We write $L^{(N)}$ for this random variable. Now we are interested in the mean of that random variable

$$\mathbb{E}\left[L^{(N)}\right]$$

and its variance

$$\operatorname{var}\left[L^{(N)}\right] = \mathbb{E}\left[(L^{(N)})^2\right] - \mathbb{E}\left[L^{(N)}\right]^2.$$

We will assume that the variance is finite.

The Mean: Because expectations are linear, we have that

$$\begin{aligned}
\mathbb{E}\left[L^{(N)}\right] &= \frac{1}{N}\mathbb{E}\left[L^{(1)} + \cdots + L^{(1)}\right] \\
&\qquad \text{(where there are N copies of $L^{(1)}$)} \\
&= \mathbb{E}\left[L^{(1)}\right] \\
&= \mathbb{E}[L]
\end{aligned}$$

The Variance: Write L_i for the i'th sample used in computing $L^{(N)}$. We have $L^{(N)} = \frac{1}{N} \sum_i L_i$, so

$$\mathbb{E}\left[(L^{(N)})^2\right] = \frac{1}{N^2} \mathbb{E}\left[\sum_i L_i^2 + \sum_i \sum_{j \neq i} L_i L_j\right]$$

but $\mathbb{E}\left[L_i^2\right] = \mathbb{E}\left[L^2\right]$. Furthermore, L_i and L_j are independent, and $\mathbb{E}[L_i] = \mathbb{E}[L]$ so we have

$$\mathbb{E}\left[(L^{(N)})^2\right] = \frac{\left(N\mathbb{E}\left[L^2\right] + N(N-1)\mathbb{E}[L]^2\right)}{N^2} = \frac{\left(\mathbb{E}\left[L^2\right] - \mathbb{E}[L]^2\right)}{N} + \mathbb{E}\left[L^2\right]$$

In turn,

$$\mathbb{E}\left[(L^{(N)})^2\right] - \mathbb{E}\left[L^{(N)}\right]^2 = \frac{\left(\mathbb{E}\left[L^2\right] - \mathbb{E}[L]^2\right)}{N}.$$

There is an easier way to remember this. Write $\mathsf{var}[L] = \mathbb{E}\left[L^2\right] - \mathbb{E}[L]^2$ for the variance of the random variable L. Then we have

$$\mathsf{var}\left[L^{(N)}\right] = \frac{\mathsf{var}[L]}{N}.$$

This should be familiar. It is the standard error of the estimate of the mean. If it isn't, it's worth remembering. The more samples you use in computing the estimate of the mean, the better the estimate.

Useful Fact: 3.1 *Mean and Variance of an Expectation Estimated from Samples*

Write X for some random variable. Write $X^{(N)}$ for the mean of N IID samples of that random variable. We have that:

$$\mathbb{E}\left[X^{(N)}\right] = \mathbb{E}[X]$$
$$\mathsf{var}\left[X^{(N)}\right] = \frac{\mathsf{var}[X]}{N}$$

Already, we have two useful facts. First, the held-out loss is the value of a random variable whose expected value is the test loss. Second, the variance of this random variable could be quite small, if we compute the held-out loss on enough held-out examples. If a random variable has small variance, you should expect to see values that are close to the mean with high probability (otherwise the variance would be bigger). The next step is to determine how often the held-out loss will be very different from the test loss. This works for held-out error, too.

3.1.2 Using Chebyshev's Inequality

Chebyshev's inequality links the observed value, the expected value, and the variance of a random variable. You should have seen this before (if you haven't, look it up, for example, in "Probability and Statistics for Computer Science"); it appears in a box below.

Useful Fact: 3.2 *Definition: Chebyshev's Inequality*

For a random variable X with finite variance, **Chebyshev's inequality** states

$$P\left(\{|X - \mathbb{E}[X]| \geq a\}\right) \leq \frac{\mathsf{var}[X]}{a^2}.$$

Combining Chebyshev's inequality and the remarks above about sample mean, we have the result in the box below.

Useful Fact: 3.3 *Held-Out Error Predicts Test Error, from Chebyshev*

There is some constant C so that

$$P\left(\left\{|L^{(N)} - \mathbb{E}[L]| \geq a\right\}\right) \leq \frac{C}{a^2 N}.$$

3.1.3 A Generalization Bound

Most generalization bounds give a value of $\mathbb{E}[L]$ that will not be exceeded with probability $1 - \delta$. The usual form of bound states that, with probability $1 - \delta$,

$$\mathbb{E}[L] \leq L^{(N)} + g(\delta, N, \ldots).$$

Then one studies how g grows as δ shrinks or N grows. It is straightforward to rearrange the Chebyshev inequality into this form. We have

$$P\left(\left\{|L^{(N)} - \mathbb{E}[L]| \geq a\right\}\right) \leq \frac{C}{a^2 N} = \delta.$$

Now solve for a in terms of δ. This yields that, with probability $1 - \delta$,

$$\mathbb{E}[L] \leq L^{(N)} + \sqrt{\frac{C\left(\frac{1}{\delta}\right)}{N}}.$$

Notice this bound is likely rather weak, because it makes the worst case assumption that all of the probability occurs when $\mathbb{E}[L]$ is larger than $L^{(N)} + w$. It does so because we don't know where the probability is, and so have to assume it is in the worst place. For almost every practical case, one does not know

$$C = \mathbb{E}_{P(X,Y)}\left[\ell^2\right] - \left(\mathbb{E}_{P(X,Y)}[\ell]\right)^2$$

and estimating C's value is not usually helpful. I have put this bound in a box for convenience.

Remember This: *There is some constant C depending on the loss and the data distribution so that with probability $1 - \delta$,*

$$\mathbb{E}[L] \leq L^{(N)} + \sqrt{\frac{C\left(\frac{1}{\delta}\right)}{N}}.$$

This result tells us roughly what we'd expect. The held-out error is a good guide to the test error. Evaluating the held-out error on more data leads to a better estimate of the test error. The bound is very general, and applies to almost any form of pointwise loss. The "almost" here is because we must assume that the loss leads to an L with finite variance, but I have never encountered a loss that does not. This means the bound applies to both regression and classification problems, and you can use any kind of classification loss.

There are two problems. First, the bound assumes we have held-out error, but what we'd really like to do is think about test error in terms of training error. Second, the bound is quite weak, and we will do better in the next section. But our better bounds will apply only to a limited range of classification losses.

3.2   Test and Training Error for a Classifier from a Finite Family

The test error of many classifiers can be bounded using the training error. I have never encountered a practical application for such bounds, but they are reassuring. They suggest that our original approach (choose a classifier with small training error) is legitimate, and they cast some light on the way that families of classifiers behave. However, these bounds are harder to obtain than bounds based on test error, because the predictor you selected minimizes the training error (at least approximately). This means that you should expect the training error to be an estimate of the test error that is too low—the classifier was chosen to achieve low training error. Equivalently, the training error is a biased estimate of the test error.

The predictor you selected came from a family of predictors, and the bias depends very strongly on the family you used. One example of a family of predictors is all linear SVMs. If you're using a linear SVM, you chose a particular set of parameter values that yields the SVM you fitted over all the other possible parameter values. As another example, if you're using a decision tree, you chose a tree from the family of all decision trees that have the depth, etc. limits that you imposed. Rather loosely, if you choose a predictor from a "big" family, then you should expect the bias is large. You are more likely to find a predictor with low training error and high test error when you search a big family.

The problem is to distinguish in a sensible way between "big" and "small" families of predictors. A natural first step is to consider only finite collections of predictors—for example, you might choose one of 10 fixed linear SVMs. Although this isn't a realistic model of learning in practice, it sets us up to deal with more difficult cases. Some rather clever tricks will then allow us to reason about continuous families of predictors, but these families need to have important and delicate properties exposed by the analysis.

From now on, we will consider only a 0-1 loss, because this will allow us to obtain much tighter bounds. We will first construct a bound on the loss of a given predictor, then consider what happens when that predictor is chosen from a finite set.

3.2.1 Hoeffding's Inequality

Chebyshev's inequality is general, and holds for any random variable with finite variance. If we assume stronger properties of the random variable, it is possible to prove very much tighter bounds.

Useful Fact: 3.4   *Definition: Hoeffding's Inequality*

Assume that X is a Bernoulli random variable that takes the value 1 with probability θ, and otherwise the value 0. Write $X^{(N)}$ for the random variable obtained by averaging N IID samples of X. Then **Hoeffding's inequality** states

$$P\left(|\theta - X^{(N)}| \geq \epsilon\right) \leq 2e^{-2N\epsilon^2}$$

The proof is more elaborate than is really tolerable here, though not hard. Now assume that our loss is 0-1. This is fairly common for classifiers, where you lose 1 if the answer is right, and 0 otherwise. The loss at any particular example is then a Bernoulli random variable, with probability $\mathbb{E}[L]$ of taking the value 1. This means we can use Hoeffding's inequality to tighten the bound of page 53. Doing so yields

---

Remember This: *With probability $1 - \delta$,*

$$\mathbb{E}[L] \leq L^{(N)} + \sqrt{\frac{\log\left(\frac{2}{\delta}\right)}{2N}}.$$

This bound is tighter than the previous bound, because $\log\left(\frac{1}{\delta}\right) \leq \left(\frac{1}{\delta}\right)$. The difference becomes very important when δ is small, which is the interesting case.

3.2.2 Test from Training for a Finite Family of Predictors

Assume we choose a predictor from a *finite* set $\mathcal{P}$ of M different predictors. We will consider only a 0-1 loss. Write the expected loss of using predictor F as $E_F = \mathbb{E}_{P(X,Y)}[l(y, \mathbf{x}, F)]$. One useful way of thinking of this loss is that it is the probability that an example will be mislabelled. Write $L_F^{(N)}$ for the estimate of this loss obtained from the training set for predictor F. From the Hoeffding inequality, we have

$$P\left(\left\{|E_F - L_F^{(N)}| \geq \epsilon\right\}\right) \leq 2e^{-2N\epsilon^2}$$

for *any* predictor F.

What we'd like to know is the generalization error for the predictor that we pick. This will be difficult to get. Instead, we will consider the worst generalization error in all the predictors—our predictor must be at least as good as this. Now consider the event $\mathcal{G}$ that at least one predictor has generalization error greater than ϵ. We have

$$\begin{aligned} \mathcal{G} \quad = \quad & \left\{|E_{F_1} - L_{F_1}^{(N)}| \geq \epsilon\right\} \cup \\ & \left\{|E_{F_2} - L_{F_2}^{(N)}| \geq \epsilon\right\} \cup \\ & \cdots \\ & \left\{|E_{F_M} - L_{F_M}^{(N)}| \geq \epsilon\right\}. \end{aligned}$$

Recall that, for two events $\mathcal{A}$ and $\mathcal{B}$, we have $P(\mathcal{A} \cup \mathcal{B}) \leq P(\mathcal{A}) + P(\mathcal{B})$ with equality only if the events are disjoint (the events that make up $\mathcal{G}$ may not be). But we have an upper bound on $P(\mathcal{G})$. In particular,

$$\begin{aligned} P(\mathcal{G}) \quad \leq \quad & P\left(\left\{|E_{F_1} - L_{F_1}^{(N)}| \geq \epsilon\right\}\right) + \\ & P\left(\left\{|E_{F_2} - L_{F_2}^{(N)}| \geq \epsilon\right\}\right) + \\ & \cdots \\ & P\left(\left\{|E_{F_M} - L_{F_M}^{(N)}| \geq \epsilon\right\}\right) \\ \leq \quad & 2Me^{-2N\epsilon^2} \end{aligned}$$

by Hoeffding's inequality.

This is sometimes known as a **union bound**.

Now notice that $P(\mathcal{G})$ is the probability that at least one predictor F in $\mathcal{P}$ has $|E_F - L_F^{(N)}| \geq \epsilon$. Equivalently, it is the probability that the largest value of $|E_F - L_F^{(N)}|$ is greater than or equal to ϵ. So we have

$$\begin{aligned} P(\mathcal{G}) \quad = \quad & P\left(\{\text{at least one predictor has generalization error} > \epsilon\}\right) \\ = \quad & P\left(\left\{\sup_{F \in \mathcal{P}} \left[|E_F - L_F^{(N)}|\right] \geq \epsilon\right\}\right) \\ \leq \quad & 2Me^{-2N\epsilon^2}. \end{aligned}$$

It is natural to rearrange this, yielding the bound in the box below. You should notice this bound does not depend on the *way* that the predictor was chosen.

Remember This: *Assume a 0-1 loss. Choose a predictor F from M different predictors, write the expected loss of using predictor F as*

$$E_F = \mathbb{E}_{P(X,Y)}[\ell(y, \mathbf{x}, F)],$$

and write $L_F^{(N)}$ for the estimate of this loss obtained from the training set for predictor F. With probability $1 - \delta$,

$$E_F \leq L_F^{(N)} + \sqrt{\frac{\log M + \log\left(\frac{2}{\delta}\right)}{2N}}$$

for any *predictor F chosen from a set of M predictors.*

3.2.3 Number of Examples Required

Generally, we expect it is easier to find a predictor with good training error but bad test error (a bad predictor) when we search a large family of predictors, and the M in the bound reflects this. Similarly, if there are relatively few examples, it should be easy to find such a predictor as well; and if there are many examples, it should be hard to find one, so there is an N in the bound, too. We can reframe the bound to ask how many examples we need to ensure that the probability of finding a bad predictor is small.

A bad predictor F is one where $E_F - L_F^{(N)} > \epsilon$ (we're not anxious about a future loss that is better than the observed loss). The probability that at least one predictor in our collection of M predictors is bad is bounded above by $Me^{-2N\epsilon^2}$. Now assume we wish to bound the failure probability above by δ. We can bound the number of examples we need to use to achieve this, by rearranging expressions, yielding the bound in the box.

Remember This: *A predictor F is bad if $E_F - L_F^{(N)} > \epsilon$. Write P_{bad} for the probability that at least one predictor in the collection (and so perhaps the predictor we select) is bad. To ensure that $P_{bad} \leq \delta$, it is enough to use*

$$N \geq \frac{1}{2\epsilon^2}\left(\log M + \log\left(\frac{1}{\delta}\right)\right)$$

examples.

3.3 An Infinite Collection of Predictors

Mostly, we're not that interested in choosing a predictor from a small discrete set. All the predictors we have looked at in previous chapters come from infinite families. The bounds in the previous section are not very helpful in this case. With some mathematical deviousness, we can obtain bounds for infinite sets of predictors, too.

We bounded the generalization error for a finite family of predictors by bounding the worst generalization error in that family. This was straightforward, but it meant the bound had a term in the number of elements in the family. If this is infinite, we have a problem. There is an important trick we can use here. It turns out that the issue to look at is not the number of predictors in the family. Instead, we think about predictors as functions that produce binary strings (Sect. 3.3.1). This is because, at each example, the predictor either gets the example right (0) or wrong (1). Order the examples in some way; then you can think of the predictor as producing an N element binary string of 0's and 1's, where there is one bit for each of the N examples in the training set. Now if you were to use a different predictor in the family, you might get a different string. What turns out to be important is the number s of different possible strings that can appear when you try every predictor in the family. This number must be finite—there are N examples, and each is either right or wrong—but might still be as big as 2^N.

There are two crucial facts that allow us to bound generalization error. First, and surprisingly, there are families of predictors where s is small, and grows slowly with N (Sect. 3.3.1). This means that, rather than worrying about infinite collections of predictors, we can attend to small finite sets of strings. Second, it turns out that we can bound generalization error using the difference between errors for some predictor given two different training datasets (Sect. 3.3.2). Because there are relatively few strings of errors, it becomes relatively straightforward to reason about this difference. These two facts yield a crucial bound on generalization error (Sect. 3.3.3).

Most of the useful facts in this section are relatively straightforward to prove (no heavy machinery is required), but the proofs take some time and trouble. Relatively few readers will really need them, and I have omitted them.

3.3.1 Predictors and Binary Functions

A predictor is a function that takes an independent variable and produces a prediction. Because we are using a 0-1 loss, choosing a predictor F is equivalent to a choice of a binary function (i.e., a function that produces either 0 or 1). The binary function is obtained by making a prediction using F, then scoring it with the loss function. This means that the family of predictors yields a family of binary functions.

We will study binary functions briefly. Assume we have some binary function b in a family of binary functions $\mathcal{B}$. Take some sample of N points $\mathbf{x}_i$. Our function b will produce a binary string, with one bit for each point. We consider the set $\mathcal{B}_N$ which consists of all the different binary strings that are produced by functions in

$\mathcal{B}$ for our chosen set of sample points. Write $\#(\mathcal{B}_N)$ for the number of different elements in this set. We could have $\#(\mathcal{B}_N) = 2^N$, because there are 2^N strings in $\mathcal{B}_N$.

In many cases, $\#(\mathcal{B}_N)$ is much smaller than 2^N. This is a property of the family of binary functions, rather than of (say) an odd choice of data points. The thing to study is the **growth function**

$$s(\mathcal{B}, N) = \sup_{\text{sets of } N \text{ points}} \#(\mathcal{B}_N).$$

This is sometimes called the **shattering number** of $\mathcal{B}$. For some interesting cases, the growth function can be recovered with elementary methods.

Worked Example 3.1 $s(\mathcal{B}, 3)$ *for a Simple Linear Classifier on the Line*

Assume that we have a 1D independent variable x, and our family of classifiers is $\text{sign}(ax + b)$ for parameters a, b. These classifiers are equivalent to a family of binary functions $\mathcal{B}$. What is $s(\mathcal{B}, 3)$?

Solution: The predictor produces a sign at each point. Now think about the sample. It should be clear that the largest set of strings isn't going to occur unless the three points are distinct. The predictor produces a string of signs (one at each point), and the binary function is obtained by testing if the label is equal to, or different from, the sign the predictor produces. This means that the number of binary strings is the same as the number of distinct strings of signs. In particular, the actual values of the labels don't matter. Now order the three points along the line so $x_1 < x_2 < x_3$. Notice there is only one sign change at $s = -b/a$; we can have $s < x_1$, $x_1 < s < x_2$, $x_2 < s < x_3$, $x_3 < s$ (we will deal with s lying on a point later). All this means the predictor can produce only the following six sign patterns (at most):

$$- - -, - - +, - + +, + + +, + + -, + - -$$

Now imagine that s lies on a data point; the rule is to choose a sign at random. It is straightforward to check that this doesn't increase the set of sign patterns (and you should). So $s(\mathcal{B}, 3) = 6 < 2^3$.

Worked Example 3.2 $s(\mathcal{B}, 4)$ *for a Simple Linear Classifier on the Plane*

Assume that we have a 2D independent variable $\mathbf{x}$, and our family of classifiers is $\mathrm{sign}(\mathbf{a}^T\mathbf{x} + b)$ for parameters $\mathbf{a}$, b. These classifiers are equivalent to a family of binary functions $\mathcal{B}$. What is $s(\mathcal{B}, 4)$?

Solution: The predictor produces a sign at each point. Now think about the sample. The predictor produces a string of signs (one at each point), and the binary function is obtained by testing if the label is equal to, or different from, the sign the predictor produces. This means that the number of binary strings is the same as the number of distinct strings of signs. It should be clear that the largest set of strings isn't going to occur unless the points points are distinct. If they're collinear, we know how to count (Example 3.1), and obtain 10. You can check the case where three points are collinear easily, to count 12. There are two remaining cases. Either $\mathbf{x}_4$ is inside the convex hull of the other three, or it is outside (Fig. 3.1). If $\mathbf{x}_4$ is inside, then you cannot see $+ + + -$ or $- - - +$. If $\mathbf{x}_4$ is outside, a linear predictor cannot predict $+$ for $\mathbf{x}_1$ and $\mathbf{x}_3$, and $-$ for points $\mathbf{x}_2$ and $\mathbf{x}_4$. This means there are 14 strings of signs possible. So $s(\mathcal{B}, 4) = 14$.

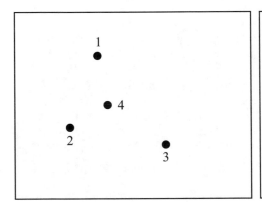 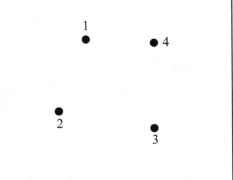

Figure 3.1:   On the **left**, four points on the plane. Because point $\mathbf{x}_4$ is inside the convex hull of $\mathbf{x}_1$, $\mathbf{x}_2$, and $\mathbf{x}_3$, a linear predictor cannot predict $+ + + -$ or $- - - +$. On the **right**, the other case: a linear predictor cannot predict $+$ for $\mathbf{x}_1$ and $\mathbf{x}_3$, and $-$ for points $\mathbf{x}_2$ and $\mathbf{x}_4$ (try drawing the decision boundary if you're uncertain)

The point of these examples is that an infinite family of predictors may yield a small family of binary functions. But $s(\mathcal{B}, N)$ can be quite hard to determine for arbitrary families of predictors. One strategy is to use what is known as the **VC dimension** of $\mathcal{P}$ (after the inventors, Vapnik and Chervonenkis).

Useful Fact: 3.5 *Definition: The VC Dimension*

The VC dimension of a class of binary functions $\mathcal{B}$ is

$$\mathsf{VC}\,(\mathcal{B}) = \sup\left\{N : s(\mathcal{B}, N) = 2^N\right\}$$

Worked Example 3.3 *The VC Dimension of the Binary Functions Produced by a Linear Classifier on the Line*

Assume that we have a 1D independent variable x, and our family of classifiers is $\mathrm{sign}(ax + b)$ for parameters a, b. These classifiers are equivalent to a family of binary functions $\mathcal{B}$. What is $\mathsf{VC}\,(\mathcal{B})$?

Solution: From Example 3.1 this number is less than three. It's easy to show that $s(\mathcal{B}, 2) = 4$, so $\mathsf{VC}\,(\mathcal{B}) = 2$.

Worked Example 3.4 *The VC Dimension of the Binary Functions Produced by a Linear Classifier on the Plane*

Assume that we have a 2D independent variable x, and our family of classifiers is $\mathrm{sign}(\mathbf{a}^T \mathbf{x} + b)$ for parameters a, b. These classifiers are equivalent to a family of binary functions $\mathcal{B}$. What is $\mathsf{VC}\,(\mathcal{B})$?

Solution: From Example 3.1 this number is less than four. It's easy to show that $s(\mathcal{B}, 3) = 8$, so $\mathsf{VC}\,(\mathcal{B}) = 3$.

Talking about the VC dimension of the binary functions produced by a family of predictors is a bit long-winded. Instead, we will refer to the VC dimension of the family of predictors. So the VC dimension of linear classifiers on the plane is 3, etc.

Remember This: *Write $\mathcal{P}$ for the family of linear classifiers on d-dimensional vectors, $\mathrm{sign}(\mathbf{a}^T \mathbf{x} + b)$. Then*

$$\mathsf{VC}\,(\mathcal{P}) = d + 1.$$

There are $d+1$ parameters in a linear classifier on d-dimensional vectors, and the VC dimension is $d+1$. Do not let this coincidence mislead you—you cannot obtain VC dimension by counting parameters. There are examples of one parameter families of predictors with infinite VC dimension. Instead, you should think of VC dimension as measuring some form of "wiggliness." For example, linear predictors aren't wiggly because if you prescribe a sign on a small set of points, you know the sign on many others. But if a family of predictors has high VC dimension, then you can take a large set of points at random, prescribe a sign at each point, and find a member of the family that takes that sign at each point.

Useful Fact: 3.6 *The Growth Number of a Family of Finite VC Dimension*

Assume $\mathsf{VC}(\mathcal{B}) = d$, which is finite. Then for all $N \geq d$, we have

$$s(\mathcal{B}, N) \leq \left(\frac{N}{d}\right)^d e^d$$

3.3.2 Symmetrization

A bad predictor F is one where $E_F - L_F^{(N)} > \epsilon$. For a finite set of predictors, we used Hoeffding's inequality to bound the probability a particular predictor was bad. We then argued that the probability that at least one predictor in our collection of M predictors is bad is bounded above by M times that bound. This won't work for an infinite set of predictors.

Assume we have a family of predictors which has finite VC dimension. Now draw a sample of N points. Associated with each predictor is a string of N binary variables (whether the predictor is right or wrong on each point). Even though there may be an infinite number of predictors, there is a finite number of distinct strings, and we can bound that number. We need a result that bounds the generalization error in terms of the behavior of these strings.

Now assume we have a second IID sample of N points, and compute the average loss over that second sample. Write $\tilde{L}_F^{(N)}$ for this new average loss. This second sample is purely an abstraction (we won't need a second training set) but it allows us to use an extremely powerful trick called symmetrization to get a bound. The result appears in two forms, in boxes, below.

Useful Fact: 3.7 *The Largest Variation of Sample Means Yields a Bound*

$$P\left(\left\{\sup_{F \in \mathcal{P}} \left[\,|\,L_F^{(N)} - L_F\,|\,\right] > \epsilon\right\}\right) \leq 2P\left(\left\{\sup_{F \in \mathcal{P}} \left[\,|\,L_F^{(N)} - \tilde{L}_F^{(N)}\,|\,\right] > \frac{\epsilon}{2}\right\}\right)$$

The proof of this result is omitted. It isn't particularly difficult, but it'll be easier to swallow with some sense of why the result is important. Notice that the right-hand side,

$$P\left(\left\{ \sup_{F \in \mathcal{P}} \left[| L_F^{(N)} - \tilde{L}_F^{(N)} | \right] > \frac{\epsilon}{2} \right\}\right)$$

is expressed entirely in terms of the values that predictors take on data. To see why this is important, remember the example of the family of linear predictors for 1D independent variables. For $N = 3$ data points, an infinite family of predictors could make only six distinct predictions. This means that the event

$$\left\{ \sup_{F \in \mathcal{P}} \left[| L_F^{(N)} - \tilde{L}_F^{(N)} | \right] > \frac{\epsilon}{2} \right\}$$

is quite easy to handle. Rather than worry about the supremum over an infinite family of predictors, we can attend to a supremum over only 36 predictions (which is six for $L_F^{(N)}$ and another six for $\tilde{L}_F^{(N)}$).

3.3.3 Bounding the Generalization Error

Useful Fact: 3.8 *Generalization Bound in Terms of VC Dimension*

Let $\mathcal{P}$ be a family of predictors with VC dimension d. With probability at least $1 - \epsilon$, we have

$$L \leq L^{(N)} + \sqrt{\frac{8}{N}\left(\log\left(\frac{4}{\epsilon}\right) + d\log\left(\frac{Ne}{d}\right)\right)}$$

Proving this fact is straightforward with the tools at hand. We start by proving

Quick statement: *Generalization Bound for an Infinite Family of Predictors*

Formal Proposition: *Let $\mathcal{P}$ be a family of predictors, and $t \geq \sqrt{\frac{2}{N}}$. We have*

$$P\left(\left\{ \sup_{F \in \mathcal{P}} \left[|\, L_F^{(N)} - L_F \,| \right] > \epsilon \right\}\right) \leq 4s(\mathcal{F}, 2N)e^{-N\epsilon^2/8}$$

Proof: Write b for a binary string obtained by computing the error for a predictor $p \in \mathcal{P}$ at $2N$ sample points, and b_i for it's i'th element. Write $\mathcal{B}$ for the set of all such strings. For a string b, write $L_b^{(N)} = \frac{1}{N}\sum_{i=1}^{N} b_i$ and $\tilde{L}_b^{(N)} = \frac{1}{N}\sum_{i=N+1}^{2N} b_i$. Now we have

$$P\left(\left\{ \sup_{F \in \mathcal{P}} |\, L_F^{(N)} - L_F \,| > \epsilon \right\}\right) \leq 2P\left(\left\{ \sup_{F \in \mathcal{P}} |\, L_F^{(N)} - \tilde{L}_F^{(N)} \,| > \epsilon/2 \right\}\right)$$

using the symmetrization idea

$$= 2P\left(\left\{ \max_{b \in \mathcal{B}} |\, L_b^{(N)} - \tilde{L}_b^{(N)} \,| > \epsilon/2 \right\}\right)$$

which is why symmetrization is useful

$$\leq 2s(\mathcal{B}, 2N)P\left(\left\{ |\, L_b^{(N)} - \tilde{L}_b^{(N)} \,| > \epsilon/2 \right\}\right)$$

union bound; $s(\mathcal{B}, 2N)$ is size of $\mathcal{B}$

$$\leq 4s(\mathcal{B}, 2N)e^{-N\epsilon^2/8}$$

Hoeffding

This yields the next step, where we bound the loss of the worst predictor in the family using VC dimension.

> **Quick statement:** *Generalization bound for family of predictors with finite VC dimension*
>
> **Formal Proposition:** *Let $\mathcal{P}$ be a family of predictors with VC dimension d. With probability at least $1 - \epsilon$, we have*
>
> $$\sup_{F \in \mathcal{P}} \mid L_F^{(N)} - L_F \mid \leq \sqrt{\frac{8}{N}\left(\log\left(\frac{4}{\epsilon}\right) + d\log\left(\frac{Ne}{d}\right)\right)}$$
>
> **Proof:** From above, we have
>
> $$P\left(\left\{\sup_{F \in \mathcal{P}} \mid L_F^{(N)} - L_F \mid > \epsilon\right\}\right) \leq 4s(\mathcal{B}, 2N)e^{-N\epsilon^2/8}$$
>
> so with probability $1 - \epsilon$,
>
> $$L_F \leq L^N + \sqrt{\frac{8}{N}\left(\log\left(\frac{4s(\mathcal{B}, N)}{\epsilon}\right)\right)}.$$
>
> But we have that $s(\mathcal{B}, N) \leq \left(\frac{N}{d}\right)^d e^d$, so
>
> $$\log\frac{4s(\mathcal{B}, N)}{\epsilon} \leq \log\frac{4}{\epsilon} + d\log\left(\frac{Ne}{d}\right).$$

The original result follows by simply rearranging terms.

3.4 You Should

3.4.1 Remember These Terms

3.4.2 Remember These Facts

3.4.3 Be Able to

- Explain why held-out loss is a good predictor of test loss.
- Remember the main content of Chebyshev's and Hoeffding's inequalities.
- Explain why training loss is a poor predictor of test loss.
- Explain roughly how one bounds test loss using training loss.

PART TWO

High Dimensional Data

High Dimensional Data

We have a dataset that is a collection of d-dimensional vectors. This chapter introduces the nasty tricks that such data can play. A dataset like this is hard to plot, though Sect. 4.1 suggests some tricks that are helpful. Most readers will already know the mean as a summary (it's an easy generalization of the 1D mean). The covariance matrix may be less familiar. This is a collection of all covariances between pairs of components. We use covariances, rather than correlations, because covariances can be represented in a matrix easily. High dimensional data has some nasty properties (it's usual to lump these under the name "the curse of dimension"). The data isn't where you think it is, and this can be a serious nuisance, making it difficult to fit complex probability models.

Natural transformations of the dataset lead to easy transformations of mean and the covariance matrix. This means we can construct a transformation that produces a new dataset, whose covariance matrix has desirable properties, from any dataset. We will exploit these methods aggressively in the next few chapters.

The main defence against the curse of dimension is to use extremely simple representations of the data. The most powerful of these is to think of a dataset as a collection of blobs of data. Each blob of data consists of points that are "reasonably close" to each other and "rather far" from other blobs. A blob can be modelled with a multivariate normal distribution. Our knowledge of what transformations do to a dataset's mean and covariance will reveal the main points about the multivariate normal distribution.

4.1 Summaries and Simple Plots

In this part, we assume that our data items are vectors. This means that we can add and subtract values and multiply values by a scalar without any distress.

For 1D data, mean and variance are a very helpful description of data that had a unimodal histogram. If there is more than one mode, one needs to be somewhat careful to interpret the mean and variance, because the mean doesn't summarize the modes particularly well, and the variance depends on how the modes are placed. In higher dimensions, the analogue of a unimodal histogram is a "blob"—a group of data points that clusters nicely together and should be understood together.

You might not believe that "blob" is a technical term, but it's quite widely used. This is because it is relatively easy to understand a single blob of data. There are good summary representations (mean and covariance, which I describe below). If a dataset forms multiple blobs, we can usually coerce it into a representation as a collection of blobs (using the methods of Chap. 8). But many datasets really are single blobs, and we concentrate on such data here. There are quite useful tricks for understanding blobs of low dimension by plotting them, which I describe in this part. To understand a high dimensional blob, we will need to think about the

© Springer Nature Switzerland AG 2019
D. Forsyth, *Applied Machine Learning*,
https://doi.org/10.1007/978-3-030-18114-7_4

coordinate transformations that places it into a particularly convenient form.

Notation: Our data items are vectors, and we write a vector as $\mathbf{x}$. The data items are d-dimensional, and there are N of them. The entire dataset is $\{\mathbf{x}\}$. When we need to refer to the i'th data item, we write $\mathbf{x}_i$. We write $\{\mathbf{x}_i\}$ for a new dataset made up of N items, where the i'th item is $\mathbf{x}_i$. If we need to refer to the j'th component of a vector $\mathbf{x}_i$, we will write $x_i^{(j)}$ (notice this isn't in bold, because it is a component not a vector, and the j is in parentheses because it isn't a power). Vectors are always column vectors.

4.1.1 The Mean

For one-dimensional data, we wrote

$$\text{mean}\,(\{x\}) = \frac{\sum_i x_i}{N}.$$

This expression is meaningful for vectors, too, because we can add vectors and divide by scalars. We write

$$\text{mean}\,(\{\mathbf{x}\}) = \frac{\sum_i \mathbf{x}_i}{N}$$

and call this the mean of the data. Notice that each component of $\text{mean}\,(\{\mathbf{x}\})$ is the mean of that component of the data. There is not an easy analogue of the median, however (how do you order high dimensional data?) and this is a nuisance. Notice that, just as for the one-dimensional mean, we have

$$\text{mean}\,(\{\mathbf{x} - \text{mean}\,(\{\mathbf{x}\})\}) = 0$$

(i.e., if you subtract the mean from a dataset, the resulting dataset has zero mean).

4.1.2 Stem Plots and Scatterplot Matrices

Plotting high dimensional data is tricky. If there are relatively few dimensions, you could just choose two (or three) of them and produce a 2D (or 3D) scatterplot. Figure 4.1 shows such a scatterplot, for data that was originally four dimensional. This is the famous iris dataset (it has to do with the botanical classification of irises), which was collected by Edgar Anderson in 1936, and made popular among statisticians by Ronald Fisher in that year. I found a copy at the UC Irvine repository of datasets that are important in machine learning (at http://archive.ics.uci.edu/ml/index.html). I will show several plots of this dataset.

Another simple but useful plotting mechanism is the stem plot. This can be a useful way to plot a few high dimensional data points. One plots each component of the vector as a vertical line, typically with a circle on the end (easier seen than said; look at Fig. 4.2). The dataset I used for this is the wine dataset, from the UC Irvine machine learning data repository. You can find this dataset at http://archive.ics.uci.edu/ml/datasets/Wine. For each of three types of wine, the data records the values of 13 different attributes. In the figure, I show the overall mean of the dataset, and also the mean of each type of wine (also known as the class means, or class-conditional means). A natural way to compare class means is to plot them on top of one another in a stem plot (Fig. 4.2).

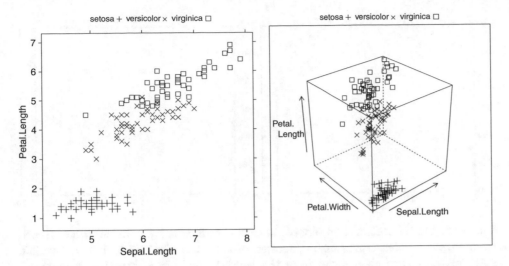

Figure 4.1: **Left:** a 2D scatterplot for the iris data. I have chosen two variables from the four, and have plotted each species with a different marker. **Right:** a 3D scatterplot for the same data. You can see from the plots that the species cluster quite tightly, and are different from one another. If you compare the two plots, you can see how suppressing a variable leads to a loss of structure. Notice that, on the left, some "x"s lie on top of boxes; you can see that this is an effect of projection by looking at the 3D picture (for each of these data points, the petal widths are quite different). You should worry that leaving out the last variable might have suppressed something important like this

Another strategy that is very useful when there aren't too many dimensions is to use a scatterplot matrix. To build one, you lay out scatterplots for each pair of variables in a matrix. On the diagonal, you name the variable that is the vertical axis for each plot in the row, and the horizontal axis in the column. This sounds more complicated than it is; look at the example of Fig. 4.3, which shows both a 3D scatterplot and a scatterplot matrix for the same dataset.

Figure 4.4 shows a scatterplot matrix for four of the variables in the height weight dataset of http://www2.stetson.edu/~jrasp/data.htm; look for bodyfat.xls at that URL). This is originally a 16-dimensional dataset, but a 16 by 16 scatterplot matrix is squashed and hard to interpret. For Fig. 4.4, you can see that weight and adiposity appear to show quite strong correlations, but weight and age are pretty weakly correlated. Height and age seem to have a low correlation. It is also easy to visualize unusual data points. Usually one has an interactive process to do so—you can move a "brush" over the plot to change the color of data points under the brush.

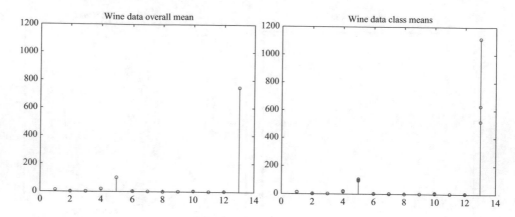

Figure 4.2: On the **left**, a stem plot of the mean of all data items in the wine dataset, from http://archive.ics.uci.edu/ml/datasets/Wine. On the **right**, I have overlaid stem plots of each class mean from the wine dataset, from http://archive.ics.uci.edu/ml/datasets/Wine, so that you can see the differences between class means

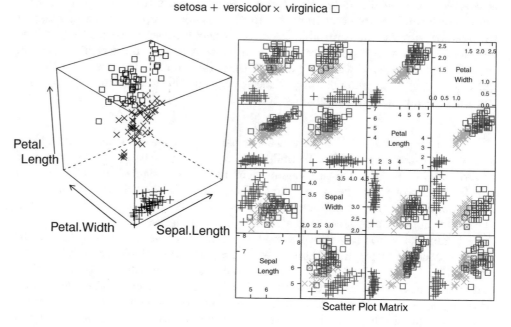

Figure 4.3: **Left:** the 3D scatterplot of the iris data of Fig. 4.1, for comparison. **Right:** a scatterplot matrix for the iris data. There are four variables, measured for each of three species of iris. I have plotted each species with a different marker. You can see from the plot that the species cluster quite tightly, and are different from one another

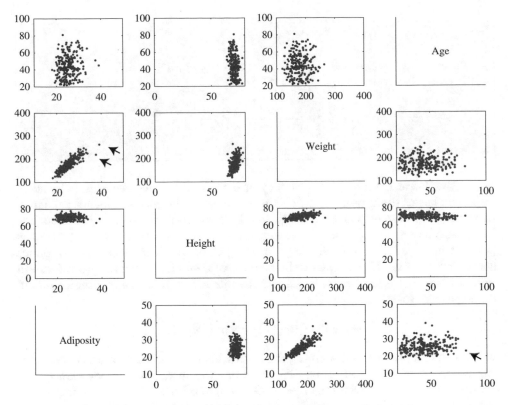

Figure 4.4: This is a scatterplot matrix for four of the variables in the height weight dataset of http://www2.stetson.edu/~jrasp/data.htm. Each plot is a scatterplot of a pair of variables. The name of the variable for the horizontal axis is obtained by running your eye down the column; for the vertical axis, along the row. Although this plot is redundant (half of the plots are just flipped versions of the other half), that redundancy makes it easier to follow points by eye. You can look at a column, move down to a row, move across to a column, etc. Notice how you can spot correlations between variables and outliers (the arrows)

4.1.3 Covariance

Variance, standard deviation, and correlation can each be seen as an instance of a more general operation on data. Extract two components from each vector of a dataset of vectors, yielding two 1D datasets of N items; write $\{x\}$ for one and $\{y\}$ for the other. The i'th element of $\{x\}$ corresponds to the i'th element of $\{y\}$ (the i'th element of $\{x\}$ is one component of some bigger vector $\mathbf{x}_i$ and the i'th element of $\{y\}$ is another component of this vector). We can define the covariance of $\{x\}$ and $\{y\}$.

Useful Fact: 4.1 *Definition: Covariance*

Assume we have two sets of N data items, $\{x\}$ and $\{y\}$. We compute the covariance by

$$\text{cov}\left(\{x\},\{y\}\right) = \frac{\sum_i (x_i - \text{mean}\left(\{x\}\right))(y_i - \text{mean}\left(\{y\}\right))}{N}$$

Covariance measures the tendency of corresponding elements of $\{x\}$ and of $\{y\}$ to be larger than (resp. smaller than) the mean. The correspondence is defined by the order of elements in the dataset, so that x_1 corresponds to y_1, x_2 corresponds to y_2, and so on. If $\{x\}$ tends to be larger (resp. smaller) than its mean for data points where $\{y\}$ is also larger (resp. smaller) than its mean, then the covariance should be positive. If $\{x\}$ tends to be larger (resp. smaller) than its mean for data points where $\{y\}$ is smaller (resp. larger) than its mean, then the covariance should be negative. Notice that

$$\text{std}\left(x\right)^2 = \text{var}\left(\{x\}\right) = \text{cov}\left(\{x\},\{x\}\right)$$

which you can prove by substituting the expressions. Recall that variance measures the tendency of a dataset to be different from the mean, so the covariance of a dataset with itself is a measure of its tendency not to be constant. More important is the relationship between covariance and correlation, in the box below.

Remember This:

$$\text{corr}\left(\{(x,y)\}\right) = \frac{\text{cov}\left(\{x\},\{y\}\right)}{\sqrt{\text{cov}\left(\{x\},\{x\}\right)}\sqrt{\text{cov}\left(\{y\},\{y\}\right)}}.$$

This is occasionally a useful way to think about correlation. It says that the correlation measures the tendency of $\{x\}$ and $\{y\}$ to be larger (resp. smaller) than their means for the same data points, *compared to* how much they change on their own.

4.1.4 The Covariance Matrix

Working with covariance (rather than correlation) allows us to unify some ideas. In particular, for data items which are d-dimensional vectors, it is straightforward to compute a single matrix that captures all covariances between all pairs of components—this is the covariance matrix.

Useful Fact: 4.2 *Definition: Covariance Matrix*

The covariance matrix is:

$$\text{Covmat}(\{\mathbf{x}\}) = \frac{\sum_i (\mathbf{x}_i - \text{mean}(\{\mathbf{x}\}))(\mathbf{x}_i - \text{mean}(\{\mathbf{x}\}))^T}{N}$$

Notice that it is quite usual to write a covariance matrix as Σ, and we will follow this convention.

Covariance matrices are often written as Σ, whatever the dataset (you get to figure out precisely which dataset is intended, from context). Generally, when we want to refer to the j, k'th entry of a matrix $\mathcal{A}$, we will write $\mathcal{A}_{jk}$, so Σ_{jk} is the covariance between the j'th and k'th components of the data.

Useful Facts: 4.3 *Properties of the Covariance Matrix*

- The j, k'th entry of the covariance matrix is the covariance of the j'th and the k'th components of $\mathbf{x}$, which we write $\text{cov}\left(\left\{x^{(j)}\right\}, \left\{x^{(k)}\right\}\right)$.
- The j, j'th entry of the covariance matrix is the variance of the j'th component of $\mathbf{x}$.
- The covariance matrix is symmetric.
- The covariance matrix is always positive semidefinite; it is positive definite, *unless* there is some vector $\mathbf{a}$ such that $\mathbf{a}^T(\mathbf{x}_i - \text{mean}(\{\mathbf{x}_i\})) = 0$ for all i.

Proposition:

$$\text{Covmat}(\{\mathbf{x}\})_{jk} = \text{cov}\left(\left\{x^{(j)}\right\}, \left\{x^{(k)}\right\}\right)$$

Proof: Recall

$$\text{Covmat}(\{\mathbf{x}\}) = \frac{\sum_i (\mathbf{x}_i - \text{mean}(\{\mathbf{x}\}))(\mathbf{x}_i - \text{mean}(\{\mathbf{x}\}))^T}{N}$$

and the j, k'th entry in this matrix will be

$$\frac{\sum_i (x_i^{(j)} - \text{mean}(\{x^{(j)}\}))(x_i^{(k)} - \text{mean}(\{x^{(k)}\}))^T}{N}$$

which is $\text{cov}\left(\left\{x^{(j)}\right\}, \left\{x^{(k)}\right\}\right)$.

Proposition:

$$\mathsf{Covmat}\left(\{\mathbf{x}\}\right)_{jj} = \Sigma_{jj} = \mathsf{var}\left(\left\{x^{(j)}\right\}\right)$$

Proof:

$$
\begin{aligned}
\mathsf{Covmat}\left(\{\mathbf{x}\}\right)_{jj} &= \mathsf{cov}\left(\left\{x^{(j)}\right\}, \left\{x^{(j)}\right\}\right) \\
&= \mathsf{var}\left(\left\{x^{(j)}\right\}\right)
\end{aligned}
$$

Proposition:

$$\mathsf{Covmat}\left(\{\mathbf{x}\}\right) = \mathsf{Covmat}\left(\{\mathbf{x}\}\right)^T$$

Proof: We have

$$
\begin{aligned}
\mathsf{Covmat}\left(\{\mathbf{x}\}\right)_{jk} &= \mathsf{cov}\left(\left\{x^{(j)}\right\}, \left\{x^{(k)}\right\}\right) \\
&= \mathsf{cov}\left(\left\{x^{(k)}\right\}, \left\{x^{(j)}\right\}\right) \\
&= \mathsf{Covmat}\left(\{\mathbf{x}\}\right)_{kj}
\end{aligned}
$$

Proposition: *Write $\Sigma = \mathsf{Covmat}\left(\{\mathbf{x}\}\right)$. If there is no vector $\mathbf{a}$ such that $\mathbf{a}^T\left(\mathbf{x}_i - \mathsf{mean}\left(\{\mathbf{x}\}\right)\right) = 0$ for all i, then for any vector $\mathbf{u}$, such that $\|\mathbf{u}\| > 0$,*

$$\mathbf{u}^T \Sigma \mathbf{u} > 0.$$

If there is such a vector $\mathbf{a}$, then

$$\mathbf{u}^T \Sigma \mathbf{u} \geq 0.$$

Proof: We have

$$
\begin{aligned}
\mathbf{u}^T \Sigma \mathbf{u} &= \frac{1}{N}\sum_i \left[\mathbf{u}^T\left(\mathbf{x}_i - \mathsf{mean}\left(\{\mathbf{x}\}\right)\right)\right]\left[\left(\mathbf{x}_i - \mathsf{mean}\left(\{\mathbf{x}\}\right)\right)^T\mathbf{u}\right] \\
&= \frac{1}{N}\sum_i \left[\mathbf{u}^T\left(\mathbf{x}_i - \mathsf{mean}\left(\{\mathbf{x}\}\right)\right)\right]^2.
\end{aligned}
$$

Now this is a sum of squares. If there is some $\mathbf{a}$ such that $\mathbf{a}^T(\mathbf{x}_i - \mathsf{mean}\left(\{\mathbf{x}\}\right)) = 0$ for every i, then the covariance matrix must be positive semidefinite (because the sum of squares could be zero in this case). Otherwise, it is positive definite, because the sum of squares will always be positive.

4.2   The Curse of Dimension

High dimensional models display unintuitive behavior (or, rather, it can take years to make your intuition see the true behavior of high dimensional models as natural). In these models, most data lies in places you don't expect. We will do several simple calculations with an easy high dimensional distribution to build some intuition.

4.2.1   The Curse: Data Isn't Where You Think It Is

Assume our data lies within a cube, with edge length two, centered on the origin. This means that each component of $\mathbf{x}_i$ lies in the range $[-1, 1]$. One simple model for such data is to assume that each dimension has uniform probability density in this range. In turn, this means that $P(x) = \frac{1}{2^d}$. The mean of this model is at the origin, which we write as $\mathbf{0}$.

The first surprising fact about high dimensional data is that most of the data can lie quite far away from the mean. For example, we can divide our dataset into two pieces. $\mathcal{A}(\epsilon)$ consists of all data items where *every* component of the data has a value in the range $[-(1 - \epsilon), (1 - \epsilon)]$. $\mathcal{B}(\epsilon)$ consists of all the rest of the data. If you think of the dataset as forming a cubical orange, then $\mathcal{B}(\epsilon)$ is the rind (which has thickness ϵ) and $\mathcal{A}(\epsilon)$ is the fruit.

Your intuition will tell you that there is more fruit than rind. This is true, for three-dimensional oranges but not true in high dimensions. The fact that the orange is cubical simplifies the calculations, but has nothing to do with the real problem.

We can compute $P(\{\mathbf{x} \in \mathcal{A}(\epsilon)\})$ and $P(\{\mathbf{x} \in \mathcal{A}(\epsilon)\})$. These probabilities tell us the probability a data item lies in the fruit (resp. rind). $P(\{\mathbf{x} \in \mathcal{A}(\epsilon)\})$ is easy to compute as

$$P(\{\mathbf{x} \in \mathcal{A}(\epsilon)\}) = (2(1 - \epsilon))^d \left(\frac{1}{2^d}\right) = (1 - \epsilon)^d$$

and

$$P(\{\mathbf{x} \in \mathcal{B}(\epsilon)\}) = 1 - P(\{\mathbf{x} \in \mathcal{A}(\epsilon)\}) = 1 - (1 - \epsilon)^d.$$

But notice that, as $d \to \infty$,

$$P(\{\mathbf{x} \in \mathcal{A}(\epsilon)\}) \to 0.$$

This means that, for large d, we expect most of the data to be in $\mathcal{B}(\epsilon)$. Equivalently, for large d, we expect that at least one component of each data item is close to either 1 or -1.

This suggests (correctly) that much data is quite far from the origin. It is easy to compute the average of the squared distance of data from the origin. We want

$$\mathbb{E}\left[\mathbf{x}^T\mathbf{x}\right] = \int_{\text{box}} \left(\sum_i x_i^2\right) P(\mathbf{x}) d\mathbf{x}$$

but we can rearrange, so that

$$\mathbb{E}\left[\mathbf{x}^T\mathbf{x}\right] = \sum_i \mathbb{E}\left[x_i^2\right] = \sum_i \int_{\text{box}} x_i^2 P(\mathbf{x}) d\mathbf{x}.$$

Now each component of $\mathbf{x}$ is independent, so that $P(\mathbf{x}) = P(x_1)P(x_2)\ldots P(x_d)$. Now we substitute, to get

$$\mathbb{E}\left[\mathbf{x}^T\mathbf{x}\right] = \sum_i \mathbb{E}\left[x_i^2\right] = \sum_i \int_{-1}^{1} x_i^2 P(x_i) dx_i = \sum_i \frac{1}{2}\int_{-1}^{1} x_i^2 dx_i = \frac{d}{3},$$

so as d gets bigger, most data points will be further and further from the origin. Worse, as d gets bigger, data points tend to get further and further from one another. We can see this by computing the average of the squared distance of data points from one another. Write $\mathbf{u}$ for one data point and $\mathbf{v}$; we can compute

$$\mathbb{E}\left[d(\mathbf{u},\mathbf{v})^2\right] = \int_{\text{box}}\int_{\text{box}} \sum_i (u_i - v_i)^2 d\mathbf{u}d\mathbf{v} = \mathbb{E}\left[\mathbf{u}^T\mathbf{u}\right] + \mathbb{E}\left[\mathbf{v}^T\mathbf{v}\right] - 2\mathbb{E}\left[\mathbf{u}^T\mathbf{v}\right]$$

but since $\mathbf{u}$ and $\mathbf{v}$ are independent, we have $\mathbb{E}\left[\mathbf{u}^T\mathbf{v}\right] = \mathbb{E}[\mathbf{u}]^T\mathbb{E}[\mathbf{v}] = 0$. This yields

$$\mathbb{E}\left[d(\mathbf{u},\mathbf{v})^2\right] = 2\frac{d}{3}.$$

This means that, for large d, we expect our data points to be quite far apart.

The nasty facts—in high dimension, data tends to be on the "outside" of a dataset, and data points tend to be unreasonably far apart—are usually true. I chose to use a uniform distribution example because the integrals are easy. If you remember how to look up integrals (or can do them yourself!), it's straightforward to reproduce these examples for a multivariate normal distribution (Sect. 4.4). One very important caveat is that the data needs to actually occupy all the dimensions. With that said, practical high dimensional data tends to allow very accurate low dimensional representations (the subject of the next chapter), and this can improve the situation somewhat.

4.2.2 Minor Banes of Dimension

High dimensional data presents a variety of important practical nuisances which follow from the curse of dimension. It is hard to estimate covariance matrices, and it is hard to build histograms.

Covariance matrices are hard to work with for two reasons. The number of entries in the matrix grows as the square of the dimension, so the matrix can get big and so difficult to store. More important, the amount of data we need to get an accurate estimate of all the entries in the matrix grows fast. As we are estimating more numbers, we need more data to be confident that our estimates are reasonable. There are a variety of straightforward work-arounds for this effect. In some cases, we have so much data there is no need to worry. In other cases, we assume that the covariance matrix has a particular form, and just estimate those parameters. There are two strategies that are usual. In one, we assume that the covariance matrix is diagonal, and estimate only the diagonal entries. In the other, we assume that the covariance matrix is a scaled version of the identity, and just estimate this scale. You should see these strategies as acts of desperation, to be used only when computing the full covariance matrix seems to produce more problems than using these approaches.

It is difficult to build histogram representations for high dimensional data. The strategy of dividing the domain into boxes, then counting data into them, fails miserably because there are too many boxes. In the case of our cube, imagine we wish to divide each dimension in half (i.e., between $[-1, 0]$ and between $[0, 1]$). Then we must have 2^d boxes. This presents two problems. First, we will have difficulty representing this number of boxes. Second, unless we are exceptionally lucky, most boxes must be empty because we will not have 2^d data items.

Instead, high dimensional data is typically represented in terms of **clusters**—coherent blobs of similar datapoints that could, under appropriate circumstances, be regarded as the same. We could then represent the dataset by, for example, the center of each cluster and the number of data items in each cluster. Since each cluster is a blob, we could also report the covariance of each cluster, if we can compute it. This representation is explored in Chaps. 8 and 9.

Remember This: *High dimensional data does not behave in a way that is consistent with most people's intuition. Points are always close to the boundary and further apart than you think. This property makes a nuisance of itself in a variety of ways. The most important is that only the simplest models work well in high dimensions.*

4.3 Using Mean and Covariance to Understand High Dimensional Data

The trick to interpreting high dimensional data is to use the mean and covariance to understand the blob. Figure 4.5 shows a two-dimensional dataset. Notice that there is obviously some correlation between the x and y coordinates (it's a diagonal blob), and that neither x nor y has zero mean. We can easily compute the mean and subtract it from the data points, and this translates the blob so that the origin is at the mean (Fig. 4.5). The mean of the new, translated dataset is zero.

Notice this blob is diagonal. We know what that means from our study of correlation—the two measurements are correlated. Now consider *rotating* the blob of data about the origin. This doesn't change the distance between any pair of points, but it does change the overall appearance of the blob of data. We can choose a rotation that means the blob looks (roughly!) like an axis aligned ellipse. *In these coordinates* there is no correlation between the horizontal and vertical components. But one direction has more variance than the other.

It turns out we can extend this approach to high dimensional blobs. We will translate their mean to the origin, then rotate the blob so that there is no correlation between any pair of distinct components (this turns out to be straightforward, which may not be obvious to you). Now the blob looks like an axis aligned ellipsoid, and we can reason about (a) what axes are "big" and (b) what that means about the original dataset.

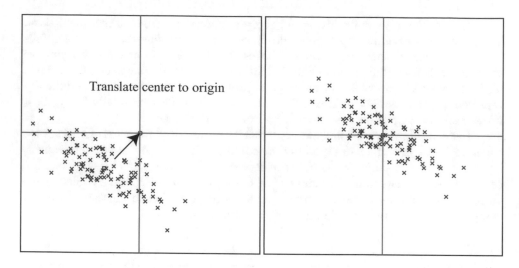

Figure 4.5: On the **left**, a "blob" in two dimensions. This is a set of data points that lie somewhat clustered around a single center, given by the mean. I have plotted the mean of these data points with a hollow square (it's easier to see when there is a lot of data). To translate the blob to the origin, we just subtract the mean from each datapoint, yielding the blob on the **right**

4.3.1 Mean and Covariance Under Affine Transformations

We have a d-dimensional dataset $\{\mathbf{x}\}$. An **affine transformation** of this data is obtained by choosing some matrix $\mathcal{A}$ and vector $\mathbf{b}$, then forming a new dataset $\{\mathbf{m}\}$, where $\mathbf{m}_i = \mathcal{A}\mathbf{x}_i + \mathbf{b}$. Here $\mathcal{A}$ doesn't have to be square, or symmetric, or anything else; it just has to have second dimension d.

It is easy to compute the mean and covariance of $\{\mathbf{m}\}$. We have

$$
\begin{aligned}
\text{mean}\left(\{\mathbf{m}\}\right) &= \text{mean}\left(\{\mathcal{A}\mathbf{x} + \mathbf{b}\}\right) \\
&= \mathcal{A}\,\text{mean}\left(\{\mathbf{x}\}\right) + \mathbf{b},
\end{aligned}
$$

so you get the new mean by multiplying the original mean by $\mathcal{A}$ and adding $\mathbf{b}$; equivalently, by transforming the old mean the same way you transformed the points.

The new covariance matrix is easy to compute as well. We have

$$
\begin{aligned}
\text{Covmat}\left(\{\mathbf{m}\}\right) &= \text{Covmat}\left(\{\mathcal{A}\mathbf{x} + \mathbf{b}\}\right) \\
&= \frac{\sum_i (\mathbf{m}_i - \text{mean}\left(\{\mathbf{m}\}\right))(\mathbf{m}_i - \text{mean}\left(\{\mathbf{m}\}\right))^T}{N} \\
&= \frac{\sum_i (\mathcal{A}\mathbf{x}_i + \mathbf{b} - \mathcal{A}\,\text{mean}\left(\{\mathbf{x}\}\right) - \mathbf{b})(\mathcal{A}\mathbf{x}_i + \mathbf{b} - \mathcal{A}\,\text{mean}\left(\{\mathbf{x}\}\right) - \mathbf{b})^T}{N} \\
&= \frac{\mathcal{A}\left[\sum_i (\mathbf{x}_i - \text{mean}\left(\{\mathbf{x}\}\right))(\mathbf{x}_i - \text{mean}\left(\{\mathbf{x}\}\right))^T\right]\mathcal{A}^T}{N} \\
&= \mathcal{A}\,\text{Covmat}\left(\{\mathbf{x}\}\right)\mathcal{A}^T.
\end{aligned}
$$

All this means that we can try and choose affine transformations that yield "good" means and covariance matrices. It is natural to choose **b** so that the mean of the new dataset is zero. An appropriate choice of $\mathcal{A}$ can reveal a lot of information about the dataset.

Remember This: *Transform a dataset $\{\mathbf{x}\}$ into a new dataset $\{\mathbf{m}\}$, where $\mathbf{m}_i = \mathcal{A}\mathbf{x}_i + \mathbf{b}$. Then*

$$\text{mean}\,(\{\mathbf{m}\}) \;=\; \mathcal{A}\text{mean}\,(\{\mathbf{x}\}) + \mathbf{b}$$
$$\text{Covmat}\,(\{\mathbf{m}\}) \;=\; \mathcal{A}\text{Covmat}\,(\{\mathbf{x}\})\mathcal{A}^T.$$

4.3.2 Eigenvectors and Diagonalization

Recall a matrix $\mathcal{M}$ is **symmetric** if $\mathcal{M} = \mathcal{M}^T$. A symmetric matrix is necessarily square. Assume $\mathcal{S}$ is a $d \times d$ symmetric matrix, $\mathbf{u}$ is a $d \times 1$ vector, and λ is a scalar. If we have

$$\mathcal{S}\mathbf{u} = \lambda \mathbf{u}$$

then $\mathbf{u}$ is referred to as an **eigenvector** of $\mathcal{S}$ and λ is the corresponding **eigenvalue**. Matrices don't have to be symmetric to have eigenvectors and eigenvalues, but the symmetric case is the only one of interest to us.

In the case of a symmetric matrix, the eigenvalues are real numbers, and there are d distinct eigenvectors that are normal to one another, and can be scaled to have unit length. They can be stacked into a matrix $\mathcal{U} = [\mathbf{u}_1, \dots, \mathbf{u}_d]$. This matrix is orthonormal, meaning that $\mathcal{U}^T\mathcal{U} = \mathcal{I}$.

This means that there is a diagonal matrix Λ and an orthonormal matrix $\mathcal{U}$ such that

$$\mathcal{S}\mathcal{U} = \mathcal{U}\Lambda.$$

In fact, there is a large number of such matrices, because we can reorder the eigenvectors in the matrix $\mathcal{U}$, and the equation still holds with a new Λ, obtained by reordering the diagonal elements of the original Λ. There is no reason to keep track of this complexity. Instead, we adopt the convention that the elements of $\mathcal{U}$ are always ordered so that the elements of Λ are sorted along the diagonal, with the largest value coming first. This gives us a particularly important procedure.

Procedure: 4.1   *Diagonalizing a Symmetric Matrix*

We can convert any symmetric matrix $\mathcal{S}$ to a diagonal form by computing

$$\mathcal{U}^T\mathcal{S}\mathcal{U} = \Lambda.$$

Numerical and statistical programming environments have procedures to compute $\mathcal{U}$ and Λ for you. We assume that the elements of $\mathcal{U}$ are always ordered so that the elements of Λ are sorted along the diagonal, with the largest value coming first.

Useful Facts: 4.4 *Orthonormal Matrices Are Rotations*

You should think of orthonormal matrices as rotations, because they do not change lengths or angles. For $\mathbf{x}$ a vector, $\mathcal{R}$ an orthonormal matrix, and $\mathbf{m} = \mathcal{R}\mathbf{x}$, we have

$$\mathbf{u}^T\mathbf{u} = \mathbf{x}^T\mathcal{R}^T\mathcal{R}\mathbf{x} = \mathbf{x}^T\mathcal{I}\mathbf{x} = \mathbf{x}^T\mathbf{x}.$$

This means that $\mathcal{R}$ doesn't change lengths. For $\mathbf{y}$, $\mathbf{z}$ both unit vectors, we have that the cosine of the angle between them is

$$\mathbf{y}^T\mathbf{x}.$$

By the argument above, the inner product of $\mathcal{R}\mathbf{y}$ and $\mathcal{R}\mathbf{x}$ is the same as $\mathbf{y}^T\mathbf{x}$. This means that $\mathcal{R}$ doesn't change angles, either.

4.3.3 Diagonalizing Covariance by Rotating Blobs

We start with a dataset of N d-dimensional vectors $\{\mathbf{x}\}$. We can translate this dataset to have zero mean, forming a new dataset $\{\mathbf{m}\}$ where $\mathbf{m}_i = \mathbf{x}_i - \text{mean}(\{\mathbf{x}\})$. Now recall that, if we were to form a new dataset $\{\mathbf{a}\}$ where

$$\mathbf{a}_i = \mathcal{A}\mathbf{m}_i$$

the covariance matrix of $\{\mathbf{a}\}$ would be

$$\text{Covmat}(\{\mathbf{a}\}) = \mathcal{A}\text{Covmat}(\{\mathbf{m}\})\mathcal{A}^T = \mathcal{A}\text{Covmat}(\{\mathbf{x}\})\mathcal{A}^T.$$

Recall also we can diagonalize $\text{Covmat}(\{\mathbf{m}\}) = \text{Covmat}(\{\mathbf{x}\})$ to get

$$\mathcal{U}^T\text{Covmat}(\{\mathbf{x}\})\mathcal{U} = \Lambda.$$

But this means we could form the dataset $\{\mathbf{r}\}$, using the rule

$$\mathbf{r}_i = \mathcal{U}^T\mathbf{m}_i = \mathcal{U}^T(\mathbf{x}_i - \text{mean}(\{\mathbf{x}\})).$$

The mean of this new dataset is clearly $\mathbf{0}$. The covariance of this dataset is

$$
\begin{aligned}
\text{Covmat}(\{\mathbf{r}\}) &= \text{Covmat}(\{\mathcal{U}^T\mathbf{x}\}) \\
&= \mathcal{U}^T\text{Covmat}(\{\mathbf{x}\})\mathcal{U} \\
&= \Lambda,
\end{aligned}
$$

where Λ is a diagonal matrix of eigenvalues of $\text{Covmat}(\{\mathbf{x}\})$ that we obtained by diagonalization. We now have a very useful fact about $\{\mathbf{r}\}$: its covariance matrix is diagonal. This means that every pair of distinct components has covariance zero, and so has correlation zero. Remember that, in describing diagonalization, we adopted the convention that the eigenvectors of the matrix being diagonalized

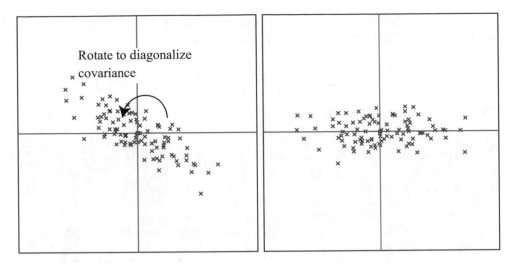

Figure 4.6: On the **left**, the translated blob of Fig. 4.5. This blob lies somewhat diagonally, because the vertical and horizontal components are correlated. On the **right**, that blob of data rotated so that there is no correlation between these components. We can now describe the blob by the vertical and horizontal variances alone, as long as we do so in the new coordinate system. In this coordinate system, the vertical variance is significantly larger than the horizontal variance—the blob is short and wide

were ordered so that the eigenvalues are sorted in the descending order along the diagonal of Λ. Our choice of ordering means that the first component of $\mathbf{r}$ has the highest variance, the second component has the second highest variance, and so on.

The transformation from $\{\mathbf{x}\}$ to $\{\mathbf{r}\}$ is a translation followed by a rotation (remember $\mathcal{U}$ is orthonormal, and so a rotation). So this transformation is a high dimensional version of what I showed in Figs. 4.5 and 4.6.

Useful Fact: 4.5 *You Can Transform Data to Zero Mean and Diagonal Covariance*

We can translate and rotate *any* blob of data into a coordinate system where it has (a) zero mean and (b) diagonal covariance matrix.

4.4 The Multivariate Normal Distribution

All the nasty facts about high dimensional data, above, suggest that we need to use quite simple probability models. By far the most important model is the **multivariate normal distribution**, which is quite often known as the **Gaussian distribution**. There are two sets of parameters in this model, the mean μ and the covariance Σ. For a d-dimensional model, the mean is a d-dimensional column vector and the covariance is a $d \times d$-dimensional matrix. The covariance is a symmetric matrix. For our definitions to be meaningful, the covariance matrix must be positive definite. The form of the distribution $p(\mathbf{x}|\mu, \Sigma)$ is

$$p(\mathbf{x}|\mu, \Sigma) = \frac{1}{\sqrt{(2\pi)^d \det(\Sigma)}} \exp\left(-\frac{1}{2}(\mathbf{x} - \mu)^T \Sigma^{-1}(\mathbf{x} - \mu)\right).$$

The following facts explain the names of the parameters:

Useful Facts: 4.6 *Parameters of a Multivariate Normal Distribution*

Assuming a multivariate normal distribution, we have

- $\mathbb{E}[\mathbf{x}] = \mu$, meaning that the mean of the distribution is μ.
- $\mathbb{E}[(\mathbf{x} - \mu)(\mathbf{x} - \mu)^T] = \Sigma$, meaning that the entries in Σ represent covariances.

Assume I now have a dataset of items $\mathbf{x}_i$, where i runs from 1 to N, and we wish to model this data with a multivariate normal distribution. The maximum likelihood estimate of the mean, $\hat{\mu}$, is

$$\hat{\mu} = \frac{\sum_i \mathbf{x}_i}{N}$$

(which is quite easy to show). The maximum likelihood estimate of the covariance, $\hat{\Sigma}$, is

$$\hat{\Sigma} = \frac{\sum_i (\mathbf{x}_i - \hat{\mu})(\mathbf{x}_i - \hat{\mu})^T}{N}$$

(which is rather a nuisance to show, because you need to know how to differentiate a determinant). These facts mean that we already know most of what is interesting about multivariate normal distributions (or Gaussians).

4.4.1 Affine Transformations and Gaussians

Gaussians behave very well under affine transformations. In fact, we've already worked out all the math. Assume I have a dataset $\mathbf{x}_i$. The mean of the maximum likelihood Gaussian model is mean $(\{\mathbf{x}_i\})$, and the covariance is Covmat $(\{\mathbf{x}_i\})$. I can now transform the data with an affine transformation, to get $\mathbf{y}_i = \mathcal{A}\mathbf{x}_i + \mathbf{b}$. The mean of the maximum likelihood Gaussian model for the transformed dataset is mean $(\{\mathbf{y}_i\})$, and we've dealt with this; similarly, the covariance is Covmat $(\{\mathbf{y}_i\})$, and we've dealt with this, too.

A very important point follows in an obvious way. I can apply an affine transformation to any multivariate Gaussian to obtain one with (a) zero mean and (b) independent components. In turn, this means that, *in the right coordinate system*, any Gaussian is a product of zero mean one-dimensional normal distributions. This fact is quite useful. For example, it means that simulating multivariate normal distributions is quite straightforward—you could simulate a standard normal distribution for each component, then apply an affine transformation.

4.4.2 Plotting a 2D Gaussian: Covariance Ellipses

There are some useful tricks for plotting a 2D Gaussian, which are worth knowing both because they're useful, and they help to understand Gaussians. Assume we are working in 2D; we have a Gaussian with mean μ (which is a 2D vector) and covariance Σ (which is a 2×2 matrix). We could plot the collection of points $\mathbf{x}$ that has some fixed value of $p(\mathbf{x}|\mu, \Sigma)$. This set of points is given by:

$$\frac{1}{2}\left((\mathbf{x} - \mu)^T \Sigma^{-1} (\mathbf{x} - \mu)\right) = c^2$$

where c is some constant. I will choose $c^2 = \frac{1}{2}$, because the choice doesn't matter, and this choice simplifies some algebra. You might recall that a set of points $\mathbf{x}$ that satisfies a quadratic like this is a conic section. Because Σ (and so Σ^{-1}) is positive definite, the curve is an ellipse. There is a useful relationship between the geometry of this ellipse and the Gaussian.

This ellipse—like all ellipses—has a major axis and a minor axis. These are at right angles, and meet at the center of the ellipse. We can determine the properties of the ellipse in terms of the Gaussian quite easily. The geometry of the ellipse isn't affected by rotation or translation, so we will translate the ellipse so that $\mu = \mathbf{0}$ (i.e., the mean is at the origin) and rotate it so that Σ^{-1} is diagonal. Writing $\mathbf{x} = [x, y]$ we get that the set of points on the ellipse satisfies

$$\frac{1}{2}\left(\frac{1}{k_1^2}x^2 + \frac{1}{k_2^2}y^2\right) = \frac{1}{2}$$

where $\frac{1}{k_1^2}$ and $\frac{1}{k_2^2}$ are the diagonal elements of Σ^{-1}. We will assume that the ellipse has been rotated so that $k_1 > k_2$. The points $(k_1, 0)$ and $(-k_1, 0)$ lie on the ellipse, as do the points $(0, k_2)$ and $(0, -k_2)$. The major axis of the ellipse, in this coordinate system, is the x-axis, and the minor axis is the y-axis. In this coordinate system, x and y are independent. If you do a little algebra, you will see that the standard deviation of x is $\mathsf{abs}(k_1)$ and the standard deviation of y is $\mathsf{abs}(k_2)$. So the ellipse is longer in the direction of largest standard deviation and shorter in the direction of smallest standard deviation.

Now rotating the ellipse means we will pre- and post-multiply the covariance matrix with some rotation matrix. Translating it will move the origin to the mean. As a result, the ellipse has its center at the mean, its major axis is in the direction of the eigenvector of the covariance with largest eigenvalue, and its minor axis is in the direction of the eigenvector with smallest eigenvalue. A plot of this ellipse, which can be coaxed out of most programming environments with relatively little effort, gives us a great deal of information about the underlying Gaussian. These ellipses are known as **covariance ellipses**.

4.4.3 Descriptive Statistics and Expectations

You might have noticed a sleight of hand in the description above. I used each of the terms mean, variance, covariance, and standard deviation in two slightly different ways. This is quite usual. One sense of each term, as in the description of covariance above, describes a property of a dataset. Terms used in this sense are known as **descriptive statistics**. The other sense is a property of probability distributions; so mean, for example, means $\mathbb{E}[X]$; variance means $\mathbb{E}\big[(X-\mathbb{E}[X])^2\big]$; and so on. Terms used in this sense are known as **expectations**. The reason we use one name for two notions is that the notions are not really all that different.

Here is a useful construction to illustrate the point. Imagine we have a dataset $\{\mathbf{x}\}$ of N items, where the i'th item is $\mathbf{x}_i$. Build a random variable X using this dataset by placing the same probability on each data item. This means that each data item has probability $1/N$. Write $\mathbb{E}[X]$ for the mean of this distribution. We have

$$\mathbb{E}[X] = \sum_i x_i P(x_i) = \frac{1}{N}\sum_i x_i = \text{mean}\,(\{x\})$$

and, by the same reasoning,

$$\text{var}[X] = \text{var}\,(\{x\}).$$

This construction works for standard deviation and covariance, too. For this particular distribution (sometimes called the **empirical distribution**), the expectations have the same value as the descriptive statistics.

There is a form of converse to this fact, which you should have seen already, and which we shall see on and off later. Imagine we have a dataset that consists of independent, identically distributed samples from a probability distribution (i.e., we know that each data item was obtained independently from the distribution). For example, we might have a count of heads in each of a number of coin flip experiments. The **weak law of large numbers** says the descriptive statistics will turn out to be accurate estimates of the expectations.

In particular, assume we have a random variable X with distribution $P(X)$ which has finite variance. We want to estimate $\mathbb{E}[X]$. Now if we have a set of IID samples of X, which we write x_i, write

$$X_N = \frac{\sum_{i=1}^N x_i}{N}.$$

This is a random variable (different sets of samples yield different values of X_N), and the weak law of large numbers gives that, for any positive number ϵ

$$\lim_{N\to\infty} P(\{\|X_N - \mathbb{E}[X]\| > \epsilon\}) = 0.$$

You can interpret this as saying that, for a set of IID random samples x_i, the probability that

$$\frac{\sum_{i=1}^{N} X_i}{N}$$

is very close to $\mathbb{E}[X]$ for large N

Useful Facts: 4.7 *Weak Law of Large Numbers*

Given a random variable X with distribution $P(X)$ which has finite variance, and a set of N IID samples $\mathbf{x}_i$ from $P(X)$, write

$$X_N = \frac{\sum_{i=1}^{N} x_i}{N}.$$

Then for any positive number ϵ

$$\lim_{N\to\infty} P(\{\|X_N - \mathbb{E}[X]\| > \epsilon\}) = 0.$$

Remember This: *Mean, variance, covariance, and standard deviation can refer either to properties of a dataset or to expectations. The context usually tells you which. There is a strong relationship between these senses. Given a dataset, you can construct an empirical distribution, whose mean, variance, and covariances (interpreted as expectations) have the same values as the mean, variance, and covariances (interpreted as descriptive statistics). If a dataset is an IID sample of a probability distribution, the mean, Variance, and covariances (interpreted as descriptive statistics) are usually very good estimates of the values of the mean, variance, and covariances (interpreted as expectations).*

4.4.4 More from the Curse of Dimension

It can be hard to get accurate estimates of the mean of a high dimensional normal distribution (and so of any other). This is mostly a minor nuisance, but it's worth understanding what is happening. The data is a set of N IID samples of a normal distribution with mean μ and covariance Σ in d- dimensional space. These points will tend to lie far away from one another. But they may not be evenly spread out, so there may be slightly more points on one side of the true mean than on the other, and so the estimate of the mean is likely noisy. It's tough to be crisp about what it means to be on one side of the true mean in high dimensions, so I'll do this in algebra, too. The estimate of the mean is

$$X^N = \frac{\sum_i \mathbf{x}_i}{N}$$

which is a random variable, because different draws of data will give different values of X^N. In the exercises, you will show that $\mathbb{E}\left[X^N\right]$ is μ (so the estimate is reasonable). One reasonable measure of the total error in estimating the mean is $(X^N - \mu)^T (X^N - \mu)$. In the exercises, you will show that the expected value of this error is

$$\frac{\mathrm{Trace}(\Sigma)}{N}$$

which may grow with d unless Σ has some strong properties. Likely, your estimate of the mean for a high dimensional distribution is poor.

4.5 You Should

4.5.1 Remember These Terms

4.5.2 Remember These Facts

Problems

   Summaries

4.1. You have a dataset $\{\mathbf{x}\}$ of N vectors, $\mathbf{x}_i$, each of which is d-dimensional. We will consider a linear function of this dataset. Write $\mathbf{a}$ for a constant vector; then the value of this linear function evaluated on the i'th data item is $\mathbf{a}^T\mathbf{x}_i$. Write $f_i = \mathbf{a}^T\mathbf{x}_i$. We can make a new dataset $\{f\}$ out of the values of this linear function.
 (a) Show that $\mathsf{mean}\,(\{f\}) = \mathbf{a}^T\mathsf{mean}\,(\{\mathbf{x}\})$ (easy).
 (b) Show that $\mathsf{var}\,(\{f\}) = \mathbf{a}^T\mathsf{Covmat}\,(\{\mathbf{x}\})\mathbf{a}$ (harder, but just push it through the definition).
 (c) Assume the dataset has the special property that there exists some $\mathbf{a}$ so that $\mathbf{a}^T\mathsf{Covmat}\,(\{\mathbf{x}\})\mathbf{a}$. Show that this means that the dataset lies on a hyperplane.

4.2. You have a dataset $\{\mathbf{x}\}$ of N vectors, $\mathbf{x}_i$, each of which is d-dimensional. Assume that $\mathsf{Covmat}\,(\{\mathbf{x}\})$ has one non-zero eigenvalue. Assume that $\mathbf{x}_1$ and $\mathbf{x}_2$ do not have the same value.
 (a) Show that you can choose a set of t_i so that you can represent *every* data item $\mathbf{x}_i$ *exactly* as
$$\mathbf{x}_i = \mathbf{x}_1 + t_i(\mathbf{x}_2 - \mathbf{x}_1).$$
 (b) Now consider the dataset of these t values. What is the relationship between (a) $\mathsf{std}\,(t)$ and (b) the non-zero eigenvalue of $\mathsf{Covmat}\,(\{\mathbf{x}\})$? Why?

4.3. You have a dataset $\{\mathbf{x}\}$ of N vectors, $\mathbf{x}_i$, each of which is d-dimensional. Assume $\mathsf{mean}\,(\{\mathbf{x}\}) = 0$. We will consider a linear function of this dataset. Write $\mathbf{a}$ for some vector; then the value of this linear function evaluated on the i'th data item is $\mathbf{a}^T\mathbf{x}_i$. Write $f_i(\mathbf{a}) = \mathbf{a}^T\mathbf{x}_i$. We can make a new dataset $\{f(\mathbf{a})\}$ out of these f_i (the notation is to remind you that this dataset depends on the choice of vector $\mathbf{a}$).
 (a) Show that $\mathsf{var}\,(\{f(s\mathbf{a})\}) = s^2\mathsf{var}\,(\{f(\mathbf{a})\})$.
 (b) The previous subexercise means that, to choose $\mathbf{a}$ to obtain a dataset with large variance in any kind of sensible way, we need to insist that $\mathbf{a}^T\mathbf{a}$ is kept constant. Show that
$$\text{Maximize } \mathsf{var}\,(\{f\})(\mathbf{a}) \text{ subject to } \mathbf{a}^T\mathbf{a} = 1$$
is solved by the eigenvector of $\mathsf{Covmat}\,(\{x\})$ corresponding to the largest eigenvalue. (You need to know Lagrange multipliers to do this, but you should.)

4.4. You have a dataset $\{\mathbf{x}\}$ of N vectors, $\mathbf{x}_i$, each of which is d-dimensional. We will consider two linear functions of this dataset, given by two vectors $\mathbf{a}$, $\mathbf{b}$.
 (a) Show that $\mathsf{cov}\,(\{\mathbf{a}^T\mathbf{x}\}, \{\mathbf{b}^T\mathbf{x}\}) = \mathbf{a}^T\mathsf{Covmat}\,(\{\mathbf{x}\})\mathbf{b}$. This is easier to do if you show that the mean has no effect on covariance, and then do the math assuming $\mathbf{x}$ has zero mean.
 (b) Show that the correlation between $\mathbf{a}^T\mathbf{x}$ and $\mathbf{b}^T\mathbf{x}$ is given by
$$\frac{\mathbf{a}^T\mathsf{Covmat}\,(\{\mathbf{x}\})\mathbf{b}}{\sqrt{\mathbf{a}^T\mathsf{Covmat}\,(\{\mathbf{x}\})\mathbf{a}}\sqrt{\mathbf{b}^T\mathsf{Covmat}\,(\{\mathbf{x}\})\mathbf{b}}}.$$

4.5. It is sometimes useful to map a dataset to have zero mean and unit covariance. Doing so is known as whitening the data (for reasons I find obscure). This can be a sensible thing to do when we don't have a clear sense of the relative scales of the components of each data vector or whiten the data might be that we know relatively little about the meaning of each component. You have a dataset $\{\mathbf{x}\}$ of N vectors, $\mathbf{x}_i$, each of which is d-dimensional. Write $\mathcal{U}$, Λ for the eigenvectors and eigenvalues of $\mathsf{Covmat}\left(\{\mathbf{x}\}\right)$.

(a) Show that $\Lambda \geq 0$

(b) Assume that some diagonal element of Λ is zero. How do you interpret this?

(c) Assume that all diagonal elements of Λ are greater than zero. Write $\Lambda^{1/2}$ for the matrix whose diagonal is the non-negative square roots of the diagonal of Λ. Write $\{\mathbf{y}\}$ for the dataset of vectors where $\mathbf{y}_i = (\Lambda^{1/2})^{-1}\mathcal{U}^T\left(\mathbf{x}_i - \mathsf{mean}\left(\{x\}\right)\right)$. Show that $\mathsf{Covmat}\left(\{y\}\right)$ is the identity matrix.

(d) Write $\mathcal{O}$ for some orthonormal matrix. Using the notation of the previous subexercise, and writing $\mathbf{z}_i = \mathcal{O}\mathbf{y}_i$, show that $\mathsf{Covmat}\left(\{z\}\right)$ is the identity matrix. Use this information to argue that there is not a unique version of a whitened dataset.

The Multivariate Normal Distribution

4.6. A dataset of points (x, y) has zero mean and covariance

$$\Sigma = \left(\begin{array}{cc} k_1^2 & 0 \\ 0 & k_2^2 \end{array} \right)$$

with $k_1 > k_2$.

(a) Show that the standard deviation of the x coordinate is $\mathsf{abs}\left(k_1\right)$ and of the y coordinate is $\mathsf{abs}\left(k_2\right)$.

(b) Show that the set of points that satisfies

$$\frac{1}{2}\left(\frac{1}{k_1^2}x^2 + \frac{1}{k_2^2}y^2 \right) = \frac{1}{2}$$

is an ellipse.

(c) Show that the major axis of the ellipse is the x axis, the minor axis of the ellipse is the y axis, and the center of the ellipse is at $(0, 0)$.

(d) Show that the height of the ellipse is $2k_1$ and the width of the ellipse is $2k_2$.

4.7. For Σ a positive definite matrix, μ some two-dimensional vector, show that the family of points that satisfies

$$\frac{1}{2}\left((\mathbf{x} - \mu)^T\Sigma^{-1}(\mathbf{x} - \mu) \right) = c^2$$

is an ellipse. An easy way to do this is to notice that ellipses remain ellipses when rotated and translated, and exploit the previous exercise.

The Curse of Dimension

4.8. A dataset consists of N IID samples from a multivariate normal distribution with dimension d. The mean of this distribution is zero, and its covariance matrix is the identity. You compute

$$X^N = \frac{1}{N} \sum_i \mathbf{x}_i.$$

The number you compute is a random variable, because you will compute a slightly different number for each different sample you draw. It turns out that the distribution of X^N is normal because the sum of normally distributed random variables is normal. You should remember (or, if you don't, memorize) the fact that

- a sample of a (1D) normal random variable is within one standard deviation of its mean about 66% of the time;
- a sample of a (1D) normal random variable is within two standard deviations of its mean about 95% of the time;
- a sample of a (1D) normal random variable is within three standard deviations of its mean about 99% of the time.

(a) Show that each component of X^N has expected value zero and variance $1/N$.

(b) Argue that about $d/3$ of the components have absolute value greater than $1/N$.

(c) Argue that about $d/20$ of the components have absolute value greater than $2/N$.

(d) Argue that about $d/100$ of the components have absolute value greater than $3/N$.

(e) What happens when d is very large compared to N?

4.9. For a dataset that consists of N IID samples $\mathbf{x}_i$ from a multivariate normal distribution with mean μ and covariance Σ, you compute

$$X^N = \frac{1}{N} \sum_i \mathbf{x}_i.$$

The number you compute is a random variable, because you will compute a slightly different number for each different sample you draw.

(a) Show that $\mathbb{E}\left[X^N\right] = \mu$. You can do this by noticing that, if $N = 1$, $\mathbb{E}\left[X^1\right] = \mu$ fairly obviously. Now use the fact that each of the samples is independent.

(b) The random variable $T^N = (X^N - \mu)^T (X^N - \mu)$ is one reasonable measure of how well X^N approximates μ. Show that

$$\mathbb{E}\left[T^N\right] = \frac{\text{Trace}(\Sigma)}{N}.$$

Do this by noticing that $\mathbb{E}\left[T^N\right]$ is the sum of the variances of the components of X^N. This exercise is much easier if you notice that translating the normal distribution to have zero mean doesn't change anything (so it's enough to work out the case where $\mu = \mathbf{0}$).

(c) Use the previous subexercise to identify situations where estimates of the mean of a normal distribution might be poor.

C H A P T E R 5

Principal Component Analysis

We have seen that a blob of data can be translated so that it has zero mean, then rotated so the covariance matrix is diagonal. In this coordinate system, we can set some components to zero, and get a representation of the data that is still accurate. The rotation and translation can be undone, yielding a dataset that is in the same coordinates as the original, but lower dimensional. The new dataset is a good approximation to the old dataset. All this yields a really powerful idea: we can choose a small set of vectors, so that each item in the original dataset can be represented as the mean vector plus a weighted sum of this set. This representation means we can think of the dataset as lying on a low dimensional space inside the original space. It's an experimental fact that this model of a dataset is usually accurate for real high dimensional data, and it is often an extremely convenient model. Furthermore, representing a dataset like this very often suppresses noise—if the original measurements in your vectors are noisy, the low dimensional representation may be closer to the true data than the measurements are.

5.1 Representing Data on Principal Components

We start with a dataset of N d-dimensional vectors $\{\mathbf{x}\}$. We translate this dataset to have zero mean, forming a new dataset $\{\mathbf{m}\}$ where $\mathbf{m}_i = \mathbf{x}_i - \text{mean}(\{\mathbf{x}\})$. We diagonalize $\text{Covmat}(\{\mathbf{m}\}) = \text{Covmat}(\{\mathbf{x}\})$ to get

$$\mathcal{U}^T \text{Covmat}(\{\mathbf{x}\})\mathcal{U} = \Lambda$$

and form the dataset $\{\mathbf{r}\}$, using the rule

$$\mathbf{r}_i = \mathcal{U}^T \mathbf{m}_i = \mathcal{U}^T(\mathbf{x}_i - \text{mean}(\{\mathbf{x}\})).$$

We saw the mean of this dataset is zero, and the covariance is diagonal. Most high dimensional datasets display another important property: many, or most, of the diagonal entries of the covariance matrix are very small. This means we can build a low dimensional representation of the high dimensional dataset that is quite accurate.

5.1.1 Approximating Blobs

The covariance matrix of $\{\mathbf{r}\}$ is diagonal, and the values on the diagonal are interesting. It is quite usual for high dimensional datasets to have a small number of large values on the diagonal, and a lot of small values. This means that the blob of data is really a low dimensional blob in a high dimensional space. For example, think about a line segment (a 1D blob) in 3D. As another example, look at Fig. 4.3; the scatterplot matrix strongly suggests that the blob of data is flattened (e.g., look at the petal width vs petal length plot).

© Springer Nature Switzerland AG 2019
D. Forsyth, *Applied Machine Learning*,
https://doi.org/10.1007/978-3-030-18114-7_5

Now assume that Covmat ($\{\mathbf{r}\}$) has many small and few large diagonal entries. In this case, the blob of data represented by $\{\mathbf{r}\}$ admits an accurate low dimensional representation. The dataset $\{\mathbf{r}\}$ is d-dimensional. We will try to represent it with an s-dimensional dataset, and see what error we incur. Choose some $s<d$. Now take each data point $\mathbf{r}_i$ and replace the last $d-s$ components with 0. Call the resulting data item $\mathbf{p}_i$. We should like to know the average error in representing $\mathbf{r}_i$ with $\mathbf{p}_i$.

This error is

$$\frac{1}{N}\sum_i\left[(\mathbf{r}_i-\mathbf{p}_i)^T(\mathbf{r}_i-\mathbf{p}_i)\right].$$

Write $r_i^{(j)}$ for the j' component of $\mathbf{r}_i$, and so on. Remember that $\mathbf{p}_i$ is zero in the last $d-s$ components. The mean error is then

$$\frac{1}{N}\sum_i\left[\sum_{j=s+1}^{j=d}\left(r_i^{(j)}\right)^2\right].$$

But we know this number, because we know that $\{\mathbf{r}\}$ has zero mean. The error is

$$\sum_{j=s+1}^{j=d}\left[\frac{1}{N}\sum_i\left(r_i^{(j)}\right)^2\right]=\sum_{j=s+1}^{j=d}\mathrm{var}\left(\left\{r^{(j)}\right\}\right)$$

which is the sum of the diagonal elements of the covariance matrix from r,r to d,d. Equivalently, writing λ_i for the ith eigenvalue of Covmat ($\{x\}$) and assuming the eigenvalues are sorted in the descending order, the error is

$$\sum_{j=s+1}^{j=d}\lambda_j$$

If this sum is small compared to the sum of the first s components, then dropping the last $d-s$ components results in a small error. In that case, we could think about the data as being s-dimensional. Figure 5.1 shows the result of using this approach to represent the blob I've used as a running example as a 1D dataset.

This is an observation of great practical importance. As a matter of experimental fact, a great deal of high dimensional data produces relatively low dimensional blobs. We can identify the main directions of variation in these blobs, and use them to understand and to represent the dataset.

5.1.2 Example: Transforming the Height–Weight Blob

Translating a blob of data doesn't change the scatterplot matrix in any interesting way (the axes change, but the picture doesn't). Rotating a blob produces really interesting results, however. Figure 5.2 shows the dataset of Fig. 4.4, translated to the origin and rotated to diagonalize it. Now we do not have names for each component of the data (they're linear combinations of the original components), but each pair is now not correlated. This blob has some interesting shape features. Figure 5.2 shows the gross shape of the blob best. Each panel of this figure has

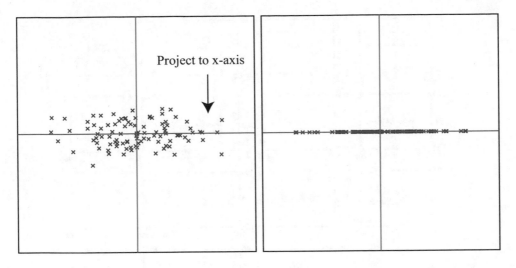

Figure 5.1: On the **left**, the translated and rotated blob of Fig. 4.6. This blob is stretched—one direction has more variance than another. Setting the y coordinate to zero for each of these datapoints results in a representation that has relatively low error, because there isn't much variance in these values. This results in the blob on the **right**. The text shows how the error that results from this projection is computed

the same scale in each direction. You can see the blob extends about 80 units in direction 1, but only about 15 units in direction 2, and much less in the other two directions. You should think of this blob as being rather cigar-shaped; it's long in one direction, but there isn't much in the others. The cigar metaphor isn't perfect (have you seen a four-dimensional cigar recently?), but it's helpful. You can think of each panel of this figure as showing views down each of the four axes of the cigar.

Now look at Fig. 5.3. This shows the same rotation of the same blob of data, but now the scales on the axis have changed to get the best look at the detailed shape of the blob. First, you can see that blob is a little curved (look at the projection onto direction 2 and direction 4). There might be some effect here worth studying. Second, you can see that some points seem to lie away from the main blob. I have plotted each data point with a dot, and the interesting points with a number. These points are clearly special in some way.

The problem with these figures is that the axes are meaningless. The components are weighted combinations of components of the original data, so they don't have any units, etc. This is annoying, and often inconvenient. But I obtained Fig. 5.2 by translating, rotating, and projecting data. It's straightforward to undo the rotation and the translation—this takes the projected blob (which we know to be a good approximation of the rotated and translated blob) back to where the original blob was. Rotation and translation don't change distances, so the result is a good approximation of the original blob, but now in the original blob's coordinates. Figure 5.4 shows what happens to the data of Fig. 4.4. This is

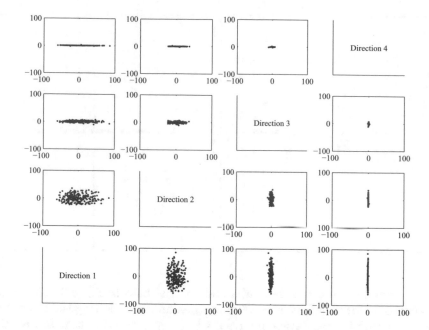

Figure 5.2: A panel plot of the bodyfat dataset of Fig. 4.4, now rotated so that the covariance between all pairs of distinct dimensions is zero. Now we do not know names for the directions—they're linear combinations of the original variables. Each scatterplot is on the same set of axes, so you can see that the dataset extends more in some directions than in others. You should notice that, in some directions, there is very little variance. This suggests that replacing the coefficient in those directions with zero (as in Fig. 5.1) should result in a representation of the data that has very little error

a two-dimensional version of the original dataset, embedded like a thin pancake of data in a four-dimensional space. Crucially, it represents the original dataset quite accurately.

5.1.3 Representing Data on Principal Components

Now consider undoing the rotation and translation for our projected dataset $\{\mathbf{p}\}$. We would form a new dataset $\{\hat{\mathbf{x}}\}$, with the ith element given by

$$\hat{\mathbf{x}}_i = \mathcal{U}\mathbf{p}_i + \mathsf{mean}\,(\{\mathbf{x}\})$$

(you should check this expression). But this expression says that $\hat{\mathbf{x}}_i$ is constructed by forming a weighted sum of the first s columns of $\mathcal{U}$ (because all the other components of $\mathbf{p}_i$ are zero), then adding $\mathsf{mean}\,(\{\mathbf{x}\})$. If we write $\mathbf{u}_j$ for the jth column of $\mathcal{U}$ and w_{ij} for a weight value, we have

$$\hat{\mathbf{x}}_i = \sum_{j=1}^{s} w_{ij}\mathbf{u}_j + \mathsf{mean}\left(\{\mathbf{x}\}\right).$$

What is important about this sum is that s is usually a lot less than d. In turn, this means that we are representing the dataset using a lower dimensional dataset. We choose an s-dimensional flat subspace of d-dimensional space, and represent each data item with a point that lies on in that subset. The $\mathbf{u}_j$ are known as **principal components** (sometimes **loadings**) of the dataset; the $r_i^{(j)}$ are sometimes known as **scores**, but are usually just called **coefficients**. Forming the representation is called **principal components analysis** or **PCA**. The weights w_{ij} are actually easy to evaluate. We have that

$$w_{ij} = r_i^{(j)} = (\mathbf{x}_i - \mathsf{mean}\left(\{\mathbf{x}\}\right))^T\mathbf{u}_j.$$

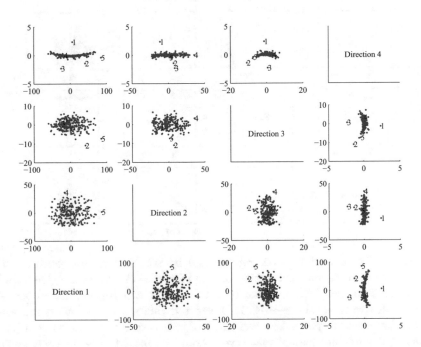

Figure 5.3: A panel plot of the bodyfat dataset of Fig. 4.4, now rotated so that the covariance between all pairs of distinct dimensions is zero. Now we do not know names for the directions—they're linear combinations of the original variables. Compare this figure with Fig. 5.2; in that figure, the axes were the same, but in this figure I have scaled the axes so you can see details. Notice that the blob is a little curved, and there are several data points that seem to lie some way away from the blob, which I have numbered

> **Remember This:**   *Data items in a d-dimensional dataset can usually be represented with good accuracy as a weighted sum of a small number s of d-dimensional vectors, together with the mean. This means that the dataset lies on an s-dimensional subspace of the d-dimensional space. The subspace is spanned by the principal components of the data.*

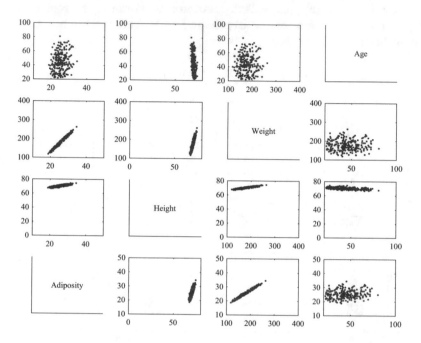

Figure 5.4:   The data of Fig. 4.4, represented by translating and rotating so that the covariance is diagonal, projecting off the two smallest directions, then undoing the rotation and translation. This blob of data is two-dimensional (because we projected off two dimensions—Fig. 5.2 suggested this was safe), but is represented in a four-dimensional space. You can think of it as a thin two-dimensional pancake of data in the four-dimensional space (you should compare to Fig. 4.4 on page 73). It is a good representation of the original data. Notice that it looks slightly thickened on edge, because it isn't aligned with the coordinate system—think of a view of a flat plate at a slight slant

5.1.4   The Error in a Low Dimensional Representation

We can easily determine the error in approximating $\{\mathbf{x}\}$ with $\{\hat{\mathbf{x}}\}$. The error in representing $\{\mathbf{r}\}$ by $\{\mathbf{p}\}$ was easy to compute. We had

$$\frac{1}{N} \sum_i \left[(\mathbf{r}_i - \mathbf{p}_i)^T (\mathbf{r}_i - \mathbf{p}_i) \right] = \sum_{j=s+1}^{j=d} \mathrm{var}\left(\left\{ r^{(j)} \right\} \right) = \sum_{j=s+1}^{j=d} \lambda_j$$

If this sum is small compared to the sum of the first s components, then dropping the last $d - s$ components results in a small error.

The average error in representing $\{\mathbf{x}\}$ with $\{\hat{\mathbf{x}}\}$ is now easy to get. Rotations and translations do not change lengths. This means that

$$\frac{1}{N}\sum_i \|\mathbf{x}_i - \hat{\mathbf{x}}_i\|^2 = \frac{1}{N}\sum_i \|\mathbf{r}_i - \mathbf{p}_i\|^2 = \sum_{j=s+1}^{j=d} \lambda_j$$

which is easy to evaluate, because these are the values of the $d - s$ eigenvalues of $\mathsf{Covmat}(\{\mathbf{x}\})$ that we decided to ignore. Now we could choose s by identifying how much error we can tolerate. More usual is to plot the eigenvalues of the covariance matrix, and look for a "knee," like that in Fig. 5.5. You can see that the sum of remaining eigenvalues is small.

Procedure: 5.1 *Principal Components Analysis*

Assume we have a general dataset $\mathbf{x}_i$, consisting of N d-dimensional vectors. Now write $\Sigma = \mathsf{Covmat}(\{\mathbf{x}\})$ for the covariance matrix. Form $\mathcal{U}$, Λ, such that

$$\Sigma\mathcal{U} = \mathcal{U}\Lambda$$

(these are the eigenvectors and eigenvalues of Σ). Ensure that the entries of Λ are sorted in the decreasing order. Choose r, the number of dimensions you wish to represent. Typically, we do this by plotting the eigenvalues and looking for a "knee" (Fig. 5.5). It is quite usual to do this by hand.

Constructing a Low Dimensional Representation: For $1 \leq j \leq s$, write $\mathbf{u}_i$ for the ith column of $\mathcal{U}$. Represent the data point $\mathbf{x}_i$ as

$$\hat{\mathbf{x}}_i = \mathsf{mean}(\{\mathbf{x}\}) + \sum_{j=1}^{s} \left[\mathbf{u}_j^T(\mathbf{x}_i - \mathsf{mean}(\{\mathbf{x}\}))\right]\mathbf{u}_j$$

The error in this representation is

$$\frac{1}{N}\sum_i \|\mathbf{x}_i - \hat{\mathbf{x}}_i\|^2 = \sum_{j=s+1}^{j=d} \lambda_j$$

5.1.5 Extracting a Few Principal Components with NIPALS

If you remember the curse of dimension, you should have noticed something of a problem in my account of PCA. When I described the curse, I said one consequence was that forming a covariance matrix for high dimensional data is hard or impossible. Then I described PCA as a method to understand the important dimensions in high dimensional datasets. But PCA appears to rely on covariance, so I should not be able to form the principal components in the first place. In fact, we can form principal components without computing a covariance matrix.

I will now assume the dataset has zero mean, to simplify notation. This is easily achieved. You subtract the mean from each data item at the start, and add the mean back once you've finished. As usual, we have N data items, each a d-dimensional column vector. We will now arrange these into a matrix,

$$\mathcal{X} = \begin{pmatrix} \mathbf{x}_1^T \\ \mathbf{x}_2^T \\ \dots \\ \mathbf{x}_N^T \end{pmatrix}$$

where each *row* of the matrix is a data vector. Now assume we wish to recover the first principal component. This means we are seeking a vector $\mathbf{u}$ and a set of N numbers w_i such that $w_i \mathbf{u}$ is a good approximation to $\mathbf{x}_i$. Now we can stack the w_i into a column vector $\mathbf{w}$. We are asking that the matrix $\mathbf{w}\mathbf{u}^T$ be a good approximation to $\mathcal{X}$, in the sense that $\mathbf{w}\mathbf{u}^T$ encodes as much of the variance of $\mathcal{X}$ as possible.

The **Frobenius norm** is a term for the matrix norm obtained by summing squared entries of the matrix. We write

$$\|\mathcal{A}\|_F^2 = \sum_{i,j} a_{ij}^2.$$

In the exercises, you will show that the right choice of $\mathbf{w}$ and $\mathbf{u}$ minimizes the cost

$$\|\mathcal{X} - \mathbf{w}\mathbf{u}^T\|_F^2$$

which we can write as

$$C(\mathbf{w}, \mathbf{u}) = \sum_{ij} (x_{ij} - w_i u_j)^2.$$

Now we need to *find* the relevant $\mathbf{w}$ and $\mathbf{u}$. Notice there is not a unique choice, because the pair $(s\mathbf{w}, (1/s)\mathbf{u})$ works as well as the pair $(\mathbf{w}, \mathbf{u})$. We will choose $\mathbf{u}$ such that $\|\mathbf{u}\| = 1$. There is still not a unique choice, because you can flip the signs in $\mathbf{u}$ and $\mathbf{w}$, but this doesn't matter. At the right $\mathbf{w}$ and $\mathbf{u}$, the gradient of the cost function will be zero.

The gradient of the cost function is a set of partial derivatives with respect to components of $\mathbf{w}$ and $\mathbf{u}$. The partial with respect to w_k is

$$\frac{\partial C}{\partial w_k} = \sum_j (x_{kj} - w_k u_j) u_j$$

which can be written in matrix vector form as

$$\nabla_\mathbf{w} C = \left(\mathcal{X} - \mathbf{w}\mathbf{u}^T\right) \mathbf{u}.$$

Similarly, the partial with respect to u_l is

$$\frac{\partial C}{\partial u_l} = \sum_i \left(x_{il} - w_i u_l\right) w_i$$

which can be written in matrix vector form as

$$\nabla_{\mathbf{u}} C = \left(\mathcal{X}^T - \mathbf{u}\mathbf{w}^T\right)\mathbf{w}.$$

At the solution, these partial derivatives are zero. Notice that, if we know the right $\mathbf{u}$, then the equation $\nabla_{\mathbf{w}} C = 0$ is linear in $\mathbf{w}$. Similarly, if we know the right $\mathbf{w}$, then the equation $\nabla_{\mathbf{u}} C = 0$ is linear in $\mathbf{u}$. This suggests an algorithm. First, assume we have an estimate of $\mathbf{u}$, say $\mathbf{u}^{(n)}$. Then we could choose the $\mathbf{w}$ that makes the partial wrt $\mathbf{w}$ zero, so

$$\hat{\mathbf{w}} = \frac{\mathcal{X}\mathbf{u}^{(n)}}{\left(\mathbf{u}^{(n)}\right)^T \mathbf{u}^{(n)}}.$$

Now we can update the estimate of $\mathbf{u}$ by choosing a value that makes the partial wrt $\mathbf{u}$ zero, using our estimate $\hat{\mathbf{w}}$, to get

$$\hat{\mathbf{u}} = \frac{\mathcal{X}^T \hat{\mathbf{w}}}{(\hat{\mathbf{w}})^T \hat{\mathbf{w}}}.$$

We need to rescale to ensure that our estimate of $\mathbf{u}$ has unit length. Write $s = \sqrt{(\hat{\mathbf{u}})^T \hat{\mathbf{u}}}$ We get

$$\mathbf{u}^{(n+1)} = \frac{\hat{\mathbf{u}}}{s}$$

and

$$\mathbf{w}^{(n+1)} = s\hat{\mathbf{w}}.$$

This iteration can be started by choosing some row of $\mathcal{X}$ as $\mathbf{u}^{(0)}$. You can test for convergence by checking $\|\mathbf{u}^{(n+1)} - \mathbf{u}^{(n)}\|$. If this is small enough, then the algorithm has converged.

To obtain a second principal component, you form $\mathcal{X}^{(1)} = \mathcal{X} - \mathbf{w}\mathbf{u}^T$ and apply the algorithm to that. You can get many principal components like this, but it's not a good way to get all of them (eventually numerical issues mean the estimates are poor). The algorithm is widely known as NIPALS (for non-linear iterative partial least squares).

5.1.6 Principal Components and Missing Values

Now imagine our dataset has missing values. We assume that the values are not missing in inconvenient patterns—if, for example, the kth component was missing for every vector, then we'd have to drop it—but don't go into what precise kind of pattern is a problem. Your intuition should suggest that we can estimate a few principal components of the dataset without particular problems. The argument is as follows. Each entry of a covariance matrix is a form of average; estimating averages in the presence of missing values is straightforward; and, when we estimate a few principal components, we are estimating far fewer numbers than when we are estimating a whole covariance matrix, so we should be able to make something work. This argument is sound, if vague.

The whole point of NIPALS is that, if you want a few principal components, you don't need to use a covariance matrix. This simplifies thinking about missing values. NIPALS is quite forgiving of missing values, though missing values make it hard to use matrix notation. Recall I wrote the cost function as $C(\mathbf{w}, \mathbf{u}) = \sum_{ij}(x_{ij} - w_i u_j)^2$. Notice that missing data occurs in $\mathcal{X}$ because there are x_{ij} whose values we don't know, but there is no missing data in $\mathbf{w}$ or $\mathbf{u}$ (we're estimating the values, and we always have *some* estimate). We change the sum so that it ranges over only the known values, to get

$$C(\mathbf{w}, \mathbf{u}) = \sum_{ij \in \text{known values}} (x_{ij} - w_i u_j)^2.$$

Now we need a shorthand to ensure that sums run over only known values. Write $\mathcal{V}(k)$ for the set of column (resp. row) indices of known values for a given row (resp. column index) k. So $i \in \mathcal{V}(k)$ means all i such that x_{ik} is known *or* all i such that x_{ki} is known (the context will tell you which). We have

$$\frac{\partial C}{\partial w_k} = \sum_{j \in \mathcal{V}(k)} (x_{kj} - w_k u_j) u_j$$

and

$$\frac{\partial C}{\partial u_l} = \sum_{i \in \mathcal{V}(l)} (x_{il} - w_i u_l) w_i.$$

These partial derivatives must be zero at the solution. This means we can use $\mathbf{u}^{(n)}$, $\mathbf{w}^{(n)}$ to estimate

$$\hat{w}_k = \frac{\sum_{j \in \mathcal{V}(k)} x_{kj} u_j^{(n)}}{\sum_j u_j^{(n)} u_j^{(n)}}$$

and

$$\hat{u}_l = \frac{\sum_{i \in \mathcal{V}(l)} x_{il} \hat{w}_l}{\sum_i \hat{w}_i \hat{w}_i}$$

We then normalize as before to get $\mathbf{u}^{(n+1)}$, $\mathbf{w}^{(n+1)}$.

Procedure: 5.2   *Obtaining Some Principal Components with NIPALS*

We assume that $\mathcal{X}$ has zero mean. Each row is a data item. Start with $\mathbf{u}^0$ as some row of $\mathcal{X}$. Write $\mathcal{V}(k)$ for the set of indices of known values for a given row or column index k. Now iterate

- compute

$$\hat{w}_k = \frac{\sum_{j \in \mathcal{V}(k)} x_{kj} u^{(n)}{}_j}{\sum_j u_j^{(n)} u_j^{(n)}}$$

  and

$$\hat{u}_l = \frac{\sum_{i \in \mathcal{V}(l)} x_{il} \hat{w}_l}{\sum_i \hat{w}_i \hat{w}_i};$$

- compute $s = \sqrt{(\hat{\mathbf{u}})^T \hat{\mathbf{u}}}$, and

$$\mathbf{u}^{(n+1)} = \frac{\hat{\mathbf{u}}}{s}$$

  and

$$\mathbf{w}^{(n+1)} = s\hat{\mathbf{w}};$$

- Check for convergence by checking that $\| \mathbf{u}^{(n+1)} - \mathbf{u}^{(n)} \|$ is small.

This procedure yields a single principal component representing the highest variance in the dataset. To obtain the next principal component, replace $\mathcal{X}$ with $\mathcal{X} - \mathbf{w}\mathbf{u}^T$ and repeat the procedure. This process will yield good estimates of the first few principal components, but as you generate more principal components, numerical errors will become more significant.

5.1.7   PCA as Smoothing

Assume that each data item $\mathbf{x}_i$ is noisy. We use a simple noise model. Write $\tilde{\mathbf{x}}_i$ for the true underlying value of the data item, and ξ_i for the value of a normal random variable with zero mean and covariance $\sigma^2 \mathcal{I}$. Then we use the model

$$\mathbf{x}_i = \tilde{\mathbf{x}}_i + \xi_i$$

(so the noise in each component is independent, has zero mean, and has variance σ^2; this is known as **additive, zero mean, independent Gaussian noise**). You should think of the measurement $\mathbf{x}_i$ as an estimate of $\tilde{\mathbf{x}}_i$. A principal component analysis of $\mathbf{x}_i$ can produce an estimate of $\tilde{\mathbf{x}}_i$ that is closer than the measurements are.

There is a subtlety here, because the noise is random, but we see the values of the noise. This means that $\mathsf{Covmat}\,(\{\xi\})$ (i.e., the covariance of the observed numbers) is the value of a random variable (because the noise is random) whose mean is $\sigma^2 \mathcal{I}$ (because that's the model). The subtlety is that $\mathsf{mean}\,(\{\xi\})$ will not necessarily be exactly $\mathbf{0}$ and $\mathsf{Covmat}\,(\{\xi\})$ will not necessarily be exactly $\sigma^2\mathcal{I}$. The weak law of large numbers tells us that $\mathsf{Covmat}\,(\{\xi\})$ will be extremely close to its expected value (which is $\sigma^2\mathcal{I}$) for a large enough dataset. We will assume that $\mathsf{mean}\,(\{\xi\}) = \mathbf{0}$ and $\mathsf{Covmat}\,(\{\xi\}) = \sigma^2\mathcal{I}$.

The first step is to write Σ for the covariance matrix of the true underlying values of the data, and $\mathsf{Covmat}\,(\{\mathbf{x}\})$ for the covariance of the observed data. Then it is straightforward that

$$\mathsf{Covmat}\,(\{\mathbf{x}\}) = \tilde{\Sigma} + \sigma^2\mathcal{I}$$

because the noise is independent of the measurements. Notice that if $\mathcal{U}$ diagonalizes $\mathsf{Covmat}\,(\{\mathbf{x}\})$, it will also diagonalize $\tilde{\Sigma}$. Write $\tilde{\Lambda} = \mathcal{U}^T\tilde{\Sigma}\mathcal{U}$. We have

$$\mathcal{U}^T\mathsf{Covmat}\,(\{\mathbf{x}\})\mathcal{U} = \Lambda = \tilde{\Lambda} + \sigma^2\mathcal{I}.$$

Now think about the diagonal entries of Λ. If they are large, then they are quite close to the corresponding components of $\tilde{\Lambda}$, but if they are small, it is quite likely they are the result of noise. But these eigenvalues are tightly linked to error in a PCA representation.

In PCA (Procedure 5.1), the d-dimensional data point $\mathbf{x}_i$ is represented by

$$\hat{\mathbf{x}}_i = \mathsf{mean}\,(\{\mathbf{x}\}) + \sum_{j=1}^{s}\left[\mathbf{u}_j^T\left(\mathbf{x}_i - \mathsf{mean}\,(\{\mathbf{x}\})\right)\right]\mathbf{u}_j$$

where $\mathbf{u}_j$ are the principal components. This representation is obtained by setting the coefficients of the $d-s$ principal components with small variance to zero. The error in representing $\{\mathbf{x}\}$ with $\{\hat{\mathbf{x}}\}$ follows from Sect. 5.1.4 and is

$$\frac{1}{N}\sum_i \|\mathbf{x}_i - \hat{\mathbf{x}}_i\|^2 = \sum_{j=s+1}^{j=d}\lambda_j.$$

Now consider the error in representing $\tilde{\mathbf{x}}_i$ (which we don't know) by $\mathbf{x}_i$ (which we do). The average error over the whole dataset is

$$\frac{1}{N}\sum_i \|\mathbf{x}_i - \tilde{\mathbf{x}}_i\|^2.$$

Because the variance of the noise is $\sigma^2\mathcal{I}$, this error must be $d\sigma^2$. Alternatively, we could represent $\tilde{\mathbf{x}}_i$ by $\hat{\mathbf{x}}_i$. The average error of this representation over the whole dataset will be

$$\frac{1}{N}\sum_i \|\hat{\mathbf{x}}_i - \tilde{\mathbf{x}}_i\|^2 = \text{Error in components that are preserved}$$
$$+\text{Error in components that are zeroed}$$
$$= s\sigma^2 + \sum_{j=s+1}^{d}\tilde{\lambda}_u.$$

Now if, for $j > s$, $\tilde{\lambda}_j < \sigma^2$, this error is smaller than $d\sigma^2$. We don't know which s guarantees this unless we know σ^2 and $\tilde{\lambda}_j$ which often doesn't happen. But it's usually possible to make a safe choice, and so **smooth** the data by reducing noise. This smoothing works because the components of the data are correlated. So the best estimate of each component of a high dimensional data item is likely not the measurement—it's a prediction obtained from all measurements. The projection onto principal components is such a prediction.

> **Remember This:** *Given a d-dimensional dataset where data items have had independent random noise added to them, representing each data item on $s < d$ principal components can result in a representation which is on average closer to the true underlying data than the original data items. The choice of s is application dependent.*

5.2 Example: Representing Colors with Principal Components

Diffuse surfaces reflect light uniformly in all directions. Examples of diffuse surfaces include matte paint, many styles of cloth, many rough materials (bark, cement, stone, etc.). One way to tell a diffuse surface is that it does not look brighter (or darker) when you look at it along different directions. Diffuse surfaces can be colored, because the surface reflects different fractions of the light falling on it at different wavelengths. This effect can be represented by measuring the spectral reflectance of a surface, which is the fraction of light the surface reflects as a function of wavelength. This is usually measured in the visual range of wavelengths (about 380 nm to about 770 nm). Typical measurements are every few nm, depending on the measurement device. I obtained data for 1995 different surfaces from http://www.cs.sfu.ca/~colour/data/ (there are a variety of great datasets here, from Kobus Barnard).

Each spectrum has 101 measurements, which are spaced 4 nm apart. This represents surface properties to far greater precision than is really useful. Physical properties of surfaces suggest that the reflectance can't change too fast from wavelength to wavelength. It turns out that very few principal components are sufficient to describe almost any spectral reflectance function. Figure 5.5 shows the mean spectral reflectance of this dataset, and Fig. 5.5 shows the eigenvalues of the covariance matrix.

This is tremendously useful in practice. One should think of a spectral reflectance as a function, usually written $\rho(\lambda)$. What the principal components analysis tells us is that we can represent this function rather accurately on a (really small) finite dimensional basis. This basis is shown in Fig. 5.5. This means that there is a mean function $r(\lambda)$ and k functions $\phi_m(\lambda)$ such that, for any $\rho(\lambda)$,

$$\rho(\lambda) = r(\lambda) + \sum_{i=1}^{k} c_i \phi_i(\lambda) + e(\lambda)$$

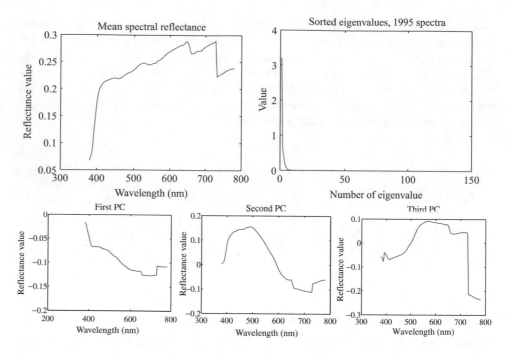

Figure 5.5: On the **top left**, the mean spectral reflectance of a dataset of 1995 spectral reflectances, collected by Kobus Barnard (at http://www.cs.sfu.ca/~colour/data/). On the **top right**, eigenvalues of the covariance matrix of spectral reflectance data, from a dataset of 1995 spectral reflectances, collected by Kobus Barnard (at http://www.cs.sfu.ca/~colour/data/). Notice how the first few eigenvalues are large, but most are very small; this suggests that a good representation using few principal components is available. The **bottom row** shows the first three principal components. A linear combination of these, with appropriate weights, added to the mean (**top left**), gives a good representation of the dataset

where $e(\lambda)$ is the error of the representation, which we know is small (because it consists of all the other principal components, which have tiny variance). In the case of spectral reflectances, using a value of k around 3–5 works fine for most applications (Fig. 5.6). This is useful, because when we want to predict what a particular object will look like under a particular light, we don't need to use a detailed spectral reflectance model; instead, it's enough to know the c_i for that object. This comes in useful in a variety of rendering applications in computer graphics. It is also the key step in an important computer vision problem, called **color constancy**. In this problem, we see a picture of a world of colored objects under unknown colored lights, and must determine what color the objects are. Modern color constancy systems are quite accurate, even though the problem sounds underconstrained. This is because they are able to exploit the fact that relatively few c_i are enough to accurately describe a surface reflectance.

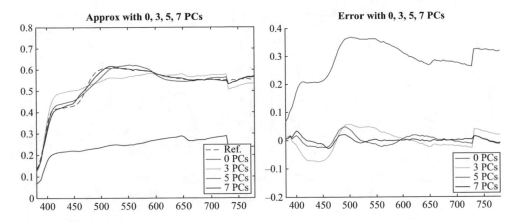

Figure 5.6: On the **left**, a spectral reflectance curve (dashed) and approximations using the mean, the mean and 3 principal components, the mean and 5 principal components, and the mean and 7 principal components. Notice the mean is a relatively poor approximation, but as the number of principal components goes up, the mean-squared distance between measurements and principal component representation falls rather quickly. On the **right** is this distance for these approximations. A projection onto very few principal components suppresses local wiggles in the data *unless* very many data items have the same wiggle in the same place. As the number of principal components increases, the representation follows the measurements more closely. The best estimate of each component of a data item is likely not the measurement—it's a prediction obtained from all measurements. The projection onto principal components is such a prediction, and you can see the smoothing effects of principal components analysis in these plots. Figure plotted from a dataset of 1995 spectral reflectances, collected by Kobus Barnard (at http://www.cs.sfu.ca/~colour/data/)

Figures 5.7 and 5.8 illustrate the smoothing process. I know neither the noise process nor the true variances (this is quite usual), so I can't say which smoothed representation is best. Each figure shows four spectral reflectances and their representation on a set of principal components. Notice how, as the number of principal components goes up, the measurements and the representation get closer together. This *doesn't* necessarily mean that more principal components are better—the measurement itself may be noisy. Notice also how representations on few principal components tend to suppress small local "wiggles" in the spectral reflectance. They are suppressed because these patterns tend not to appear in the same place in all spectral reflectances, so the most important principal components tend not to have them. The noise model tends to produce these patterns, so that the representation on a small set of principal components may well be a more accurate estimate of the spectral reflectance than the measurement is.

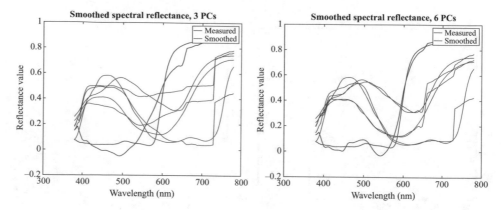

Figure 5.7: The best estimate of each component of a data item is likely not the measurement—it's a prediction obtained from all measurements. The projection onto principal components is such a prediction, and these plots show the smoothing effects of principal components analysis. Each figure shows four spectral reflectances, together with the smoothed version computed using principal components. A projection onto very few principal components suppresses local wiggles in the data *unless* very many data items have the same wiggle in the same place. As the number of principal components increases, the representation follows the measurements more closely. Figure 5.8 shows representations on more principal components

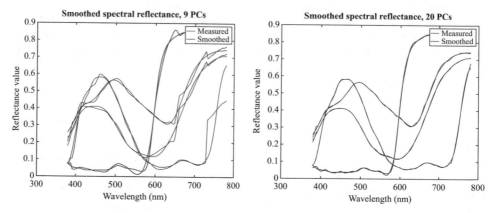

Figure 5.8: Each figure shows four spectral reflectances, together with the smoothed version computed using principal components. Compare with Fig. 5.7, and notice how few principal components are required to get a representation very close to the measurements (compare the eigenvalue plot in Fig. 5.5, which makes this point less directly). For some number of principal components, the smoothed representation is better than the measurements, though it's usually hard to be sure *which* number without knowing more about the measurement noise. If the measurement noise is very low, the wiggles are data; if it is high, the wiggles are likely noise

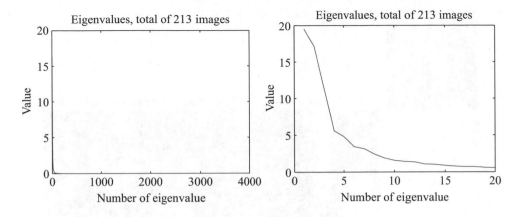

Figure 5.9:   On the **left**,the eigenvalues of the covariance of the Japanese facial expression dataset; there are 4096, so it's hard to see the curve (which is packed to the left). On the **right**, a zoomed version of the curve, showing how quickly the values of the eigenvalues get small

5.3   Example: Representing Faces with Principal Components

An image is usually represented as an array of values. We will consider intensity images, so there is a single intensity value in each cell. You can turn the image into a vector by rearranging it, for example, stacking the columns onto one another. This means you can take the principal components of a set of images. Doing so was something of a fashionable pastime in computer vision for a while, though there are some reasons that this is not a great representation of pictures. However, the representation yields pictures that can give great intuition into a dataset.

Figure 5.10 shows the mean of a set of face images encoding facial expressions of Japanese women (available at http://www.kasrl.org/jaffe.html; there are tons of face datasets at http://www.face-rec.org/databases/). I reduced the images to 64 × 64, which gives a 4096 dimensional vector. The eigenvalues of the covariance of this dataset are shown in Fig. 5.9; there are 4096 of them, so it's hard to see a trend, but the zoomed figure suggests that the first couple of hundred contain most of the variance. Once we have constructed the principal components, they can be rearranged into images; these images are shown in Fig. 5.10. Principal components give quite good approximations to real images (Fig. 5.11).

The principal components sketch out the main kinds of variation in facial expression. Notice how the mean face in Fig. 5.10 looks like a relaxed face, but with fuzzy boundaries. This is because the faces can't be precisely aligned, because each face has a slightly different shape. The way to interpret the components is to remember one adjusts the mean toward a data point by adding (or subtracting) some scale times the component. So the first few principal components have to do with the shape of the haircut; by the fourth, we are dealing with taller/shorter faces; then several components have to do with the height of the eyebrows, the shape of the chin, and the position of the mouth; and so on. These are all images of women who are not wearing spectacles. In face pictures taken from a wider set of

Mean image from Japanese Facial Expression dataset

First sixteen principal components of the Japanese Facial Expression dat

Figure 5.10: The mean and first 16 principal components of the Japanese facial expression dataset

models, moustaches, beards, and spectacles all typically appear in the first couple of dozen principal components.

A representation on enough principal components results in pixel values that are closer to the true values than the measurements (this is one sense of the word "smoothing"). Another sense of the word is blurring. Irritatingly, blurring reduces noise, and some methods for reducing noise, like principal components, also blur (Fig. 5.11). But this doesn't mean the resulting images are better *as images*. In fact, you don't have to blur an image to smooth it. Producing images that are both accurate estimates of the true values and look like sharp, realistic images require quite substantial technology, beyond our current scope.

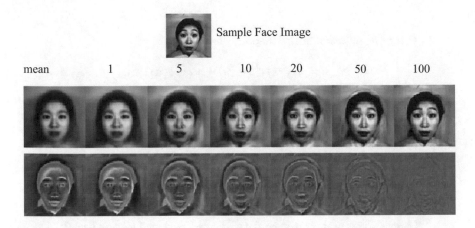

Figure 5.11: Approximating a face image by the mean and some principal components; notice how good the approximation becomes with relatively few components

5.4 You Should

5.4.1 Remember These Terms

5.4.2 Remember These Facts

5.4.3 Remember These Procedures

5.4.4 Be Able to

- Create, plot, and interpret the first few principal components of a dataset.
- Compute the error resulting from ignoring some principal components.
- Interpret the principal components of a dataset.

Problems

5.1. Using the notation of the chapter, show that

$$w_{ij} = r_i^{(j)} = (\mathbf{x}_i - \mathsf{mean}\,(\{\mathbf{x}\}))^T \mathbf{u}_j.$$

5.2. We have N d-dimensional data items forming a dataset $\{\mathbf{x}\}$. We translate this dataset to have zero mean, compute

$$\mathcal{U}^T \mathsf{Covmat}\,(\{\mathbf{x}\})\mathcal{U} = \Lambda$$

and form the dataset $\{\mathbf{r}\}$, using the rule

$$\mathbf{r}_i = \mathcal{U}^T \mathbf{m}_i = \mathcal{U}^T(\mathbf{x}_i - \mathsf{mean}\,(\{\mathbf{x}\})).$$

Choose some $s<d$, take each data point $\mathbf{r}_i$, and replace the last $d-s$ components with 0. Call the resulting data item $\mathbf{p}_i$.
(a) Show that

$$\frac{1}{N}\sum_i \left[(\mathbf{r}_i - \mathbf{p}_i)^T (\mathbf{r}_i - \mathbf{p}_i)\right] = \sum_{j=s+1}^{j=d} \mathsf{var}\,\left(\left\{r^{(j)}\right\}\right).$$

(b) Sort the eigenvalues of $\mathsf{Covmat}\,(\{\mathbf{x}\})$ in the descending order, and write λ_i for the ith (so that $\lambda_1 \geq \lambda_2 \cdots \geq \lambda_N$). Show that

$$\frac{1}{N}\sum_i \left[(\mathbf{r}_i - \mathbf{p}_i)^T (\mathbf{r}_i - \mathbf{p}_i)\right] = \sum_{j=s+1}^{j=d} \lambda_j.$$

5.3. You have a dataset of N vectors $\mathbf{x}_i$ in d-dimensions, stacked into a matrix $\mathcal{X}$. This dataset has zero mean. You would like to determine the principal component of this dataset corresponding to the largest eigenvalue of its covariance. Write $\mathbf{u}$ for this principal component.
(a) The Frobenius norm is a term for the matrix norm obtained by summing squared entries of the matrix. We write

$$\|\mathcal{A}\|_F^2 = \sum_{i,j} a_{ij}^2.$$

Show that

$$\|\mathcal{A}\|_F^2 = \mathsf{Trace}(\mathcal{A}\mathcal{A}^T)$$

(b) Show that

$$\mathsf{Trace}(\mathcal{A}\mathcal{B}) = \mathsf{Trace}(\mathcal{B}\mathcal{A}).$$

I have found this fact worth remembering. It may help to remember the trace is defined only for square matrices.
(c) Show that, if $\mathbf{u}$ and $\mathbf{w}$ together minimize

$$\|\mathcal{X} - \mathbf{w}\mathbf{u}^T\|_F^2$$

then

$$\left(\mathbf{w}^T\mathbf{w}\right)\mathbf{u} = \mathcal{X}^T\mathbf{w}$$
$$\left(\mathbf{u}^T\mathbf{u}\right)\mathbf{w} = \mathcal{X}\mathbf{u}$$

Do this by differentiating and setting to zero, the text of the NIPALS section should help.

(d) $\mathbf{u}$ is a unit vector—why?

(e) Show that

$$\mathcal{X}^T\mathcal{X}\mathbf{u} = \left(\mathbf{w}^T\mathbf{w}\right)\mathbf{u}$$

and so that, if $\mathbf{u}$ minimizes the Frobenius norm as above, it must be some eigenvector of $\mathsf{Covmat}\left(\{x\}\right)$.

(f) Show that, if $\mathbf{u}$ is a unit vector, then

$$\mathrm{Trace}(\mathbf{u}\mathbf{u}^T) = 1$$

(g) Assume that $\mathbf{u}$, $\mathbf{w}$ satisfy the equations for a minimizer, above, then show

$$\|\mathcal{X} - \mathbf{w}\mathbf{u}^T\|_F^2 = \mathrm{Trace}(\mathcal{X}^T\mathcal{X} - \mathbf{u}(\mathbf{w}^T\mathbf{w})\mathbf{u}^T)$$
$$= \mathrm{Trace}(\mathcal{X}^T\mathcal{X}) - (\mathbf{w}^T\mathbf{w})$$

(h) Use the information above to argue that if $\mathbf{u}$ and $\mathbf{w}$ together minimize

$$\|\mathcal{X} - \mathbf{w}\mathbf{u}^T\|_F^2$$

then $\mathbf{u}$ is the eigenvector of $\mathcal{X}^T\mathcal{X}$ corresponding to the largest eigenvalue.

5.4. You have a dataset of N vectors $\mathbf{x}_i$ in d-dimensions, stacked into a matrix $\mathcal{X}$. This dataset has zero mean. You would like to determine the principal component of this dataset corresponding to the largest eigenvalue of its covariance. Write $\mathbf{u}$ for this principal component. Assume that each data item $\mathbf{x}_i$ is noisy. We use a simple noise model. Write $\tilde{\mathbf{x}}_i$ for the true underlying value of the data item, and ξ_i for the value of a normal random variable with zero mean and covariance $\sigma^2\mathcal{I}$. Then we use the model

$$\mathbf{x}_i = \tilde{\mathbf{x}}_i + \xi_i$$

We will assume that $\mathsf{mean}\left(\{\xi\}\right) = \mathbf{0}$ and $\mathsf{Covmat}\left(\{\xi\}\right) = \sigma^2\mathcal{I}$.

(a) Notice that the noise is independent of the dataset. This means that $\mathsf{mean}\left(\{\mathbf{x}\xi^T\}\right) = \mathsf{mean}\left(\{\mathbf{x}\}\right)\mathsf{mean}\left(\{\xi^T\}\right) = 0$. Show that

$$\mathsf{Covmat}\left(\{\mathbf{x}\}\right) = \tilde{\Sigma} + \sigma^2\mathcal{I}.$$

(b) Show that if $\mathcal{U}$ diagonalizes $\mathsf{Covmat}\left(\{\mathbf{x}\}\right)$, it will also diagonalize $\tilde{\Sigma}$.

Programming Exercises

5.5. Obtain the iris dataset from the UC Irvine machine learning data repository at http://https://archive.ics.uci.edu/ml/machine-learning-databases/iris/iris.data.

(a) Plot a scatterplot matrix of this dataset, showing each species with a different marker.

(b) Now obtain the first two principal components of the data. Plot the data on those two principal components alone, again showing each species with a different marker. Has this plot introduced significant distortions? Explain

5.6. Take the wine dataset from the UC Irvine machine learning data repository at https://archive.ics.uci.edu/ml/datasets/Wine.

(a) Plot the eigenvalues of the covariance matrix in sorted order. How many principal components should be used to represent this dataset? Why?

(b) Construct a stem plot of each of the first 3 principal components (i.e., the eigenvectors of the covariance matrix with largest eigenvalues). What do you see?

(c) Compute the first two principal components of this dataset, and project it onto those components. Now produce a scatterplot of this two-dimensional dataset, where data items of class 1 are plotted as a "1," class 2 as a "2," and so on.

5.7. Take the wheat kernel dataset from the UC Irvine machine learning data repository at http://archive.ics.uci.edu/ml/datasets/seeds. Compute the first two principal components of this dataset, and project it onto those components.

(a) Produce a scatterplot of this projection. Do you see any interesting phenomena?

(b) Plot the eigenvalues of the covariance matrix in sorted order. How many principal components should be used to represent this dataset? why?

5.8. The UC Irvine machine learning data repository hosts a collection of data on breast cancer diagnostics, donated by Olvi Mangasarian, Nick Street, and William H. Wolberg. You can find this data at http://archive.ics.uci.edu/ml/datasets/Breast+Cancer+Wisconsin+(Diagnostic). For each record, there is an id number, 10 continuous variables, and a class (benign or malignant). There are 569 examples. Separate this dataset randomly into 100 validation, 100 test, and 369 training examples. Plot this dataset on the first three principal components, using different markers for benign and malignant cases. What do you see?

5.9. The UC Irvine Machine Learning data archive hosts a dataset of measurements of abalone at http://archive.ics.uci.edu/ml/datasets/Abalone. Compute the principal components of all variables except Sex. Now produce a scatterplot of the measurements projected onto the first two principal components, plotting an "m" for male abalone, an "f" for female abalone, and an "i" for infants. What do you see?

5.10. Obtain the iris dataset from the UC Irvine machine learning data repository at http://https://archive.ics.uci.edu/ml/machine-learning-databases/iris/iris.data. We will investigate the use of principal components to smooth data.

(a) Ignore the species names, so you should have 150 data items with four measurements each. For each value in $\{0.1, 0.2, 0.5, 1\}$, form a dataset by adding an independent sample from a normal distribution with this standard deviation to *each* entry in the original dataset. Now for each value, plot the mean-squared error between the original dataset and an expansion onto 1, 2, 3, and 4 principal components. You should see that, as the noise gets larger, using fewer principal components gives a more accurate estimate of the original dataset (i.e., the one without noise).

(b) We will now try the previous subexercise with a very much different noise model. For each of $w = \{10, 20, 30, 40\}$, construct a mask matrix each of whose entries is a sample of a binomial random variable with probability $p = 1 - w/600$ of turning up 1. This matrix should have about w zeros in it. Ignore the species names, so you should have 150 data items with four measurements each. Now form a new dataset by multiplying each location in the original dataset by the corresponding mask location (so you are randomly setting a small set of measurements to zero). Now for each value of w, plot the mean-squared error between the original dataset and an expansion onto 1, 2, 3, and 4 principal components. You should see that, as the noise gets larger, using fewer principal components gives a more accurate estimate of the original dataset (i.e., the one without noise).

CHAPTER 6

Low Rank Approximations

A principal components analysis models high dimensional data points with an accurate, low dimensional, model. Now form a data matrix from the approximate points. This data matrix must have low rank (because the model is low dimensional) *and* it must be close to the original data matrix (because the model is accurate). This suggests modelling data with a low rank matrix.

Assume we have data in $\mathcal{X}$, with rank d, and we wish to produce $\mathcal{X}_s$ such that (a) the rank of $\mathcal{X}_s$ is s (which is less than d) and (b) such that $\|\mathcal{X} - \mathcal{X}_s\|^2$ is minimized. The resulting $\mathcal{X}_s$ is called a **low rank approximation** to $\mathcal{X}$. Producing a low rank approximation is a straightforward application of the singular value decomposition (SVD).

We have already seen examples of useful low rank approximations. NIPALS— which is actually a form of partial SVD—produces a rank one approximation to a matrix (check this point if you're uncertain). A new, and useful, application is to use a low rank approximation to make a low dimensional map of a high dimensional dataset (Sect. 6.2).

The link between principal components analysis and low rank approximation suggests (correctly) that you can use a low rank approximation to smooth and suppress noise. Smoothing is extremely powerful, and Sect. 6.3 describes an important application. The count of words in a document gives a rough representation of the document's meaning. But there are many different words an author could use for the same idea ("spanner" or "wrench," say), and this effect means that documents with quite similar meaning could have quite different word counts. Word counts can be smoothed very effectively with a low rank approximation to an appropriate matrix. There are two quite useful applications. First, this low rank approximation yields quite good measures of how similar documents are. Second, the approximation can yield a representation of the underlying meaning of a word which is useful in dealing with unfamiliar words.

6.1 The Singular Value Decomposition

For any $m \times p$ matrix $\mathcal{X}$, it is possible to obtain a decomposition

$$\mathcal{X} = \mathcal{U}\Sigma\mathcal{V}^T$$

where $\mathcal{U}$ is $m \times m$, $\mathcal{V}$ is $p \times p$, and Σ is $m \times p$ and is diagonal. The diagonal entries of Σ are non-negative. Both $\mathcal{U}$ and $\mathcal{V}$ are orthonormal (i.e., $\mathcal{U}\mathcal{U}^T = \mathcal{I}$ and $\mathcal{V}\mathcal{V}^T = \mathcal{I}$). This decomposition is known as the **singular value decomposition**, almost always abbreviated to **SVD**.

If you don't recall what a diagonal matrix looks like when the matrix *isn't* square, it's simple. All entries are zero, except the i, i entries for i in the range 1 to

© Springer Nature Switzerland AG 2019
D. Forsyth, *Applied Machine Learning*,
https://doi.org/10.1007/978-3-030-18114-7_6

$\min(m, p)$. So if Σ is tall and thin, the top square is diagonal and everything else is zero; if Σ is short and wide, the left square is diagonal and everything else is zero. The terms on the diagonal of Σ are usually called the **singular values**. There is a significant literature on methods to compute the SVD efficiently, accurately, and at large scale, which we ignore: any decent computing environment should do this for you if you find the right function. Read the manual for your environment.

Procedure: 6.1 *Singular Value Decomposition*

Given a matrix $\mathcal{X}$, any halfway decent numerical linear algebra package or computing environment will produce a decomposition

$$\mathcal{X} = \mathcal{U}\Sigma\mathcal{V}^T$$

and $\mathcal{U}$ and $\mathcal{V}$ are both orthonormal, Σ is diagonal with non-negative entries. Most environments that can do an SVD can be persuaded to provide the columns of $\mathcal{U}$ and rows of $\mathcal{V}^T$ corresponding to the k largest singular values.

There are many SVDs for a given matrix, because you could reorder the singular values and then reorder $\mathcal{U}$ and $\mathcal{V}$. We will always assume that the diagonal entries in Σ go from largest to smallest as one moves down the diagonal. In this case, the columns of $\mathcal{U}$ and the rows of $\mathcal{V}^T$ corresponding to non-zero diagonal elements of Σ are unique.

Notice that there is a relationship between forming an SVD and diagonalizing a matrix. In particular, $\mathcal{X}^T\mathcal{X}$ is symmetric, and it can be diagonalized as

$$\mathcal{X}^T\mathcal{X} = \mathcal{V}\Sigma^T\Sigma\mathcal{V}^T.$$

Similarly, $\mathcal{X}\mathcal{X}^T$ is symmetric, and it can be diagonalized as

$$\mathcal{X}\mathcal{X}^T = \mathcal{U}\Sigma\Sigma^T\mathcal{U}.$$

Remember This: *A singular value decomposition (SVD) decomposes a matrix $\mathcal{X}$ as $\mathcal{X} = \mathcal{U}\Sigma\mathcal{V}^T$ where $\mathcal{U}$ is $m \times m$, $\mathcal{V}$ is $p \times p$, and Σ is $m \times p$ and is diagonal. The diagonal entries of Σ are non-negative. Both $\mathcal{U}$ and $\mathcal{V}$ are orthonormal. The SVD of $\mathcal{X}$ yields the diagonalization of $\mathcal{X}^T\mathcal{X}$ and the diagonalization of $\mathcal{X}\mathcal{X}^T$.*

6.1.1 SVD and PCA

Now assume we have a dataset with zero mean. As usual, we have N data items, each a d- dimensional column vector. We will now arrange these into a matrix,

$$\mathcal{X} = \begin{pmatrix} \mathbf{x}_1^T \\ \mathbf{x}_2^T \\ \dots \\ \mathbf{x}_N^T \end{pmatrix}$$

where each *row* of the matrix is a data vector. The covariance matrix is

$$\mathsf{Covmat}\left(\{X\}\right) = \frac{1}{N}\mathcal{X}^T\mathcal{X}$$

(zero mean, remember). Form the SVD of $\mathcal{X}$, to get

$$\mathcal{X} = \mathcal{U}\Sigma\mathcal{V}^T.$$

But we have $\mathcal{X}^T\mathcal{X} = \mathcal{V}\Sigma^T\Sigma\mathcal{V}^T$ so that

$$\mathsf{Covmat}\left(\{X\}\right)\mathcal{V} = \frac{1}{N}\left(\mathcal{X}^T\mathcal{X}\right)\mathcal{V} = \mathcal{V}\frac{\Sigma^T\Sigma}{N}$$

and $\Sigma^T\Sigma$ is diagonal. By pattern matching, the columns of $\mathcal{V}$ contain the principal components of $\mathcal{X}$, and

$$\frac{\Sigma^T\Sigma}{N}$$

are the variances on each component. All this means we can read the principal components of a dataset of the SVD of that dataset, without actually forming the covariance matrix—we just form the SVD of $\mathcal{X}$, and the columns of $\mathcal{V}$ are the principal components. Remember, these are the columns of $\mathcal{V}$—it's easy to get mixed up about $\mathcal{V}$ and $\mathcal{V}^T$ here.

We have seen NIPALS as a way of extracting some principal components from a data matrix. In fact, NIPALS is a method to recover a partial SVD of $\mathcal{X}$. Recall that NIPALS produces a vector $\mathbf{u}$ and a vector $\mathbf{w}$ so that $\mathbf{wu}^T$ is as close as possible to $\mathcal{X}$, and $\mathbf{u}$ is a unit vector. By pattern matching, we have that

- $\mathbf{u}^T$ is the row of $\mathcal{V}^T$ corresponding to the largest singular value;
- $\frac{\mathbf{w}}{\|\mathbf{w}\|}$ is the column of $\mathcal{U}$ corresponding to the largest singular value;
- $\|\mathbf{w}\|$ is the largest singular value.

It is easy to show that if you use NIPALS to extract several principal components, you will get several rows of $\mathcal{V}^T$, several columns of $\mathcal{U}$, and several singular values. Be careful, however: this isn't an efficient or accurate way to extract many singular values, because numerical errors accumulate. If you want a partial SVD with many singular values, you should be searching for specialist packages, not making your own.

> **Remember This:** *Assume $\mathcal{X}$ has zero mean. Then the SVD of $\mathcal{X}$ yields the principal components of the dataset represented by this matrix. NIPALS is a method to recover a partial SVD of $\mathcal{X}$. There are other specialized methods.*

6.1.2 SVD and Low Rank Approximations

Assume we have $\mathcal{X}$, with rank d, and we wish to produce $\mathcal{X}_s$ such that (a) the rank of $\mathcal{X}_s$ is s (which is less than d) and (b) such that $\|\mathcal{X} - \mathcal{X}_s\|^2$ is minimized. An SVD will yield $\mathcal{X}_s$. Take the SVD to get $\mathcal{X} = \mathcal{U}\Sigma\mathcal{V}^T$. Now write Σ_s for the matrix obtained by setting all but the s largest singular values in Σ to 0. We have that

$$\mathcal{X}_s = \mathcal{U}\Sigma_s\mathcal{V}^T.$$

It is obvious that $\mathcal{X}_s$ has rank s. You can show (exercises) that $\|\mathcal{X} - \mathcal{X}_s\|^2$ is minimized, by noticing that $\|\mathcal{X} - \mathcal{X}_s\|^2 = \|\Sigma - \Sigma_s\|^2$.

There is one potential point of confusion. There are a lot of zeros in Σ_s, and they render most of the columns of $\mathcal{U}$ and rows of $\mathcal{V}^T$ irrelevant. In particular, write $\mathcal{U}_s$ for the $m \times s$ matrix consisting of the first s columns of $\mathcal{U}$, and so on; and write $\Sigma_s^{(s)}$ for the $s \times s$ submatrix of Σ_s with non-zero diagonal. Then we have

$$\mathcal{X}_s = \mathcal{U}\Sigma_s\mathcal{V}^T = \mathcal{U}_s\Sigma_s^{(s)}(\mathcal{V}_s)^T$$

and it is quite usual to switch from one representation to the other without comment. I try not to switch notation like this, but it's quite common practice.

6.1.3 Smoothing with the SVD

As we have seen, principal components analysis can smooth noise in the data matrix (Sect. 5.1.7). That argument was for one particular kind of noise, but experience shows that PCA can smooth other kinds of noise (there is an example in the exercises for Chap. 5). This means that the entries of a data matrix can be smoothed by computing a low rank approximation of $\mathcal{X}$.

I have already shown that PCA can smooth data. In PCA (Procedure 5.1), the d-dimensional data point $\mathbf{x}_i$ is represented by

$$\hat{\mathbf{x}}_i = \mathsf{mean}\left(\{\mathbf{x}\}\right) + \sum_{j=1}^{s}\left[\mathbf{u}_j^T\left(\mathbf{x}_i - \mathsf{mean}\left(\{\mathbf{x}\}\right)\right)\right]\mathbf{u}_j$$

where $\mathbf{u}_j$ are the principal components. A low rank approximation represents the ith row of $\mathcal{X}$ (which is $\mathbf{x}_i^T$) as

$$\hat{\mathbf{x}}_i^T = \sum_{j=1}^{r} w_{ij}\mathbf{v}_j^T$$

where $\mathbf{v}_j^T$ is a row of $\mathcal{V}^T$ (obtained from the SVD) and where w_{ij} are weights that can be computed from the SVD. In each case, the data point is represented by a projection onto a low dimensional space, so it is fair to conclude the SVD can smooth something.

Just like smoothing with a PCA, smoothing with an SVD works for a wide range of noise processes. In one very useful example, each component of the data might be a count. For concreteness, let the entries be counts of roadkill species per mile of highway. Each row would correspond to a species, each column to a particular mile. Counts like this would typically be noisy, because you see rare species only occasionally. At least for rare species, the count for most miles would be 0, but occasionally, you would count 1. The 0 is too low a per-mile estimate, and the 1 is too high, but one doesn't see a fraction of a roadkill (ideally!). Constructing a low rank approximation tends to lead to better estimates of the counts.

Missing data is a particularly interesting form of noise—the noise process deletes entries in the data matrix—and low rank approximations are quite effective in dealing with this. Assume you know most, but not all, entries of $\mathcal{X}$. You would like to build an estimate of the whole matrix. If you expect that the true whole matrix has low rank, you can compute a low rank approximation to the matrix. For example, the entries in the data matrix are scores of how well a viewer liked a film. Each row of the data matrix corresponds to one viewer; each column corresponds to one film. At useful scales, most viewers haven't seen most films, so most of the data matrix is missing data. However, there is good reason to believe that users are "like" each other—the rows are unlikely to be independent, because if two viewers both like (say) horror movies they might very well also both dislike (say) documentaries. Films are "like" each other, too. Two horror movies are quite likely to be liked by viewers who like horror movies but dislike documentaries. All this means that the rows (resp. columns) of the true data matrix are very likely to be highly dependent. More formally, the true data matrix is likely to have low rank. This suggests using an SVD to fill in the missing values.

Numerical and algorithmic questions get tricky here. If the rank is very low, you could use NIPALS to manage the question of missing entries. If you are dealing with a larger rank, or many missing values, you need to be careful about numerical error, and you should be searching for specialist packages, not making your own with NIPALS.

Remember This: *Taking an SVD of a data matrix usually produces a smoothed estimate of the data matrix. Smoothing is guaranteed to be effective if the entries are subject to additive, zero mean, independent Gaussian noise, but often works very well if the entries are noisy counts. Smoothing can be used to fill in missing values, too.*

6.2 Multidimensional Scaling

One way to get insight into a dataset is to plot it. But choosing what to plot for a high dimensional dataset could be difficult. Assume we must plot the dataset in two dimensions (by far the most common choice). We wish to build a scatterplot in two dimensions—but where should we plot each data point? One natural requirement is that the points be laid out in two dimensions in a way that reflects how they sit in many dimensions. In particular, we would like points that are far apart in the high dimensional space to be far apart in the plot, and points that are close in the high dimensional space to be close in the plot.

6.2.1 Choosing Low D Points Using High D Distances

We will plot the high dimensional point $\mathbf{x}_i$ at $\mathbf{y}_i$, which is an s-dimensional vector (almost always, s will be 2 or 3). Now the squared distance between points i and j in the high dimensional space is

$$D_{ij}^{(2)}(\mathbf{x}) = (\mathbf{x}_i - \mathbf{x}_j)^T (\mathbf{x}_i - \mathbf{x}_j)$$

(where the superscript is to remind you that this is a squared distance). We could build an $N \times N$ matrix of squared distances, which we write $\mathcal{D}^{(2)}(\mathbf{x})$. The i, jth entry in this matrix is $D_{ij}^{(2)}(\mathbf{x})$, and the $\mathbf{x}$ argument means that the distances are between points in the high dimensional space. Now we could choose the $\mathbf{y}_i$ to make

$$\sum_{ij} \left(D_{ij}^{(2)}(\mathbf{x}) - D_{ij}^{(2)}(\mathbf{y}) \right)^2$$

as small as possible. Doing so should mean that points that are far apart in the high dimensional space are far apart in the plot, and that points that are close in the high dimensional space are close in the plot.

In its current form, the expression is difficult to deal with, but we can refine it. Because translation does not change the distances between points, it cannot change either of the $\mathcal{D}^{(2)}$ matrices. So it is enough to solve the case when the mean of the points $\mathbf{x}_i$ is zero. We assume that the mean of the points is zero, so

$$\frac{1}{N} \sum_i \mathbf{x}_i = \mathbf{0}.$$

Now write $\mathbf{1}$ for the n-dimensional vector containing all ones, and $\mathcal{I}$ for the identity matrix. Notice that

$$D_{ij}^{(2)} = (\mathbf{x}_i - \mathbf{x}_j)^T (\mathbf{x}_i - \mathbf{x}_j) = \mathbf{x}_i \cdot \mathbf{x}_i - 2\mathbf{x}_i \cdot \mathbf{x}_j + \mathbf{x}_j \cdot \mathbf{x}_j.$$

Now write

$$\mathcal{A} = \left[\mathcal{I} - \frac{1}{N}\mathbf{1}\mathbf{1}^T \right].$$

Now you can show that

$$-\frac{1}{2}\mathcal{A}\mathcal{D}^{(2)}(\mathbf{x})\mathcal{A}^T = \mathcal{X}\mathcal{X}^T.$$

I now argue that, to make $\mathcal{D}^{(2)}(\mathbf{y})$ is close to $\mathcal{D}^{(2)}(\mathbf{x})$, it is enough to choose $\mathbf{y}_i$ so that $\mathcal{Y}\mathcal{Y}^T$ close to $\mathcal{X}\mathcal{X}^T$. Proving this will take us out of our way unnecessarily, so I omit a proof.

6.2.2 Using a Low Rank Approximation to Factor

We need to find a set of $\mathbf{y}_i$ so that (a) the $\mathbf{y}_i$ are s-dimensional and (b) $\mathcal{Y}$ (the matrix made by stacking the $\mathbf{y}_i$) minimizes the distance between $\mathcal{Y}\mathcal{Y}^T$ and $\mathcal{X}\mathcal{X}^T$. Notice that $\mathcal{Y}\mathcal{Y}^T$ must have rank s.

Now form an SVD of $\mathcal{X}$, to get

$$\mathcal{X} = \mathcal{U}\Sigma\mathcal{V}^T$$

Recall $\Sigma_s^{(s)}$ is the $s \times s$ submatrix of Σ_s with non-zero diagonal, $\mathcal{U}_s$ is the $m \times s$ matrix consisting of the first s columns of $\mathcal{U}$, and so on. Consider

$$\mathcal{X}_s = \mathcal{U}_s\Sigma_s\mathcal{V}_s^T.$$

We have that $\mathcal{X}_s\mathcal{X}_s^T$ is the closest rank s approximation to $\mathcal{X}\mathcal{X}^T$. The rows of $\mathcal{X}_s$ are d-dimensional, so it isn't the matrix we seek. But

$$\mathcal{X}_s\mathcal{X}_s^T = \left(\mathcal{U}_s\Sigma_s\mathcal{V}_s^T\right)\left(\mathcal{V}_s\Sigma_s\mathcal{U}_s^T\right)$$

and $\mathcal{V}_s^T\mathcal{V}_s$ is the $s \times s$ identity matrix. This means that

$$\mathcal{Y} = \mathcal{U}_s\Sigma_s$$

is the matrix we seek. We can obtain $\mathcal{Y}$ even if we don't know $\mathcal{X}$. It is enough to know $\mathcal{X}\mathcal{X}^T$. This is because

$$\mathcal{X}\mathcal{X}^T = \left(\mathcal{U}\Sigma\mathcal{V}^T\right)\left(\mathcal{V}\Sigma\mathcal{U}^T\right) = \mathcal{U}\Sigma^2\mathcal{U}^T$$

so diagonalizing $\mathcal{X}\mathcal{X}^T$ is enough. This method for constructing a plot is known as **principal coordinate analysis**.

This plot might not be perfect, because reducing the dimension of the data points should cause some distortions. In many cases, the distortions are tolerable. In other cases, we might need to use a more sophisticated scoring system that penalizes some kinds of distortion more strongly than others. There are many ways to do this; the general problem is known as **multidimensional scaling**. I pick up this theme in Sect. 19.1, which demonstrates more sophisticated methods for the problem.

Procedure: 6.2 *Principal Coordinate Analysis*

Assume we have a matrix $D^{(2)}$ consisting of the squared differences between each pair of N points. We do not need to know the points. We wish to compute a set of points in s dimensions, such that the distances between these points are as similar as possible to the distances in $D^{(2)}$.

- Form $\mathcal{A} = \left[\mathcal{I} - \frac{1}{N}\mathbf{1}\mathbf{1}^T\right]$.
- Form $\mathcal{W} = \frac{1}{2}\mathcal{A}\mathcal{D}^{(2)}\mathcal{A}^T$.
- Form $\mathcal{U}$, Λ, such that $\mathcal{W}\mathcal{U} = \mathcal{U}\Lambda$ (these are the eigenvectors and eigenvalues of $\mathcal{W}$). Ensure that the entries of Λ are sorted in the decreasing order. Notice that you need only the top s eigenvalues and their eigenvectors, and many packages can extract these rather faster than constructing all.
- Choose s, the number of dimensions you wish to represent. Form Λ_s, the top left $s \times s$ block of Λ. Form $\Lambda_s^{(1/2)}$, whose entries are the positive square roots of Λ_s. Form $\mathcal{U}_s$, the matrix consisting of the first s columns of $\mathcal{U}$.

Then

$$\mathcal{Y} = \mathcal{U}_s \Sigma_s = \begin{bmatrix} \mathbf{y}_1 \\ \dots \\ \mathbf{y}_N \end{bmatrix}$$

is the set of points to plot.

6.2.3 Example: Mapping with Multidimensional Scaling

Multidimensional scaling gets positions (the $\mathcal{Y}$ of Sect. 6.2.1) from distances (the $\mathcal{D}^{(2)}(\mathbf{x})$ of Sect. 6.2.1). This means we can use the method to build maps from distances alone. I collected distance information from the web (I used http://www.distancefromto.net, but a Google search on "city distances" yields a wide range of possible sources), then applied multidimensional scaling. I obtained distances between the South African provincial capitals, in kilometers. I then used principal coordinate analysis to find positions for each capital, and rotated, translated, and scaled the resulting plot to check it against a real map (Fig. 6.1).

One natural use of principal coordinate analysis is to see if one can spot any structure in a dataset. Does the dataset form a blob, or is it clumpy? This isn't a perfect test, but it's a good way to look and see if anything interesting is happening. In Fig. 6.2, I show a 3D plot of the spectral data, reduced to three dimensions using principal coordinate analysis. The plot is quite interesting. You should notice that the data points are spread out in 3D, but actually seem to lie on a complicated curved surface—they very clearly don't form a uniform blob. To me, the structure looks somewhat like a butterfly. I don't know why this occurs (perhaps the universe is doodling), but it certainly suggests that something worth investigating is going on. Perhaps the choice of samples that were measured is funny; perhaps

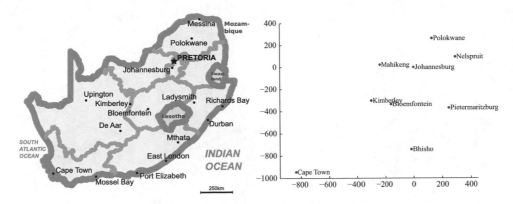

Figure 6.1: On the **left**, a public domain map of South Africa, obtained from http://commons.wikimedia.org/wiki/File:Map_of_South_Africa.svg, and edited to remove surrounding countries. On the **right**, the locations of the cities inferred by multidimensional scaling, rotated, translated, and scaled to allow a comparison to the map by eye. The map doesn't have all the provincial capitals on it, but it's easy to see that MDS has placed the ones that are there in the right places (use a piece of ruled tracing paper to check)

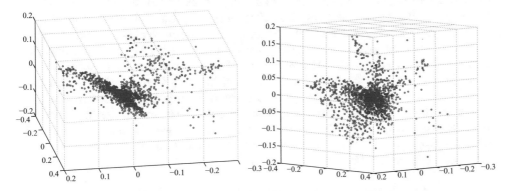

Figure 6.2: Two views of the spectral data of Sect. 5.2, plotted as a scatterplot by applying principal coordinate analysis to obtain a 3D set of points. Notice that the data spreads out in 3D, but seems to lie on some structure; it certainly isn't a single blob. This suggests that further investigation would be fruitful

the measuring instrument doesn't make certain kinds of measurement; or perhaps there are physical processes that prevent the data from spreading out over the space (Fig. 6.3).

Our algorithm has one really interesting property. In some cases, we do not actually know the data points as vectors. Instead, we *just* know distances between the data points. This happens often in the social sciences, but there are important cases in computer science as well. As a rather contrived example, one could survey people about breakfast foods (say, eggs, bacon, cereal, oatmeal, pancakes, toast, muffins, kippers, and sausages for a total of nine items). We ask each person to rate the similarity of each pair of distinct items on some scale. We advise people

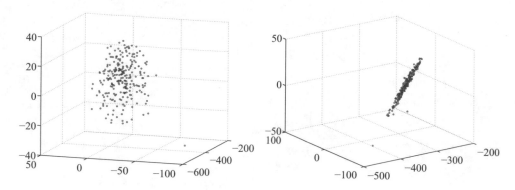

Figure 6.3: Two views of a multidimensional scaling to three dimensions of the height-weight dataset of Fig. 4.4. Notice how the data seems to lie in a flat structure in 3D, with one outlying data point. This means that the distances between data points can be (largely) explained by a 2D representation

that similar items are ones where, if they were offered both, they would have no particular preference; but, for dissimilar items, they would have a strong preference for one over the other. The scale might be "very similar," "quite similar," "similar," "quite dissimilar," and "very dissimilar" (scales like this are often called **Likert scales**). We collect these similarities from many people for each pair of distinct items, and then average the similarity over all respondents. We compute distances from the similarities in a way that makes very similar items close and very dissimilar items distant. Now we have a table of distances between items, and can compute a $\mathcal{Y}$ and produce a scatterplot. This plot is quite revealing, because items that most people think are easily substituted appear close together, and items that are hard to substitute are far apart. The neat trick here is that we did not start with a $\mathcal{X}$, but with just a set of distances; but we were able to associate a vector with "eggs," and produce a meaningful plot.

6.3 Example: Text Models and Latent Semantic Analysis

It is really useful to be able to measure the similarity between two documents, but it remains difficult to build programs that understand natural language. Experience shows that very simple models can be used to measure similarity between documents without going to the trouble of building a program that understands their content. Here is a representation that has been successful. Choose a vocabulary (a list of different words), then represent the document by a vector of word counts, where we simply ignore every word outside the vocabulary. This is a viable representation for many applications because quite often, most of the words people actually use come from a relatively short list (typically 100s to 1000s, depending on the particular application). The vector has one component for each word in the list, and that component contains the number of times that particular word is used. This model is sometimes known as a **bag-of-words** model.

Details of how you put the vocabulary together can be quite important. It is not a good idea to count extremely common words, sometimes known as **stop**

words, because every document has lots of them and the counts don't tell you very much. Typical stop words include "and," "the," "he," "she," and so on. These are left out of the vocabulary. Notice that the choice of stop words can be quite important, and depends somewhat on the application. It's often, but not always, helpful to **stem** words—a process that takes "winning" to "win," "hugely" to "huge," and so on. This isn't always helpful, and can create confusion (for example, a search for "stock" may be looking for quite different things than a search for "stocking"). We will always use datasets that have been preprocessed to produce word counts, but you should be aware that preprocessing this data is hard and involves choices that can have significant effects on the application.

Assume we have a set of N documents we wish to deal with. We have removed stop words, chosen a d-dimensional vocabulary, and counted the number of times each word appears in each document. The result is a collection of N d-dimensional vectors. Write the ith vector $\mathbf{x}_i$ (these are usually called **word vectors**). There is one minor irritation here; I have used d for the dimension of the vector $\mathbf{x}_i$ for consistency with the rest of the text, but d is the number of terms in the vocabulary *not* the number of documents.

The distance between two word vectors is usually a poor guide to the similarity of two documents. One reason is quite small, changes in word use might lead to large differences between count vectors. For example, some authors might write "car" when others write "auto." In turn, two documents might have a large (resp. small) count for "car" and a small (resp. large) count for "auto." Just looking at the counts would significantly overstate the difference between the vectors.

6.3.1 The Cosine Distance

The number of words in a document isn't particularly informative. As an extreme example, we could append a document to itself to produce a new document. The new document would have twice as many copies of each word as the old one, so the distance from the new document's word vector to other word vectors would have changed a lot. But the meaning of the new document wouldn't have changed. One way to overcome this nuisance is to normalize the vector of word counts in some way. It is usual to normalize the word counts by the magnitude of the count vector.

The distance between two word count vectors, normalized to be unit vectors, is

$$\left\| \frac{\mathbf{x}_i}{\|\mathbf{x}_i\|} - \frac{\mathbf{x}_j}{\|\mathbf{x}_j\|} \right\|^2 = 2 - 2\frac{\mathbf{x}_i^T \mathbf{x}_j}{\|\mathbf{x}_i\|\|\mathbf{x}_j\|}.$$

The expression

$$d_{ij} = \frac{\mathbf{x}_i^T \mathbf{x}_j}{\|\mathbf{x}_i\|\|\mathbf{x}_j\|}$$

is often known as the **cosine distance** between documents. While this is widely referred to as a distance, it isn't really. If two documents are very similar, their cosine distance will be close to 1; if they are really different, their cosine distance will be close to -1. Experience has shown that their cosine distance is a very effective measure of the similarity of documents i and j.

6.3.2 Smoothing Word Counts

Measuring the cosine distance for word counts has problems. We have seen one important problem already: if one document uses "car" and the other "auto," the two might be quite similar and yet have cosine distance that is close to zero. Remember, cosine distance close to zero suggests they're far apart. This is because the word counts are misleading. If you count, say, "car" once, you should have a non-zero count for "auto" as well. You could regard the zero count for "auto" as noise. This suggests smoothing word counts.

Arrange the word vectors into a matrix in the usual way, to obtain

$$\mathcal{X} = \begin{bmatrix} \mathbf{x}_1^T \\ \ldots \\ \mathbf{x}_N^T \end{bmatrix}.$$

This matrix is widely called a **document-term matrix** (its transpose is called a **term-document matrix**). This is because you can think of it as a table of counts; each row represents a document, each column represents a term from the vocabulary. We will use this object to produce a reduced dimension representation of the words in each document; this will smooth the word counts. Take an SVD of $\mathcal{X}$, yielding

$$\mathcal{X} = \mathcal{U}\Sigma\mathcal{V}^T.$$

Write Σ_r for the matrix obtained by setting all but the r largest singular values in Σ to 0, and construct

$$\mathcal{X}^{(r)} = \mathcal{U}\Sigma_r\mathcal{V}^T.$$

You should think of $\mathcal{X}^{(r)}$ as a smoothing of $\mathcal{X}$. The argument I used to justify seeing principal components as a smoothing method (Sect. 5.1.7) doesn't work here, because the noise model doesn't apply. But a qualitative argument supports the idea that we are smoothing. Each document that contains the word "car" should also have a non-zero count for the word "automobile" (and vice versa) because the two words mean about the same thing. The original matrix of word counts $\mathcal{X}$ doesn't have this information, because it relies on counting actual words. The counts in $\mathcal{X}^{(r)}$ are better estimates of what true word counts should be than one can obtain by simply counting words, because they take into account correlations between words.

Here is one way to think about this. Because word vectors in $\mathcal{X}^{(r)}$ are compelled to occupy a low dimensional space, counts "leak" between words with similar meanings. This happens because most documents that use "car" will tend to have many *other* words in common with most documents that use "auto." For example, it's highly unlikely that every document that uses "car" instead of "auto" also uses "spanner" instead of "wrench," and vice versa. A good low dimensional representation will place documents that use a large number of words with similar frequencies close together, even if they use some words with different frequencies; in turn, a document that uses "auto" will likely have the count for that word go down somewhat, and the count for "car" go up. Recovering information from the SVD of $\mathcal{X}$ is referred to as **latent semantic analysis**.

We have that

$$\left(\mathbf{x}_i^{(r)}\right)^T = \sum_{k=1}^r u_{ik}\sigma_k \mathbf{v}_k^T = \sum_{k=1}^r a_{ik}\mathbf{v}_k^T$$

so each $\mathbf{x}_i^{(r)}$ is a weighted sum of the first r rows of $\mathcal{V}^T$.

Forming a unit vector out of these smoothed word counts yields a natural representation for the ith document as

$$\mathbf{d}_i = \frac{\mathbf{x}_i^{(r)}}{\|\mathbf{x}_i^{(r)}\|}.$$

The distance between $\mathbf{d}_i$ and $\mathbf{d}_j$ is a good representation of the differences in meaning of document i and document j (it's $2 -$ cosine distance).

A key application for latent semantic analysis is in search. Assume you have a few query words, and you need to find documents that are suggested by those words. You can represent the query words as a word vector $\mathbf{x}_q$, which you can think of as a very small document. We will find nearby documents by: computing a low dimensional unit vector $\mathbf{d}_q$ for the query word vector, then finding nearby documents by an approximate nearest neighbor search on the document dataset. Computing a $\mathbf{d}_q$ for the query word vector is straightforward. We find the best representation of $\mathbf{x}_q$ on the space spanned by $\{\mathbf{v}_1, \ldots, \mathbf{v}_r\}$, then scale that to have unit norm.

Now $\mathcal{V}$ is orthonormal, so $\mathbf{v}_k^T\mathbf{v}_m$ is 1 for $k = m$, and zero otherwise. This means that

$$\left(\mathbf{x}_i^{(r)}\right)^T \left(\mathbf{x}_j^{(r)}\right) = \left(\sum_{k=1}^r a_{ik}\mathbf{v}_k^T\right)\left(\sum_{m=1}^r a_{jm}\mathbf{v}_m\right) = \sum_{k=1}^r a_{ik}a_{jk}.$$

But all the terms we are interested are inner products between document vectors. In turn, this means we could adopt a low dimensional representation for documents explicitly, and so, for example, use

$$\mathbf{d}_i = \frac{[a_{i1}, \ldots, a_{ir}]}{\sum_k a_{ik}^2}.$$

This representation has a much lower dimension than the normalized smoothed document vector, but contains exactly the same information.

Remember This: *Documents can be represented by smoothed word counts. The word counts are smoothed by constructing a low rank approximation to a document-word matrix. Each document is then represented by a unit vector proportional to the smoothed counts. The distance between these unit vectors yields the cosine distance. Documents can be retrieved by forming a unit vector representing the smoothed word counts of the query, then using nearest neighbors.*

6.3.3 Example: Mapping NIPS Documents

At https://archive.ics.uci.edu/ml/datasets/NIPS+Conference+Papers+1987-2015, you can find a dataset giving word counts for each word that appears at least 50 times in the NIPS conference proceedings from 1987 to 2015, by paper. It's big. There are 11,463 distinct words in the vocabulary, and 5811 total documents. We will use LSA to compute smoothed word counts in documents, and to map documents.

First, we need to deal with practicalities. Taking the SVD of a matrix this size will present problems, and storing the result will present quite serious problems. Storing $\mathcal{X}$ is quite easy, because most of the entries are zero, and a sparse matrix representation will work. But the whole point of the exercise is that $\mathcal{X}^{(r)}$ is *not* sparse, and this will have about 10^7 entries. Nonetheless, I was able to form an SVD in R, though it took about 30 min on my laptop. Figure 6.4 shows a multidimensional scaling of distances between normalized smoothed word counts. You should notice that documents are fairly evenly spread over the space. To give some meaning to the space, I have plotted the 10 words most strongly correlated with a document appearing in the corresponding grid block (highest correlation at top left in block, lowest in bottom right). Notice how the word clusters shade significantly across the coordinates. This is (rather rough) evidence that distances between smoothed normalized word counts do capture aspects of meaning.

6.3.4 Obtaining the Meaning of Words

It is difficult to know what a word means by looking at it, unless you have seen it before or it is an inflected version of a word you have seen before. A high percentage of readers won't have seen "peridot," "incarnadine," "whilom," or "numbat" before. If any of these are unfamiliar, simply looking at the letters isn't going to tell you what they mean. This means that unfamiliar words are quite different from unfamiliar pictures. If you look at a picture of something you haven't seen before, you're likely to be able to make some sensible guesses as to what it is like (how you do this remains very poorly understood; but that you can do this is everyday experience).

We run into unfamiliar words all the time, but the words around them seem to help us figure out what the unfamiliar words mean. As a demonstration, you should find these texts, which I modified from sentences found on the internet, helpful

- Peridot: "A sweet row of Peridot sit between golden round beads, strung from a delicate plated chain" (suggesting some form of decorative stone).
- Incarnidine: "A spreading stain incarnadined the sea" (a color description of some sort).
- Whilom: "Portions of the whilom fortifications have been converted into promenades." (a reference to the past).
- Numbat: "They fed the zoo numbats modified cat chow with crushed termite" (some form of animal, likely not vegetarian, and perhaps a picky eater).

This is a demonstration of a general point. Words near a particular word give strong and often very useful hints to that word's meaning, an effect known as

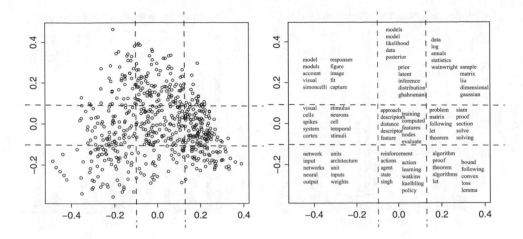

Figure 6.4: On the **left**, a multidimensional scaling mapping the NIPS documents into 2D. The distances between points represent (as well as an MDS can) the distances between normalized smoothed word counts. I have plotted every 10th document, to avoid crowding the plot. Superimposed on the figure is a grid, dividing each coordinate at the 33% (resp. 66%) quantile. On the **right**, I have plotted the 10 words most strongly correlated with a document appearing in the corresponding grid block (highest correlation at top left in block, lowest in bottom right). Each block has quite different sets of words, but there is evidence that: changes in the coordinates result in changes in document content; the dataset still has proper names in it, though insiders might notice the names are in sensible places; the horizontal coordinate seems to represent a practical–conceptual axis; and increasing values of the vertical coordinate seems to represent an increasingly statistical flavor. This is (rather rough) evidence that distances between smoothed normalized word counts do capture aspects of meaning

distributional semantics. Latent semantic analysis offers a way to exploit this effect to estimate representations of word meaning (Fig. 6.4).

Each row of $\mathcal{X}^{(r)}$ is a smoothed count of the number of times each word appears in a single document (Fig. 6.5). In contrast, each column is a smoothed count of the number of times a single word appears in each document. Imagine we wish to know the similarity in meaning between two words. Represent the ith word by the ith *column* of $\mathcal{X}^{(r)}$, which I shall write as $\mathbf{w}_i$, so that

$$\mathcal{X}^{(r)} = [\mathbf{w}_1, \ldots, \mathbf{w}_d].$$

Using a word more often (or less often) should likely not change its meaning. In turn, this means we should represent the ith word by

$$\mathbf{n}_i = \frac{\mathbf{w}_i}{\|\mathbf{w}_i\|}$$

and the distance between the ith and jth words is the distance between $\mathbf{n}_i$ and $\mathbf{n}_j$. This distance gives quite a good representation of word meaning, because two

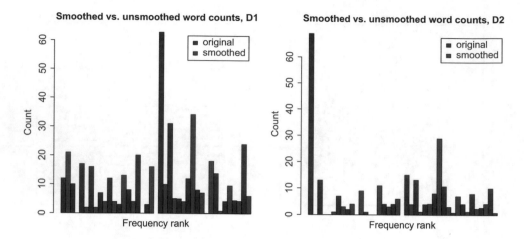

Figure 6.5: Unsmoothed and smoothed word counts for two different documents, where smoothing is by LSA to 1000 intermediate dimensions. Each figure shows one document; the blue bars are unsmoothed counts and the red bars are smoothed counts. The figure shows the counts for the 100 words that appear most frequently in the whole dataset, ordered by the rank of the word count (most common word first, etc.) Notice that generally, large counts tend to go down, and small counts tend to go up, as one would expect

words that are close in this distance will tend to appear in the same documents. For example, "auto" and "car" should be close. As we saw above, the smoothing will tend to reduce counts of "auto" and increase counts of "car" for documents that have only "auto," and so on. In turn, this means that "auto" will tend to appear in the same documents as "car," meaning that the distance between their normalized smoothed counts should be small.

We have that

$$\left(\mathbf{w}_i^{(r)}\right) = \sum_{k=1}^{r} (\sigma_k v_{ki}) \mathbf{u}_k = \sum_{k=1}^{r} b_{ik} \mathbf{u}_k$$

so each $\mathbf{w}_i^{(r)}$ is a weighted sum of the first r columns of $\mathcal{U}$.

Now $\mathcal{U}$ is orthonormal, so $\mathbf{u}_k^T \mathbf{u}_m$ is 1 for $k = m$, and zero otherwise. This means that

$$\left(\mathbf{w}_i^{(r)}\right)^T \left(\mathbf{w}_j^{(r)}\right) = \left(\sum_{k=1}^{r} b_{ik} \mathbf{u}_k^T\right) \left(\sum_{m=1}^{r} b_{jm} \mathbf{u}_m\right) = \sum_{k=1}^{r} b_{ik} b_{jk}.$$

But all the terms we are interested are inner products between word vectors. In turn, this means we could adopt a low dimensional representation for words explicitly, and so, for example, use

$$\mathbf{n}_i = \frac{[b_{i1}, \ldots, b_{ir}]}{\sum_k b_{ik}^2}.$$

This representation has a much lower dimension than the normalized smoothed word vector, but contains exactly the same information. This representation of a word is an example of a **word embedding**—a representation that maps a word to a point in some high dimensional space, where embedded points have good properties. In this case, we seek an embedding that places words with similar meanings near one another.

Remember This: *Words can be represented by smoothed counts of the documents they appear in. This works because words with similar meanings tend to appear in similar documents.*

6.3.5 Example: Mapping NIPS Words

LSA does not give a particularly strong word embedding, as this example will show. I used the dataset of Sect. 6.3.3, and computed a representation on 1000 dimensions. Figure 6.6 shows a multidimensional scaling (using the method of Sect. 6.2.3) onto two dimensions, where distances between points are given by distances between the normalized vectors of Sect. 6.3.4. I have shown only the top 80 words, so that the figures are not too cluttered to read.

Some results are natural. For example, "used" and "using" lie close to one another, as do "algorithm" and "algorithms"; "network" and "networks"; and "features" and "feature." This suggests the data wasn't stemmed or even preprocessed to remove plurals. Most of the pairs that seem to make sense (and aren't explained as plurals or inflections) seem to have more to do with phrases than with meaning. For example, "work" and "well" are close ("work well"); "problem" and "case" ("problem case"); "probability" and "distribution" ("probability distribution"). Some pairs are close because the words likely appear near one another in common phrases. So "classification" and "feature" suggest "feature based classification" or "classification by feature."

This tendency can be seen in the k-nearest neighbors of embedded words, too. Table 6.1 shows the six nearest neighbors for the 20 most frequent words. But there is evidence that the embedding is catching some kind of semantics, too. Notice that "network,", "neural," "units," and "weights" are close, as they should be. Similarly, "distribution," "distributions," and "probability" are close, and so are "algorithm" and "problem."

Embedding words in a way that captures semantics is a hard problem. Good recent algorithms use finer measures of word similarity than the pattern of documents a word appears in. Strong recent methods, like Word2Vec or Glove, pay most attention to the words that appear near the word of interest, and construct embeddings that try to explain such similarity statistics. These methods tend to be trained on very large datasets, too; much larger than this one.

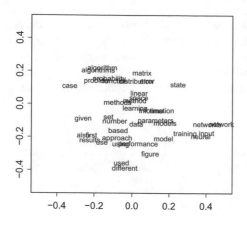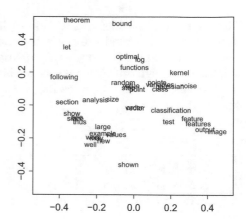

Figure 6.6: On the **left**, the 40 most frequent words in the NIPS dataset, plotted using a multidimensional scaling of the document frequencies, smoothed using latent semantic analysis. On the **right**, the next 40 most frequent words, plotted in the same way. I used 1000 dimensions for the smoothing. Words that have a similar pattern of incidence in documents appear near one another

> **Remember This:** *Strong recent word embedding methods, like Word2Vec or Glove, pay most attention to the words that appear near the word of interest, and construct embeddings that try to explain such similarity statistics.*

6.3.6 TF-IDF

The raw count of the number of times a word appears in a document may not be the best value to use in a term-document matrix. If a word is very common in all documents, then the fact that it appears often in a given document isn't that informative about what the document means. If a word appears only in a few documents, but is quite common in those documents, the number of times the word appears may understate how important it is. For example, in a set of documents about small pets, a word like "cat" is likely to appear often in each document; a word like "tularemia" is unlikely to appear often in many documents, but will tend to be repeated in a document if it appears. You can then argue that observing "cat" five times is a lot less informative about the document than observing "tularemia" five times is. Much time and trouble has been spent on making this very appealing argument more rigorous, without significant benefits that I'm aware of.

All this suggests that you might use a modified word score. The standard is known as **TF-IDF** (or, very occasionally, **term frequency-inverse document frequency**). Write c_{ij} for the number of times the ith word appears in the jth document, N for the number of documents, and N_i for the number of documents that contain at least one instance of the ith word. Then one TF-IDF score is

Model	Models	Also	Using	Used	Figure	Parameters
Learning	Also	Used	Using	Machine	Results	Use
Data	Using	Also	Used	Use	Results	Set
Algorithm	Algorithms	Problem	Also	Set	Following	Number
Set	Also	Given	Using	Results	Used	Use
Function	Functions	Also	Given	Using	Defined	Paper
Using	Used	Use	Also	Results	Given	First
Time	Also	First	Given	Used	University	Figure
Figure	Shown	Shows	Used	Using	Also	Different
Number	Also	Results	Set	Given	Using	Used
Problem	Problems	Following	Paper	Also	Set	Algorithm
Models	Model	Using	Also	Used	Parameters	Use
Used	Using	Use	Also	Results	First	University
Training	Used	Set	Using	Test	Use	Results
Given	Also	Using	Set	Results	University	First
Also	Results	Using	Use	Used	Well	First
Results	Also	Using	Used	Paper	Use	Show
Distribution	Distributions	Given	Probability	Also	University	Using
Network	Networks	Neural	Input	Output	Units	Weights
Based	Using	Also	Use	Used	Results	Given

TABLE 6.1: The leftmost column gives the top 20 words, by frequency of use in the NIPS dataset. Each row shows the seven closest words to each query word using the cosine distance applied to document counts of the word smoothed using latent semantic analysis. I used 1000 dimensions. Words that have similar patterns of use across documents do have important similarities, but these are not restricted to similarities of meaning. For example, "algorithm" is very similar to "algorithms," and also to "following" (likely because the phrase "following algorithm" is quite common) and to "problem" (likely because it's natural to have an algorithm to solve a problem)

$$c_{ij} \log \frac{N}{N_i}$$

(where we exclude cases where $N_i = 0$ because the term then doesn't appear in any document). Notice that a term appears in most documents, the score is about the same as the count; but if the term appears in few documents, the score is rather larger than the count. Using this score, rather than a count, tends to produce improved behavior from systems that use term-document matrices. There are a variety of ingenious variants of this score—the Wikipedia page lists many—each of which tends to produce changes in systems (typically, some things get better and some get worse). Don't forget the logarithm, which got dropped from the acronym for no reason I know.

Remember This: *Weighting word counts can produce significant improvements in document retrieval. Weights are chosen so that words that are common are downweighted, and words that are rare but common in a particular document are upweighted.*

6.4 You Should

6.4.1 Remember These Terms

6.4.2 Remember These Facts

6.4.3 Remember These Procedures

6.4.4 Be Able to

- Use a singular value decomposition to obtain principal components.
- Use a singular value decomposition to produce a principal coordinate analysis.

Problems

6.1. You have a dataset of N vectors $\mathbf{x}_i$ in d-dimensions, stacked into a matrix $\mathcal{X}$. This dataset *does not* have zero mean. The data $\mathbf{x}_i$ is noisy. We use a simple noise model. Write $\tilde{\mathbf{x}}_i$ for the true underlying value of the data item, and ξ_i

for the value of a normal random variable with zero mean and covariance $\sigma^2 \mathcal{I}$. Then we use the model

$$\mathbf{x}_i = \tilde{\mathbf{x}}_i + \xi_i.$$

In matrices, we write

$$\mathcal{X} = \tilde{\mathcal{X}} + \Xi.$$

We will assume that $\mathsf{mean}\,(\{\xi\}) = \mathbf{0}$ and $\mathsf{Covmat}\,(\{\xi\}) = \sigma^2 \mathcal{I}$.

(a) Show that our assumptions mean that the row rank of Ξ is d. Do this by contradiction: show if the row rank of Ξ is $r < d$, there is some rotation $\mathcal{U}$ so that each $\mathcal{U}\xi$ has zeros in the last $d - r$ components; now think about the covariance matrix of $\mathcal{U}\xi$.

(b) Assume that the row rank of $\tilde{\mathcal{X}}$ is $s \ll d$. Show that the row rank of $\mathcal{X}$ is d. Do this by noticing that the noise is independent of the dataset. This means that $\mathsf{mean}\,(\{\mathbf{x}\xi^T\}) = \mathsf{mean}\,(\{\mathbf{x}\})\mathsf{mean}\,(\{\xi^T\}) = 0$. Now show that

$$\mathsf{Covmat}\,(\{\mathbf{x}\}) = \mathsf{Covmat}\,(\{\tilde{x}\}) + \sigma^2 \mathcal{I}.$$

Now use the results of the previous exercise.

(c) We now have a geometric model for both $\tilde{\mathcal{X}}$ and $\mathcal{X}$. The points in $\tilde{\mathcal{X}}$ lie on some hyperplane that passes through the origin in d-dimensional space. This hyperplane has dimension s.

(d) The points in $\mathcal{X}$ lie on a "thickened" version of this hyperplane which has dimension d because the matrix has rank d. Show that the variance of the data in any direction normal to the original hyperplane is σ^2.

(e) Use the previous subexercises to argue that a rank s approximation of $\mathcal{X}$ lies closer to $\tilde{\mathcal{X}}$ than $\mathcal{X}$ does. Use the Frobenius norm.

6.2. Write $\mathcal{D}^{(2)}$ for the matrix whose i,jth component is

$$D^{(2)}_{ij} = \left(\mathbf{x}_i - \mathbf{x}_j\right)^T \left(\mathbf{x}_i - \mathbf{x}_j\right) = \mathbf{x}_i \cdot \mathbf{x}_i - 2\mathbf{x}_i \cdot \mathbf{x}_j + \mathbf{x}_j \cdot \mathbf{x}_j$$

where $\mathsf{mean}\,(\{x\}) = \mathbf{0}$. Now write

$$\mathcal{A} = \left[\mathcal{I} - \frac{1}{N}\mathbf{1}\mathbf{1}^T\right].$$

Show that

$$-\frac{1}{2}\mathcal{A}\mathcal{D}^{(2)}(\mathbf{x})\mathcal{A}^T = \mathcal{X}\mathcal{X}^T.$$

6.3. You have a dataset of N vectors $\mathbf{x}_i$ in d-dimensions, stacked into a matrix $\mathcal{X}$, and wish to build an s-dimensional dataset $\mathcal{Y}_s$ so that $\mathcal{Y}_s\mathcal{Y}_f^T$ minimizes $\|\mathcal{Y}_s\mathcal{Y}_f^T - \mathcal{X}\mathcal{X}^T\|_F$. Form an SVD, to get

$$\mathcal{X} = \mathcal{U}\Sigma\mathcal{V}^T$$

and write

$$\mathcal{Y} = \mathcal{U}_s\Sigma_s$$

(the subscript-s notation is in the chapter).

(a) Show that

$$\|\mathcal{Y}_s\mathcal{Y}_f^T - \mathcal{X}\mathcal{X}^T\|_F = \|\Sigma_s^2 - \Sigma^2\|_F.$$

Explain why this means that $\mathcal{Y}_s$ is a solution.

(b) For any $s \times s$ orthonormal matrix $\mathcal{R}$, show that $\mathcal{Y}_R = \mathcal{U}_s\Sigma_s\mathcal{R}$ is also a solution. Interpret this geometrically.

Programming Exercises

6.4. At https://archive.ics.uci.edu/ml/datasets/NIPS+Conference+Papers+1987-2015, you can find a dataset giving word counts for each word that appears at least 50 times in the NIPS conference proceedings from 1987 to 2015, by paper. It's big. There are 11,463 distinct words in the vocabulary, and 5811 total documents. We will investigate simple document clustering with this dataset.

   (a) Reproduce Fig. 6.5 using approximations with rank 100, 500, and 2000. Which is best, and why?

   (b) Now use a TF-IDF weight to reproduce Fig. 6.5 using approximations with rank 100, 500, and 2000. Which is best, and why?

   (c) Reproduce Fig. 6.6 using approximations with rank 100, 500, and 2000. Which is best, and why?

   (d) Now use a TF-IDF weight to reproduce Fig. 6.6 using approximations with rank 100, 500, and 2000. Which is best, and why?

6.5. Choose a state. For the 15 largest cities in your chosen state, find the distance between cities and the road mileage between cities. These differ because of the routes that roads take; you can find these distances by careful use of the internet. Prepare a map showing these cities on the plane using principal coordinate analysis for each of these two distances. How badly does using the road network distort to make a map distort the state? Does this differ from state to state? Why?

6.6. CIFAR-10 is a dataset of 32×32 images in 10 categories, collected by Alex Krizhevsky, Vinod Nair, and Geoffrey Hinton. It is often used to evaluate machine learning algorithms. You can download this dataset from https://www.cs.toronto.edu/~kriz/cifar.html.

   (a) For each category, compute the mean image and the first 20 principal components. Plot the error resulting from representing the images of each category using the first 20 principal components against the category.

   (b) Compute the distances between mean images for each pair of classes. Use principal coordinate analysis to make a 2D map of the means of each categories. For this exercise, compute distances by thinking of the images as vectors.

   (c) Here is another measure of the similarity of two classes. For class A and class B, define $E(A \to B)$ to be the average error obtained by representing all the images of class A using the mean of class A and the first 20 principal components of class B. This should tell you something about the similarity of the classes. For example, imagine images in class A consist of dark circles that are centered in a light image, but where different images have circles of different sizes; images of class B are dark on the left, light on the right, but different images change from dark to light at different vertical lines. Then the mean of class A should look like a fuzzy centered blob, and its principal components make the blob bigger or smaller. The principal components of class B will move the dark patch left or right. Encoding an image of class A with the principal components of class B should work very badly. But if class C consists of dark circles that move left or right from image to image, encoding an image of class C using A's principal components might work tolerably. Now define the similarity between classes to be $(1/2)(E(A \to B) + E(B \to A))$. Use principal coordinate analysis to make a 2D map of the classes. Compare this map to the map in the previous exercise—are they different? why?

CHAPTER 7

Canonical Correlation Analysis

In many applications, one wants to associate one kind of data with another. For example, every data item could be a video sequence together with its sound track. You might want to use this data to learn to associate sounds with video, so you can predict a sound for a new, silent, video. You might want to use this data to learn how to read the (very small) motion cues in a video that result from sounds in a scene (so you could, say, read a conversation off the tiny wiggles in the curtain caused by the sound waves). As another example, every data item could be a captioned image. You might want to predict words from pictures to label the pictures, or predict pictures from words to support image search. The important question here is: what aspects of the one kind of data can be predicted from the other kind of data?

In each case, we deal with a dataset of N pairs, $\mathbf{p}_i = [\mathbf{x}_i, \mathbf{y}_i]^T$, where $\mathbf{x}_i$ is a d_x dimensional vector representing one kind of data (e.g., words, sound, image, video) and $\mathbf{y}_i$ is a d_y dimensional vector representing the other kind. I will write $\{\mathbf{x}\}$ for the $\mathbf{x}$ part, etc., but notice that our agenda of prediction assumes that the pairing is significant—if you could shuffle one of the parts without affecting the outcome of the algorithm, then you couldn't predict one from the other.

We could do a principal components analysis on $\{\mathbf{p}\}$, but that approach misses the point. We are primarily interested in the *relationship* between $\{\mathbf{x}\}$ and $\{\mathbf{y}\}$ and the principal components capture only the major components of variance of $\{\mathbf{p}\}$. For example, imagine the $\mathbf{x}_i$ all have a very large scale, and the $\mathbf{y}_i$ all have a very small scale. Then the principal components will be determined by the $\mathbf{x}_i$. We assume that $\{\mathbf{x}\}$ and $\{\mathbf{y}\}$ have zero mean, because it will simplify the equations and is easy to achieve. There is a standard procedure for dealing with data like this. This is quite good at, say, predicting words to attach to pictures. However, it can result in a misleading analysis, and I show how to check for this.

7.1 Canonical Correlation Analysis

Canonical correlation analysis (or CCA) seeks linear projections of $\{\mathbf{x}\}$ and $\{\mathbf{y}\}$ such that one is easily predicted from the other. A projection of $\{\mathbf{x}\}$ onto one dimension can be represented by a vector $\mathbf{u}$. The projection yields a dataset $\{\mathbf{u}^T\mathbf{x}\}$ whose i'th element is $\mathbf{u}^T\mathbf{x}_i$. Assume we project $\{\mathbf{x}\}$ onto $\mathbf{u}$ and $\{\mathbf{y}\}$ onto $\mathbf{v}$. Our ability to predict one from the other is measured by the correlation of these two datasets. So we should look for $\mathbf{u}$, $\mathbf{v}$ so that

$$\mathrm{corr}\left(\left\{\mathbf{u}^T\mathbf{x}, \mathbf{v}^T\mathbf{y}\right\}\right)$$

is maximized. If you are worried that a negative correlation with a large absolute value also allows good prediction, and this isn't accounted for by the expression, you should remember that we get to choose the sign of $\mathbf{v}$.

© Springer Nature Switzerland AG 2019
D. Forsyth, *Applied Machine Learning*,
https://doi.org/10.1007/978-3-030-18114-7_7

   We need some more notation. Write Σ for the covariance matrix of $\{\mathbf{p}\}$. Recall $\mathbf{p}_i = [\mathbf{x}_i, \mathbf{y}_i]^t$. This means the covariance matrix has a block structure, where one block is covariance of x components of $\{\mathbf{p}\}$ with each other, another is covariance of y components with each other, and the third is covariance of x components with y components. We write

$$\Sigma = \begin{bmatrix} \Sigma_{xx} & \Sigma_{xy} \\ \Sigma_{yx} & \Sigma_{yy} \end{bmatrix} = \begin{bmatrix} x-x \text{ covariance} & x-y \text{ covariance} \\ y-x \text{ covariance} & y-y \text{ covariance} \end{bmatrix}.$$

We have that

$$\text{corr}\left(\{\mathbf{u}^T\mathbf{x}, \mathbf{v}^T\mathbf{y}\}\right) = \frac{\mathbf{u}^T \Sigma_{xy} \mathbf{v}}{\sqrt{\mathbf{u}^T \Sigma_{xx} \mathbf{u}} \sqrt{\mathbf{v}^T \Sigma_{yy} \mathbf{v}}}$$

and maximizing this ratio will be hard (think about what the derivatives look like). There is a useful trick. Assume $\mathbf{u}^*$, $\mathbf{v}^*$ are values at a maximum. Then they must also be solutions of the problem

$$\text{Max } \mathbf{u}^T \Sigma_{xy} \mathbf{v} \quad \text{Subject to} \quad \mathbf{u}^T \Sigma_{xx} \mathbf{u} = c_1 \text{ and } \mathbf{v}^T \Sigma_{yy} \mathbf{v} = c_2$$

(where c_1, c_2 are positive constants of no particular interest). This second problem is quite easy to solve. The Lagrangian is

$$\mathbf{u}^T \Sigma_{xy} \mathbf{v} - \lambda_1 (\mathbf{u}^T \Sigma_{xx} \mathbf{u} - c_1) - \lambda_2 (\mathbf{v}^T \Sigma_{yy} \mathbf{v} - c_2)$$

so we must solve

$$\Sigma_{xy} \mathbf{v} - \lambda_1 \Sigma_{xx} \mathbf{u} = 0$$
$$\Sigma_{xy}^T \mathbf{u} - \lambda_2 \Sigma_{yy} \mathbf{v} = 0$$

For simplicity, we assume that there are no redundant variables in $\mathbf{x}$ or $\mathbf{y}$, so that Σ_{xx} and Σ_{yy} are both invertible. We substitute $(1/\lambda_1)\Sigma_{xx}^{-1}\Sigma_{xy}\mathbf{v} = \mathbf{u}$ to get

$$\Sigma_{yy}^{-1} \Sigma_{xy}^T \Sigma_{xx}^{-1} \Sigma_{xy} \mathbf{v} = (\lambda_1 \lambda_2)\mathbf{v}.$$

Similar reasoning yields

$$\Sigma_{xx}^{-1} \Sigma_{xy} \Sigma_{yy}^{-1} \Sigma_{xy}^T \mathbf{u} = (\lambda_1 \lambda_2)\mathbf{u}.$$

So $\mathbf{u}$ and $\mathbf{v}$ are eigenvectors of the relevant matrices. But which eigenvectors? Notice that

$$\mathbf{u}^T \Sigma_{xy} \mathbf{v} = \mathbf{u}^T \left(\lambda_1 \Sigma_{xx} \mathbf{u}\right) = \left(\lambda_2 \mathbf{v}^T \Sigma_{yy}\right)\mathbf{v}$$

so that

$$\text{corr}\left(\{\mathbf{u}^T\mathbf{x}, \mathbf{v}^T\mathbf{y}\}\right) = \frac{\mathbf{u}^T \Sigma_{xy} \mathbf{v}}{\sqrt{\mathbf{u}^T \Sigma_{xx} \mathbf{u}} \sqrt{\mathbf{v}^T \Sigma_{yy} \mathbf{v}}} = \sqrt{\lambda_1}\sqrt{\lambda_2}$$

meaning that the eigenvectors corresponding to the largest eigenvalues give the largest correlation directions, to the second largest give the second largest correlation directions, and so on. There are $\min(d_x, d_y)$ directions in total. The values of $\text{corr}\left(\{\mathbf{u}^T\mathbf{x}, \mathbf{v}^T\mathbf{y}\}\right)$ for the different directions are often called **canonical correlations**. The projections are sometimes known as **canonical variables**.

Worked Example 7.1   *Anxiety and Wildness in Mice*

Compute the canonical correlations between indicators of anxiety and of wildness in mice, using the dataset at http://phenome.jax.org/db/q?rtn=projects/details&sym=Jaxpheno7

Solution: You should read the details on the web page that publishes the data. The anxiety indicators are: `transfer_arousal`, `freeze`, `activity`, `tremor`, `twitch`, `defecation_jar`, `urination_jar`, `defecation_arena`, `urination_arena`, and the wildness indicators are: `biting`, `irritability`, `aggression`, `vocal`, `finger_approach`. After this, it's just a question of finding a package and putting the data in it. I used R's `cancor`, and found the following five canonical correlations: 0.62, 0.53, 0.40, 0.35, 0.30. You shouldn't find the presence of strong correlations shocking (anxious mice should be bitey), but we don't have any evidence this isn't an accident. The example in the subsection below goes into this question in more detail.

This data was collected by The Jackson Laboratory, who ask it be cited as: Neuromuscular and behavioral testing in males of six inbred strains of mice. MPD:Jaxpheno7. Mouse Phenome Database website, The Jackson Laboratory, Bar Harbor, Maine USA. http://phenome.jax.org

---

Procedure: 7.1   *Canonical Correlation Analysis*

Given a dataset of N pairs, $\mathbf{p}_i = [\mathbf{x}_i, \mathbf{y}_i]^T$, where $\mathbf{x}_i$ is a d_x dimensional vector representing one kind of data (e.g., words, sound, image, video) and $\mathbf{y}_i$ is a d_y dimensional vector representing the other kind. Write Σ for the covariance matrix of $\{\mathbf{p}\}$. We have

$$\Sigma = \left[\begin{array}{cc} \Sigma_{xx} & \Sigma_{xy} \\ \Sigma_{yx} & \Sigma_{yy} \end{array} \right].$$

Write $\mathbf{u}_j$ for the eigenvectors of

$$\Sigma_{xx}^{-1} \Sigma_{xy} \Sigma_{yy}^{-1} \Sigma_{xy}^T$$

sorted in the descending order of eigenvalue. Write $\mathbf{v}_j$ for the eigenvectors of

$$\Sigma_{yy}^{-1} \Sigma_{xy}^T \Sigma_{xx}^{-1} \Sigma_{xy}$$

sorted in the descending order of eigenvalue. Then $\mathbf{u}_1^T \mathbf{x}_i$ is most strongly correlated with $\mathbf{v}_1 \mathbf{y}_i$; $\mathbf{u}_2^T \mathbf{x}_i$ is second most strongly correlated with $\mathbf{v}_2 \mathbf{y}_i$; and so on, up to $j = \min(d_x, d_y)$.

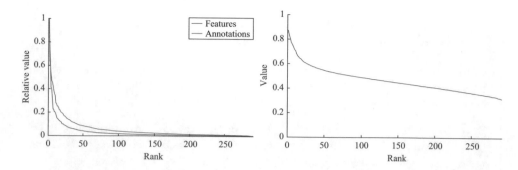

Figure 7.1: On the **left**, the 291 largest eigenvalues of the covariance for features and for word vectors, normalized by the largest eigenvalue in each case, plotted against rank. Notice in each case relatively few eigenvalues capture most of the variance. The word vectors are 291 dimensional, so this figure shows all the variances for the word vectors, but there are a total of 3000 eigenvalues for the features. On the **right**, the canonical correlations for this dataset. Notice that there are some rather large correlations, but quite quickly the values are small

7.2 Example: CCA of Words and Pictures

CCA is commonly used to find good matching spaces. Here is an example. Assume we have a set of captioned images. It is natural to want to build two systems: given an image, caption it; and given a caption, produce a good image. There is a very wide range of methods that have been deployed to attack this problem. Perhaps the simplest—which is surprisingly effective—is to use a form of nearest neighbors in a cleverly chosen space. We have N images described by feature vectors $\mathbf{x}_i$, corresponding to N captions described by word vectors $\mathbf{y}_i$. The i'th image corresponds to the i'th caption. The image features have been constructed using specialized methods (there are some constructions in Chaps. 8 and 17, but coming up with the best construction is still a topic of active research, and way outside the scope of this book). The word vectors are like those of Sect. 6.3.

We would like to map the word vectors and the image features into a new space. We will assume that the features have extracted all the useful properties of the images and captions, and so a linear map of each is sufficient. If an image and a caption correspond, we would like their feature vectors to map to points that are nearby. If a caption describes an image very poorly, we would like its feature vector to map far away from where the image's feature vector maps.

Assume we have this new space. Then we could come up with a caption for a new image by mapping the image into the space, and picking the nearest point that represents a caption. We could come up with an image for a new caption by mapping the caption into the space, then picking the nearest point that represents an image. This strategy (with some tuning, improvements, and so on) remains extremely hard to beat.

For this example, I will use a dataset called the IAPR TC-12 benchmark, which is published by ImageCLEF. A description of the dataset can be found at https://www.imageclef.org/photodata. There are 20,000 images, each of which has

Figure 7.2: Four images with true (in red) and predicted (green) label words. Words
are predicted using a CCA of image features and word vectors, as described in the
text. Images are from a test set, not used in constructing the CCA. The yellow box
gives the cosine distance between the predicted and true word vectors, smoothed by
projection to a 150- dimensional space as in Chap. 5. For these images, the cosine
distances are reasonably close to one, and the predictions are quite good

a text annotation. The annotations use a vocabulary of 291 words, and the word
vectors are binary (i.e., word is there or not). I used image features published by
Mathieu Guillaumin, at https://lear.inrialpes.fr/people/guillaumin/data.php. These
features are not the current state of the art for this problem, but they're easily
available and effective. There are many different features available at this location,
but for these figures, I used the DenseSiftV3H1 feature set. I matched test images to training captions using the 150 canonical variables with largest canonical
correlations.

The first thing you should notice (Fig. 7.1) is that both image features and
text behave as you should expect. There are a small number of large eigenvalues
in the covariance matrix, and a large number of small eigenvalues. Because the
scaling of the features is meaningless, I have plotted the eigenvalues as fractions
of the largest value. Notice also that the first few canonical correlations are large
(Fig. 7.1).

There are two ways to evaluate a system like this. The first is qualitative, and
if you're careless or optimistic, it looks rather good. Figure 7.2 shows a set of images
with true word labels and labels predicted using the nearest neighbors procedure.
Predicted labels are in green, and true labels are in red. Mostly, these labellings
should look quite good to you. Some words are missing from the predictions, true,
but most predictions are about right.

A quantitative evaluation reflects the "about right" sense. For each image, I
formed the cosine distance between the predicted word vector and the true word
vector, smoothed by projection to a 150-dimensional space. This is the number
in the yellow box. These numbers are fairly close to one for Fig. 7.2, which is a

Figure 7.3: Four images with true (in red) and predicted (green) label words. Words are predicted using a CCA of image features and word vectors, as described in the text. Images are from a test set, not used in constructing the CCA. The yellow box gives the cosine distance between the predicted and true word vectors, smoothed by projection to a 150- dimensional space as in Chap. 5. For these images, the cosine distances are rather far from one, and the predictions are not as good as those in Fig. 7.2

good sign. But Figs. 7.3 and 7.4 suggest real problems. There are clearly images for which the predictions are poor. In fact, predictions are poor for most images, as Fig. 7.5 shows. This figure gives the cosine distance between predicted and true labels (again, smoothed by projection to a 150-dimensional space), sorted from best to worst, for all test images. Most produce really bad label vectors with very low cosine distance.

Improving this is a matter of image features. The features I have used here are outdated. I used them because it was easy to get many different sets of features for the same set of images (yielding some rather interesting exercises). Modern feature constructions allow improved labelling of images, but modern systems still tend to use CCA, although often in more complex forms than we can deal with here.

7.3   Example: CCA of Albedo and Shading

Here is a classical computer vision problem. The brightness of a diffuse (=dull, not shiny or glossy) surface in an image is the product of two effects: the **albedo** (the percentage of incident light that it reflects) and the **shading** (the amount of light incident on the surface). We will observe the brightness in an image, and the problem is to recover the albedo and the shading separately. This has been an important problem in computer vision since the early 1970s, and in human vision since the mid nineteenth century. The problem still gets regular and significant

Figure 7.4: Four images with true (in red) and predicted (green) label words. Words are predicted using a CCA of image features and word vectors, as described in the text. Images are from a test set, not used in constructing the CCA. The yellow box gives the cosine distance between the predicted and true word vectors, smoothed by projection to a 150- dimensional space as in Chap. 5. For these images, the cosine distances are rather close to zero, and the predictions are bad

attention in the computer vision literature, because it's hard, and because it seems to be important.

We will confine our discussion to smooth (=not rough) surfaces, to prevent the complexity spiralling out of control. Albedo is a property of surfaces. A dark surface has low albedo (it reflects relatively little of the light that falls on it) and a light surface has high albedo (it reflects most of the light that falls on it). Shading is a property of the geometry of the light sources with respect to the surface. When you move an object around in a room, its shading may change a lot (though people are surprisingly bad at noticing this), but its albedo doesn't change at all. To change an object's albedo, you need (say) a marker or paint. All this suggests that a CCA of albedo against shading will suggest there is no correlation.

Because this is a classical problem, there are datasets one can download. There is a very good dataset giving the albedo and shading for images, collected by Roger Grosse, Micah K. Johnson, Edward H. Adelson, and William T. Freeman at http://www.cs.toronto.edu/~rgrosse/intrinsic/. These images show individual objects on black backgrounds, and there are masks identifying object pixels. For each image in the dataset, there is an albedo map (basically, an image of the albedos) and a shading map. These maps are constructed by clever photographic techniques. I constructed random 11 × 11 tiles of albedo and shading for each of the 20 objects depicted. I chose 20 tiles per image (so 400 in total), centered at random locations, but chosen so that every pixel in a tile lies on an object pixel. The albedo tiles I chose for a particular image were in the same locations in that image as the shading tiles—each pair of tiles represents a pair of albedo-shading in some image patch. I

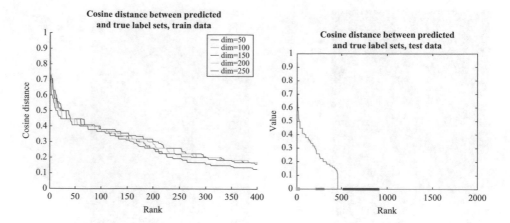

Figure 7.5: **Left:** all values of cosine distance between predicted and true word labels, sorted from best to worst, for the CCA method of the text, for different numbers of canonical variables, for the training data. The distances are fairly good, and 150 seems like a reasonable choice of dimension. On the **right**, cosine distances between predicted and true for *test* data; this looks much worse. I have marked the regions where the "good," "medium," and "bad" figures come from. Note that most values are bad—predicting words from images is hard. Accurate predictions require considerably more sophisticated feature constructions than we have used here

then reshaped each tile into a 121-dimensional vector, and computed a CCA. The top 10 values of canonical correlations I obtained were: 0.96, 0.94, 0.93, 0.93, 0.92, 0.92, 0.91, 0.91, 0.90, 0.88.

If this doesn't strike you as ridiculous, then you should check you understand the definitions of albedo and shading. How could albedo and shading be correlated? Do people put dark objects in light places, and light objects in dark places? The correct answer is that they are not correlated, but that this analysis has missed one important, nasty point. The objective function we are maximizing is a ratio

$$\text{corr}\left(\left\{\mathbf{u}^T\mathbf{x}, \mathbf{v}^T\mathbf{y}\right\}\right) = \frac{\mathbf{u}^T\Sigma_{xy}\mathbf{v}}{\sqrt{\mathbf{u}^T\Sigma_{xx}\mathbf{u}}\sqrt{\mathbf{v}^T\Sigma_{yy}\mathbf{v}}}.$$

Now look at the denominator of this fraction, and recall our work on PCA. The whole point of PCA is that there are many directions $\mathbf{u}$ such that $\mathbf{u}^T\Sigma_{xx}\mathbf{u}$ is small— these are the directions that we can drop in building low dimensional models. But now they have a potential to be a significant nuisance. We could have the objective function take a large value simply because the terms in the denominator are very small. This is what happens in the case of albedo and shading. You can check this by looking at Fig. 7.6, or by actually looking at the size of the canonical correlation directions (the $\mathbf{u}$'s and $\mathbf{v}$'s). You will find that, if you compute $\mathbf{u}$ and $\mathbf{v}$ using the procedure I described, these vectors have large magnitude (I found magnitudes of the order of 100). This suggests, correctly, that they're associated with small eigenvalues in Σ_{xx} and Σ_{yy}.

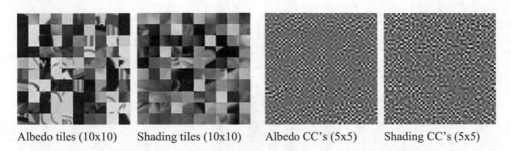

Albedo tiles (10x10)   Shading tiles (10x10)   Albedo CC's (5x5)   Shading CC's (5x5)

Figure 7.6: On the **left**, a 10×10 grid of tiles of albedo (**far left**) and shading (**center left**), taken from Grosse et al.'s dataset. The position of the tiles is keyed, so (for example) the albedo tile at 3, 5 corresponds to the shading tile at 3, 5. On the **right**, the first 25 canonical correlation directions for albedo (**center right**) and shading (**far right**). I have reshaped these into tiles and zoomed them. The scale of smallest value is black, and largest white. These are ordered so the pair with highest correlation is at the top left, next highest is one step to the right, etc. You should notice that these directions do not look even slightly like the patterns in the original tiles, or like any pattern you expect to encounter in a real image. This is because they're not: these are directions that have very small variance

Just a quick check with intuition and an image tells us that these canonical correlations don't mean what we think. But this works only for a situation where we have intuition, etc. We need a test that tells whether the large correlation values have arisen by accident.

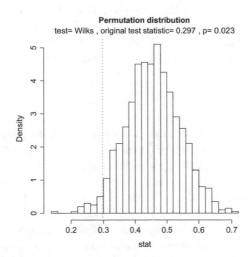

Figure 7.7: A histogram of values of Wilks' lambda obtained from permuted versions of the mouse dataset of example 7.1. The value obtained for the original dataset is shown by the vertical line. Notice that most values are larger (about 97% of values), meaning that we would see the canonical correlation values we see only about once in 30 experiments if they were purely a chance effect. There is very likely a real effect here

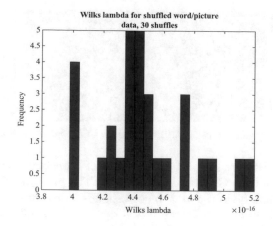

Figure 7.8: A histogram of values of Wilks' lambda obtained from permuted versions of the word and picture dataset of Sect. 7.2. I computed the value for the first 150 canonical variates, and used 30 shuffles (which takes quite a long time). The value for the true dataset is $9.92e{-}25$, which suggests very strongly that the correlations are not accidental

7.3.1 Are Correlations Significant?

There is an easy and useful strategy for testing this. If there really are meaningful correlations between the $\{\mathbf{x}\}$ and $\{\mathbf{y}\}$, they should be disrupted if we reorder the datasets. So if something important is changed by reordering one dataset, there is evidence that there is a meaningful correlation. The recipe is straightforward. We choose a method to summarize the canonical correlations in a number (this is a statistic, a term you should remember). In the case of canonical correlations, the usual choice is **Wilks' lambda** (or Wilks' λ if you're fussy).

Write ρ_i for the i'th canonical correlation. For $r \leq \min(d_x, d_y)$, Wilks' lambda is

$$\Lambda(r) = \prod_{i=1}^{i=r}(1 - \rho_i^2).$$

Notice if there are a lot of strong correlations in the first r, we should get a small value of $\Lambda(r)$, and as r increases, $\Lambda(r)$ gets smaller. We can now use the following procedure. For each r, compute $\Lambda(r)$. Now construct a collection of new datasets by randomly reordering the items in $\{\mathbf{y}\}$, and for each we compute the value of $\Lambda(r)$. This gives an estimate of the distribution of values of $\Lambda(r)$ available *if there is no correlation*. We then ask what fraction of the reordered datasets have an even smaller value of $\Lambda(r)$ than the observed value. If this fraction is small for a given r, then it is unlikely that the correlations we observed arose by accident. Write $f(r)$ for the fraction of randomly reordered datasets that have $\Lambda(r)$ smaller than that observed for the real dataset. Generally, $f(r)$ grows as r increases, and you can use this to decide what number of canonical correlations are not accidental. All this is fairly easily done using a package (I used CCP in R).

Figure 7.7 shows what happens for the mouse canonical correlation of example 7.1. You should notice that this is a significance test, and follows the usual

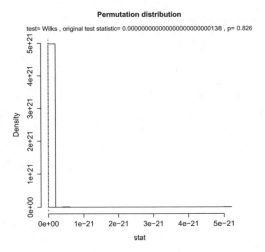

Figure 7.9: A histogram of values of Wilks' lambda obtained from permuted versions of the 400 tile albedo-shading dataset discussed in the text. The value obtained for the original dataset is shown by the vertical line, and is really tiny (rather less than $1e-21$). But rather more than four-fifths (82.6%) of the values obtained by permuting the data are even tinier, meaning that we would see the canonical correlation values we see or smaller about 4 in every 5 experiments if they were purely a chance effect. There is no reason to believe the two have a correlation

recipe for such tests except that we estimate the distribution of the statistic empirically. Here about 97% of random permutations have a larger value of the Wilks' lambda than that of the original data, which means that we would see the canonical correlation values we see only about once in 30 experiments if they were purely a chance effect. You should read this as quite good evidence there is a correlation. As Fig. 7.8 shows, there is good evidence that the correlations for the words and pictures data of Sect. 7.2 are not accidental, either.

But the albedo-shading correlations really are accidental. Figure 7.9 shows what happens for albedo and shading. The figure is annoying to interpret, because the value of the Wilks' lambda is extremely small; the big point is that almost every permutation of the data has an even smaller value of the Wilks' lambda—the correlations are entirely an accident, and are of no statistical significance.

> **Remember This:** *A canonical correlation analysis can mislead you. The problem is the division in the objective function. If you're working with data where many principal components have very small variances, you can get large correlations as a result. You should always check whether the CCA is actually telling you something useful. A natural check is the Wilks' lambda procedure.*

CCA: CCA can mislead you

7.4 You Should

7.4.1 Remember These Terms

7.4.2 Remember These Facts

7.4.3 Remember These Procedures

7.4.4 Be Able to

- Use a canonical correlation analysis to investigate correlations between two types of data.
- Use Wilks' lambda to determine whether correlations are the result of real effects.

Programming Exercises

7.1. We investigate CCA to predict words from pictures using Mathieu Guillaumin's published features, available at https://lear.inrialpes.fr/people/guillaumin/data. php.

    (a) Reproduce Figs. 7.1 and 7.5 of Sect. 7.2, using the same features and the same number of canonical variables.

    (b) One reasonable summary of performance is the mean of the cosine distance between true and predicted label vectors over all test images. This number will vary depending on how many of the canonical variables you use to match. Plot this number over the range $[1 \ldots 291]$, using at least 10 points.

    (c) Based on the results of the previous subexercise, choose a good number of canonical variables. For the 30 most common words in the vocabulary, compute the total error rate, the false positive rate, and the false negative rate for predictions over the whole test set. Does this suggest any way to improve the method?

7.2. We investigate image features using Mathieu Guillaumin's published features, available at https://lear.inrialpes.fr/people/guillaumin/data.php.

    (a) Compute a CCA of the GIST features against the DenseSiftV3H1 features, and plot the sorted canonical correlations. You should get a figure like Fig. 7.1. Does this suggest that different feature sets encode different aspects of the image?

    (b) If you concatenate GIST features with DenseSiftV3H1 features, do you get improved word predictions?

7.3. Here is a much more elaborate exercise investigating CCA to predict words from pictures using Mathieu Guillaumin's published features, available at https://lear.inrialpes.fr/people/guillaumin/data.php.

    (a) Reproduce Fig. 7.5 of Sect. 7.2, for *each* of the available image feature sets. Is any particular feature set better overall?

    (b) Now take the top 50 canonical variables of each feature set for the images, and concatenate them. This should yield 750 variables that you can use as image features. Reproduce Fig. 7.5 for this set of features. Was there an improvement in performance?

    (c) Finally, if you can get your hands on some hefty linear algebra software, concatenate all the image feature sets. Reproduce Fig. 7.5 for this set of features. Was there an improvement in performance?

PART THREE

Clustering

CHAPTER 8

Clustering

One very good, very simple, model for data is to assume that it consists of multiple blobs. To build models like this, we must determine (a) what the blob parameters are and (b) which data points belong to which blob. Generally, we will collect together data points that are close and form blobs out of them. The blobs are usually called **clusters**, and the process is known as **clustering**.

Clustering is a somewhat puzzling activity. It is extremely useful to cluster data, and it seems to be quite important to do it reasonably well. But it surprisingly hard to give crisp criteria for a good (resp. bad) clustering of a dataset. Typically, one evaluates clustering by seeing how well it supports an application.

There are many applications of clustering. You can summarize a dataset by clustering it, then reporting a summary of each cluster. Summaries might be either a typical element of each cluster or (say) the mean of each cluster. Clusters can help expose structure in a dataset that is otherwise quite difficult to see. For example, in Sect. 8.2.5, I show ways of visualizing the difference between sets of grocery stores by clustering customer records. It turns out that different sets of stores get different types of customer, but you can't see that by looking directly at the customer records. Instead, you can assign customers to types by clustering the records, then look at what types of customer go to what set of store. This observation yields a quite general procedure for building features for complex signals (images, sound, accelerometer data). The method can take signals of varying size and produce a fixed size feature vector, which is then used for classification.

8.1 Agglomerative and Divisive Clustering

There are two natural recipes you can use to produce clustering algorithms. In **agglomerative clustering**, you start with each data item being a cluster, and then merge clusters recursively to yield a good clustering (Procedure 8.1). The difficulty here is that we need to know a good way to measure the distance between clusters, which can be somewhat harder than the distance between points. In **divisive clustering**, you start with the entire dataset being a cluster, and then split clusters recursively to yield a good clustering (Procedure 8.2). The difficulty here is we need to know some criterion for splitting clusters.

© Springer Nature Switzerland AG 2019
D. Forsyth, *Applied Machine Learning*,
https://doi.org/10.1007/978-3-030-18114-7_8

> **Procedure: 8.1** *Agglomerative Clustering*
>
> Choose an inter-cluster distance. Make each point a separate cluster. Now, until the clustering is satisfactory,
>
> - Merge the two clusters with the smallest inter-cluster distance.

> **Procedure: 8.2** *Divisive Clustering*
>
> Choose a splitting criterion. Regard the entire dataset as a single cluster. Now, until the clustering is satisfactory,
>
> - choose a cluster to split;
> - then split this cluster into two parts.

To turn these recipes into algorithms requires some more detail. For agglomerative clustering, we need to choose a good inter-cluster distance to fuse nearby clusters. Even if a natural distance between data points is available, there is no canonical inter-cluster distance. Generally, one chooses a distance that seems appropriate for the dataset. For example, one might choose the distance between the closest elements as the inter-cluster distance, which tends to yield extended clusters (statisticians call this method **single-link clustering**). Another natural choice is the maximum distance between an element of the first cluster and one of the second, which tends to yield rounded clusters (statisticians call this method **complete-link clustering**). Finally, one could use an average of distances between elements in the cluster, which also tends to yield rounded clusters (statisticians call this method **group average clustering**).

For divisive clustering, we need a splitting method. This tends to be something that follows from the logic of the application, because the ideal is an efficient method to find a natural split in a large dataset. We won't pursue this question further.

Finally, we need to know when to stop. This is an intrinsically difficult task if there is no model for the process that generated the clusters. The recipes I have described generate a hierarchy of clusters. Usually, this hierarchy is displayed to a user in the form of a **dendrogram**—a representation of the structure of the hierarchy of clusters that displays inter-cluster distances—and an appropriate choice of clusters is made from the dendrogram (see the example in Fig. 8.1).

Another important thing to notice about clustering from the example of Fig. 8.1 is that there is no right answer. There are a variety of different clusterings of the same data. For example, depending on what scales in that figure mean, it might be right to zoom out and regard all of the data as a single cluster, or to zoom in and regard each data point as a cluster. Each of these representations may be useful.

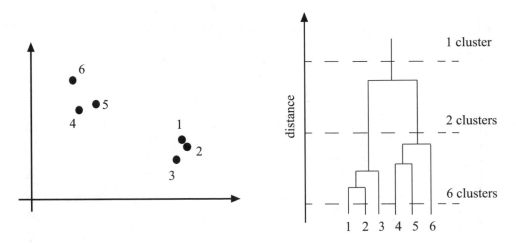

Figure 8.1: **Left**, a dataset; **right**, a dendrogram obtained by agglomerative clustering using single-link clustering. If one selects a particular value of distance, then a horizontal line at that distance splits the dendrogram into clusters. This representation makes it possible to guess how many clusters there are and to get some insight into how good the clusters are

8.1.1 Clustering and Distance

In the algorithms above, and in what follows, we assume that the features are scaled so that distances (measured in the usual way) between data points are a good representation of their similarity. This is quite an important point. For example, imagine we are clustering data representing brick walls. The features might contain several distances: the spacing between the bricks, the length of the wall, the height of the wall, and so on. If these distances are given in the same set of units, we could have real trouble. For example, assume that the units are centimeters. Then the spacing between bricks is of the order of one or two centimeters, but the heights of the walls will be in the hundreds of centimeters. In turn, this means that the distance between two data points is likely to be completely dominated by the height and length data. This could be what we want, but it might also not be a good thing.

There are some ways to manage this issue. One is to know what the features measure, and know how they should be scaled. Usually, this happens because you have a deep understanding of your data. If you don't (which happens!), then it is often a good idea to try and normalize the scale of the dataset. There are two good strategies. The simplest is to translate the data so that it has zero mean (this is just for neatness—translation doesn't change distances), then scale each direction so that it has unit variance. More sophisticated is to translate the data so that it has zero mean, then transform it so that each direction is independent and has unit variance. Doing so is sometimes referred to as **decorrelation** or **whitening**; I described how to do this in the exercises (p. 90).

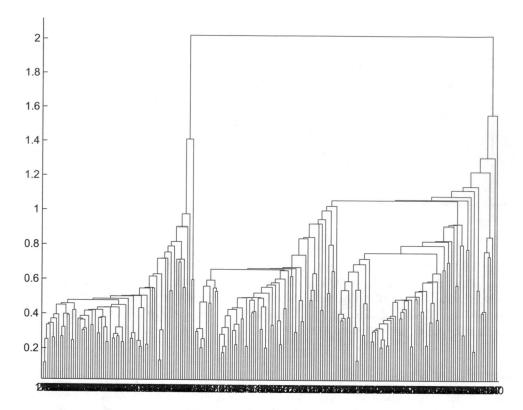

Figure 8.2:   A dendrogram obtained from the seed dataset, using single-link clustering. Recall that the data points are on the horizontal axis, and that the vertical axis is distance; there is a horizontal line linking two clusters that get merged, established at the height at which they're merged. I have plotted the entire dendrogram, despite the fact it's a bit crowded at the bottom, because you can now see how clearly the dataset clusters into a small set of clusters—there are a small number of vertical "runs"

Worked Example 8.1   *Agglomerative Clustering*

Cluster the seed dataset from the UC Irvine Machine Learning Dataset Repository (you can find it at http://archive.ics.uci.edu/ml/datasets/seeds).

Solution: Each item consists of seven measurements of a wheat kernel; there are three types of wheat represented in this dataset. For this example, I used Matlab, but many programming environments will provide tools that are useful for agglomerative clustering. I show a dendrogram in Fig. 8.2). I deliberately forced Matlab to plot the whole dendrogram, which accounts for the crowded look of the figure. As you can see from the dendrogram and from Fig. 8.3, this data clusters rather well.

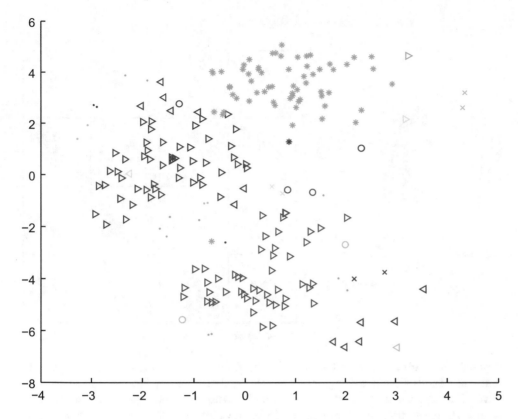

Figure 8.3: A clustering of the seed dataset, using agglomerative clustering, single-link distance, and requiring a maximum of 30 clusters. I have plotted each cluster with a distinct marker (though some markers differ only by color). Notice that there are a set of fairly natural isolated clusters. The original data is eight-dimensional, which presents plotting problems; I show a scatterplot on the first two principal components (though I computed distances for clustering in the original eight-dimensional space)

Remember This: *Agglomerative clustering starts with each data point a cluster, then recursively merges. There are three main ways to compute the distance between clusters. Divisive clustering starts with all in one cluster, then recursively splits. Choosing a split can be tricky.*

8.2 The k-Means Algorithm and Variants

Assume we have a dataset that, we believe, forms many clusters that look like blobs. If we knew where the center of each of the clusters was, it would be easy to tell which cluster each data item belonged to—it would belong to the cluster with

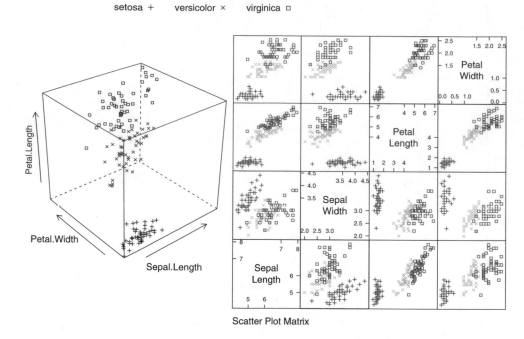

Figure 8.4: **Left:** a 3D scatterplot for the famous Iris data, collected by Edgar Anderson in 1936, and made popular among statisticians by Ronald Fisher in that year. I have chosen three variables from the four, and have plotted each species with a different marker. You can see from the plot that the species cluster quite tightly, and are different from one another. **Right:** a scatterplot matrix for the Iris data. There are four variables, measured for each of three species of iris. I have plotted each species with a different marker. You can see from the plot that the species cluster quite tightly, and are different from one another

the closest center. Similarly, if we knew which cluster each data item belonged to, it would be easy to tell where the cluster centers were—they'd be the mean of the data items in the cluster. This is the point closest to every point in the cluster.

We can formalize this fairly easily by writing an expression for the squared distance between data points and their cluster centers. Assume that we know how many clusters there are in the data, and write k for this number. There are N data items. The ith data item to be clustered is described by a feature vector $\mathbf{x}_i$. We write $\mathbf{c}_j$ for the center of the jth cluster. We write $\delta_{i,j}$ for a discrete variable that records which cluster a data item belongs to, so

$$\delta_{i,j} = \begin{cases} 1 & \text{if } \mathbf{x}_i \text{ belongs to cluster } j \\ 0 & \text{otherwise.} \end{cases}$$

We require that every data item belongs to exactly one cluster, so that $\sum_j \delta_{i,j} = 1$. We require that every cluster contains at least one point, because we assumed we knew how many clusters there were, so we must have that $\sum_i \delta_{i,j} > 0$ for every j. We can now write the sum of squared distances from data points to cluster centers as

$$\Phi(\delta, \mathbf{c}) = \sum_{i,j} \delta_{i,j} \left[(\mathbf{x}_i - \mathbf{c}_j)^T (\mathbf{x}_i - \mathbf{c}_j) \right].$$

Notice how the $\delta_{i,j}$ are acting as "switches." For the i'th data point, there is only one non-zero $\delta_{i,j}$ which selects the distance from that data point to the appropriate cluster center. It is natural to want to cluster the data by choosing the δ and $\mathbf{c}$ that minimizes $\Phi(\delta, \mathbf{c})$. This would yield the set of k clusters and their cluster centers such that the sum of distances from points to their cluster centers is minimized.

There is no known algorithm that can minimize Φ exactly in reasonable time. The $\delta_{i,j}$ are the problem: it turns out to be hard to choose the best allocation of points to clusters. The algorithm we guessed above is a remarkably effective approximate solution. Notice that if we know the $\mathbf{c}$'s, getting the δ's is easy—for the i'th data point, set the $\delta_{i,j}$ corresponding to the closest $\mathbf{c}_j$ to one and the others to zero. Similarly, if the $\delta_{i,j}$ are known, it is easy to compute the best center for each cluster—just average the points in the cluster. So we iterate:

- Assume the cluster centers are known and allocate each point to the closest cluster center.
- Replace each center with the mean of the points allocated to that cluster.

We choose a start point by randomly choosing cluster centers, and then iterate these stages alternately. This process eventually converges to a local minimum of the objective function (the value either goes down or is fixed at each step, and it is bounded below). It is not guaranteed to converge to the global minimum of the objective function, however. It is also not guaranteed to produce k clusters, unless we modify the allocation phase to ensure that each cluster has some non-zero number of points. This algorithm is usually referred to as k-**means** (summarized in Algorithm 8.3).

Procedure: 8.3 *k-Means Clustering*

Choose k. Now choose k data points $\mathbf{c}_j$ to act as cluster centers. Until the cluster centers change very little

- Allocate each data point to cluster whose center is nearest.
- Now ensure that every cluster has at least one data point; one way to do this is by supplying empty clusters with a point chosen at random from points far from their cluster center.
- Replace the cluster centers with the mean of the elements in their clusters.

Usually, we are clustering high dimensional data, so that visualizing clusters can present a challenge. If the dimension isn't too high, then we can use panel plots.

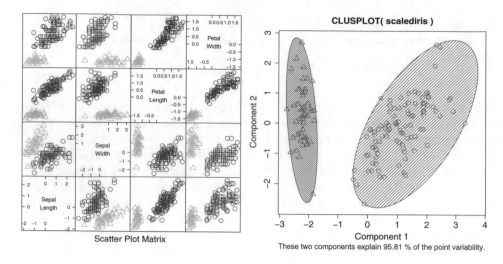

Figure 8.5: On the **left**, a panel plot of the iris data clustered using k-means with $k = 2$. By comparison with Fig. 8.4, notice how the *versicolor* and *verginica* clusters appear to have been merged. On the **right**, this dataset projected onto the first two principal components, with one blob drawn over each cluster

An alternative is to project the data onto two principal components, and plot the clusters there; the process for plotting 2D covariance ellipses from Sect. 4.4.2 comes in useful here. A natural dataset to use to explore k-means is the iris data, where we know that the data should form three clusters (because there are three species). Recall this dataset from Sect. 4.1. I reproduce Fig. 4.3 from that section as Fig. 8.4, for comparison. Figures 8.5, 8.6, and 8.7 show different k-means clusterings of that data.

One natural strategy for initializing k-means is to choose k data items at random, then use each as an initial cluster center. This approach is widely used, but has some difficulties. The quality of the clustering can depend quite a lot on initialization, and an unlucky choice of initial points might result in a poor clustering. One (again quite widely adopted) strategy for managing this is to initialize several times, and choose the clustering that performs best in your application. Another strategy, which has quite good theoretical properties and a good reputation, is known as k-**means++**. You choose a point $\mathbf{x}$ uniformly and at random from the dataset to be the first cluster center. Then you compute the squared distance between that point and each other point; write $d_i^2(\mathbf{x})$ for the distance from the i'th point to the first center. You now choose the other $k - 1$ cluster centers as IID draws from the probability distribution

$$\frac{d_i^2(\mathbf{x})}{\sum_u d_u^2(\mathbf{x})}.$$

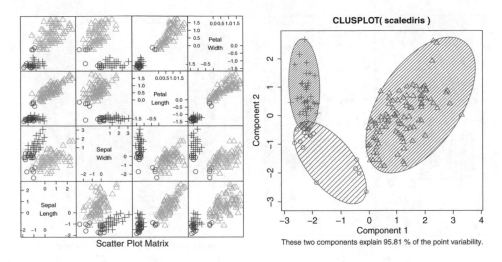

Figure 8.6: On the **left**, a panel plot of the iris data clustered using k-means with $k = 3$. By comparison with Fig. 8.4, notice how the clusters appear to follow the species labels. On the **right**, this dataset projected onto the first two principal components, with one blob drawn over each cluster

8.2.1 How to Choose k

The iris data is just a simple example. We know that the data forms clean clusters, and we know there should be three of them. Usually, we don't know how many clusters there should be, and we need to choose this by experiment. One strategy is to cluster for a variety of different values of k, then look at the value of the cost function for each. If there are more centers, each data point can find a center that is close to it, so we expect the value to go down as k goes up. This means that looking for the k that gives the smallest value of the cost function is not helpful, because that k is always the same as the number of data points (and the value is then zero). However, it can be very helpful to plot the value as a function of k, then look at the "knee" of the curve. Figure 8.8 shows this plot for the iris data. Notice that $k = 3$—the "true" answer—doesn't look particularly special, but $k = 2$, $k = 3$, or $k = 4$ all seem like reasonable choices. It is possible to come up with a procedure that makes a more precise recommendation by penalizing clusterings that use a large k, because they may represent inefficient encodings of the data. However, this is often not worth the bother.

In some special cases (like the iris example), we might know the right answer to check our clustering against. In such cases, one can evaluate the clustering by looking at the number of different labels in a cluster (sometimes called the purity), and the number of clusters. A good solution will have few clusters, all of which have high purity. Mostly, we don't have a right answer to check against. An alternative strategy, which might seem crude to you, for choosing k is extremely important in practice. Usually, one clusters data to use the clusters in an application (one of the most important, vector quantization, is described in Sect. 8.3). There are usually natural ways to evaluate this application. For example, vector quantization

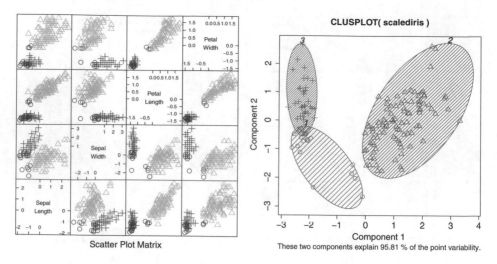

Figure 8.7: On the **left**, a panel plot of the iris data clustered using k-means with $k = 5$. By comparison with Fig. 8.4, notice how *setosa* seems to have been broken in two groups, and *versicolor* and *verginica* into a total of three. On the **right**, this dataset projected onto the first two principal components, with one blob drawn over each cluster

is often used as an early step in texture recognition or in image matching; here one can evaluate the error rate of the recognizer, or the accuracy of the image matcher. One then chooses the k that gets the best evaluation score on validation data. In this view, the issue is not how good the clustering is; it's how well the system that uses the clustering works.

8.2.2 Soft Assignment

One difficulty with k-means is that each point must belong to exactly one cluster. But, given we don't know how many clusters there are, this seems wrong. If a point is close to more than one cluster, why should it be forced to choose? This reasoning suggests we assign points to cluster centers with weights. These weights are different from the original $\delta_{i,j}$ because they are not forced to be either zero or one, however. Write $w_{i,j}$ for the weight connecting point i to cluster center j. Weights should be non-negative (i.e., $w_{i,j} \geq 0$), and each point should carry a total weight of 1 (i.e., $\sum_j w_{i,j} = 1$), so that if the i'th point contributes more to one cluster center, it is forced to contribute less to all others. You should see $w_{i,j}$ as a simplification of the $\delta_{i,j}$ in the original cost function. We can write a new cost function

$$\Phi(w, \mathbf{c}) = \sum_{i,j} w_{i,j} \left[(\mathbf{x}_i - \mathbf{c}_j)^T (\mathbf{x}_i - \mathbf{c}_j) \right],$$

which we would like to minimize by choice of w and $\mathbf{c}$. There isn't any improvement in the problem, because for any choice of $\mathbf{c}$, the best choice of w is to allocate each point to its closest cluster center. This is because we have not specified any relationship between w and $\mathbf{c}$.

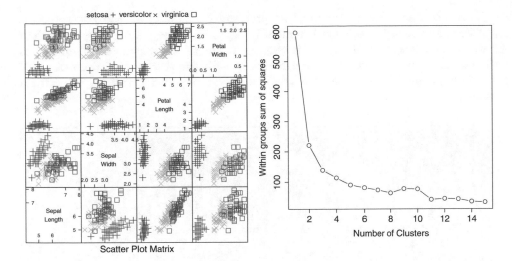

Figure 8.8: On the **left**, the scatterplot matrix for the Iris data, for reference. On the **right**, a plot of the value of the cost function for each of several different values of k. Notice how there is a sharp drop in cost going from $k = 1$ to $k = 2$, and again at $k = 4$; after that, the cost falls off slowly. This suggests using $k = 2$, $k = 3$, or $k = 4$, depending on the precise application

But w and $\mathbf{c}$ should be coupled. We would like $w_{i,j}$ to be large when $\mathbf{x}_i$ is close to $\mathbf{c}_j$, and small otherwise. Write $d_{i,j}$ for the distance $\|\mathbf{x}_i - \mathbf{c}_j\|$, choose a scaling parameter $\sigma > 0$, and write

$$s_{i,j} = e^{\frac{-d_{i,j}^2}{2\sigma^2}}.$$

This $s_{i,j}$ is often called the **affinity** between the point i and the center j; it is large when they are close in σ units, and small when they are far apart. Now a natural choice of weights is

$$w_{i,j} = \frac{s_{i,j}}{\sum_{l=1}^k s_{i,l}}.$$

All these weights are non-negative, and sum to one. The weight linking a point and a cluster center is large if the point is much closer to one center than to any other. The scaling parameter σ sets the meaning of "much closer"—we measure distance in units of σ.

Once we have weights, re-estimating the cluster centers is easy. We use the weights to compute a weighted average of the points. In particular, we re-estimate the j'th cluster center by

$$\frac{\sum_i w_{i,j} \mathbf{x}_i}{\sum_i w_{i,j}}.$$

Notice that k-means is a special case of this algorithm where σ limits to zero. In this case, each point has a weight of one for some cluster, and zero for all others, and the weighted mean becomes an ordinary mean. I have collected the description into a box (Procedure 8.4) for convenience.

Notice one other feature of this procedure. As long as you use sufficient precision for the arithmetic (which might be a problem), $w_{i,j}$ is *always* greater than zero. This means that no cluster is empty. In practice, if σ is small compared to the distances between points, you can end up with empty clusters. You can tell if this is happening by looking at $\sum_i w_{i,j}$; if this is very small or zero, you have a problem.

Procedure: 8.4 *k-Means with Soft Weights*

Choose k. Choose k data points $\mathbf{c}_j$ to act as initial cluster centers. Choose a scale, σ. Until the cluster centers change very little:

- First, we estimate the weights. For each pair of a data point $\mathbf{x}_i$ and a cluster $\mathbf{c}_j$, compute the affinity

$$s_{i,j} = e^{\dfrac{-\|\mathbf{x}_i - \mathbf{c}_j\|^2}{2\sigma^2}}.$$

- Now for each pair of a data point $\mathbf{x}_i$ and a cluster $\mathbf{c}_j$ compute the soft weight linking the data point to the center

$$w_{i,j} = s_{i,j} / \sum_{l=1}^{k} s_{i,l}.$$

- For each cluster, compute a new center

$$\mathbf{c}_j = \frac{\sum_i w_{i,j}\mathbf{x}_i}{\sum_i w_{i,j}}.$$

8.2.3 Efficient Clustering and Hierarchical k-Means

One important difficulty occurs in applications. We might need to have an enormous dataset (millions of items is a real possibility), and so a very large k. In this case, k-means clustering becomes difficult because identifying which cluster center is closest to a particular data point scales linearly with k (and we have to do this for every data point at every iteration). There are two useful strategies for dealing with this problem.

The first is to notice that, if we can be reasonably confident that each cluster contains many data points, some of the data is redundant. We could randomly subsample the data, cluster that, then keep the cluster centers. This helps rather a lot, but not enough if you expect the data will contain many clusters.

A more effective strategy is to build a hierarchy of k-means clusters. We randomly subsample the data (typically quite aggressively), then cluster this with a small value of k. Each data item is then allocated to the closest cluster center, and the data in each cluster is clustered again with k-means. We now have something that looks like a two-level tree of clusters. Of course, this process can be repeated to produce a multi-level tree of clusters.

8.2.4 *k*-Medoids

In some cases, we want to cluster objects that can't be averaged. One case where this happens is when you have a table of distances between objects, but do not know vectors representing the objects. For example, you could collect data giving the distances between cities, without knowing where the cities are (as in Sect. 6.2.3, particularly Fig. 6.1), then try and cluster using this data. As another example, you could collect data giving similarities between breakfast items as in Sect. 6.2.3, then turn the similarities into distances by taking the negative logarithm. This gives a usable table of distances. You still can't average kippers with oatmeal, so you couldn't use *k*-means to cluster this data.

A variant of *k*-means, known as *k*-medoids, applies to this case. In *k*-medoids, the cluster centers are data items rather than averages, and so are called "medoids." The rest of the algorithm has a familiar form. We assume *k*, the number of cluster centers, is known. We initialize the cluster centers by choosing examples at random. We then iterate two procedures. In the first, we allocate each data point to the closest medoid. In the second, we choose the best medoid for each cluster by finding the data point that minimizes the sum of distances of points in the cluster to that medoid. This point can be found by simply searching all the points in the cluster.

8.2.5 Example: Groceries in Portugal

Clustering can be used to expose structure in datasets that isn't visible with simple tools. Here is an example. At http://archive.ics.uci.edu/ml/datasets/Wholesale+ customers, you will find a dataset giving sums of money spent annually on different commodities by customers in Portugal. The commodities are divided into a set of categories (fresh; milk; grocery; frozen; detergents and paper; and delicatessen) relevant for the study. These customers are divided by channel (two channels, corresponding to different types of shop) and by region (three regions). You can think of the data as being divided into six blocks (one for each pair of channel and region). There are 440 customer records, and there are many customers in each block. The data was provided by M. G. M. S. Cardoso.

Figure 8.9 shows a panel plot of the customer data; the data has been clustered, and I gave each of 10 clusters its own marker. You (or at least, I) can't see any evidence of the six blocks here. This is due to the form of the visualization, rather than a true property of the data. People tend to like to live near people who are "like" them, so you could expect people in a region to be somewhat similar; you could reasonably expect differences between blocks (regional preferences; differences in wealth; and so on). Retailers have different channels to appeal to different people, so you could expect people using different channels to be different. But you don't see this in the plot of clusters. In fact, the plot doesn't really show much structure at all, and is basically unhelpful.

Here is a way to think about structure in the data. There are likely to be different "types" of customer. For example, customers who prepare food at home might spend more money on fresh or on grocery, and those who mainly buy prepared

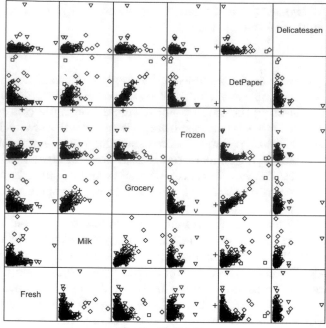

Scatter Plot Matrix

Figure 8.9: A panel plot of the wholesale customer data of http://archive.ics.uci. edu/ml/datasets/Wholesale+customers, which records sums of money spent annually on different commodities by customers in Portugal. This data is recorded for six different blocks (two channels each within three regions). I have plotted each block with a different marker, but you can't really see much structure here, for reasons explained in the text

food might spend more money on delicatessen; similarly, coffee drinkers with cats or with children might spend more on milk than the lactose-intolerant, and so on. So we can expect customers to cluster in types. An effect like this is hard to see on a panel plot of the clustered data (Fig. 8.9). The plot for this dataset is hard to read, because the dimension is fairly high for a panel plot and the data is squashed together in the bottom left corner. However, you can see the effect when you cluster the data and look at the cost function in representing the data with different values of k—quite a small set of clusters gives quite a good representation of the customers (Fig. 8.10). The panel plot of cluster membership (also in that figure) isn't particularly informative. The dimension is quite high, and clusters get squashed together.

There is an important effect which isn't apparent in the panel plots. Some of what cause customers to cluster in types are driven by things like wealth and the tendency of people to have neighbors who are similar to them. This means that different blocks should have different fractions of each type of customer. There might be more deli-spenders in wealthier regions; more milk-spenders and detergent-spenders in regions where it is customary to have many children; and so on. This sort of

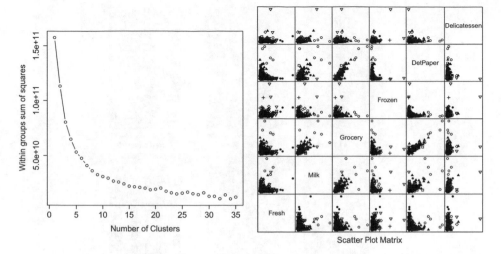

Figure 8.10:   On the **left**, the cost function (of Sect. 8.2) for clusterings of the customer data with k-means for k running from 2 to 35. This suggests using a k somewhere in the range 10–30; I chose 10. On the **right**, I have clustered this data to 10 cluster centers with k-means. The clusters seem to be squashed together, but the plot on the left suggests that clusters do capture some important information. Using too few clusters will clearly lead to problems. Notice that I did not scale the data, because each of the measurements is in a comparable unit. For example, it wouldn't make sense to scale expenditures on fresh and expenditures on grocery with a different scale. The main point of the plot on the right is that it's hard to interpret, and that we need a better way to represent the underlying data

structure will not be apparent in a panel plot. A block of a few milk-spenders and many detergent-spenders will have a few data points with high milk expenditure values (and low other values) and also many data points with high detergent expenditure values (and low other values). In a panel plot, this will look like two blobs; but if there is a second block with many milk-spenders and few detergent-spenders it will also look like two blobs, lying roughly on top of the first set of blobs. It will be hard to spot the difference between the blocks.

An easy way to see this difference is to look at histograms of the types of customer *within each block*. Figure 8.11 shows this representation for the shopper dataset. The figure shows the histogram of customer types that appears in each block. The blocks do appear to contain quite different distributions of customer type, as you would expect. It looks as though the channels (rows in this figure) are more different than the regions (columns in this figure). Again, you might expect this: regions might contain slightly different customers (e.g., as a result of regional food preferences), but different channels are intended to cater to different customers.

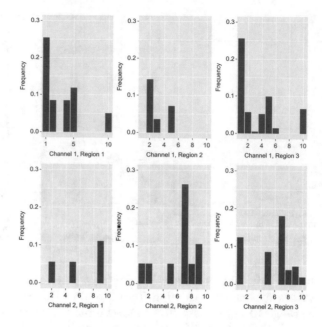

Figure 8.11:   The histogram of different types of customer, by block, for the customer data. Notice how the distinction between the blocks is now apparent—the blocks do appear to contain quite different distributions of customer type. It looks as though the channels (rows in this figure) are more different than the regions (columns in this figure)

8.2.6   General Comments on k-Means

If you experiment with k-means, you will notice one irritating habit of the algorithm. It almost always produces either some rather spread out clusters or some single element clusters. Most clusters are usually rather tight and blobby clusters, but there is usually one or more bad cluster. This is fairly easily explained. Because every data point must belong to some cluster, data points that are far from all others (a) belong to some cluster and (b) very likely "drag" the cluster center into a poor location. This applies even if you use soft assignment, because every point must have total weight one. If the point is far from all others, then it will be assigned to the closest with a weight very close to one, and so may drag it into a poor location, or it will be in a cluster on its own.

There are ways to deal with this. If k is very big, the problem is often not significant, because then you simply have many single element clusters that you can ignore. It isn't always a good idea to have too large a k, because then some larger clusters might break up. An alternative is to have a junk cluster. Any point that is too far from the closest true cluster center is assigned to the junk cluster, and the center of the junk cluster is not estimated. Notice that points should not be assigned to the junk cluster permanently; they should be able to move in and out of the junk cluster as the cluster centers move.

> **Remember This:** *k-Means clustering is the "go-to" clustering algorithm. You should see it as a basic recipe from which many algorithms can be concocted. The recipe is: iterate: allocate each data point to the closest cluster center; re-estimate cluster centers from their data points. There are many variations, improvements, etc., that are possible on this recipe. We have seen soft weights and k-medoids. k-Means is not usually best implemented with the method I described (which isn't particularly efficient, but gets to the heart of what is going on). Implementations of k-means differ in important ways from my rather high-level description of the algorithm; for any but tiny problems, you should use a package, and you should look for a package that uses the Lloyd–Hartigan method.*

8.3 Describing Repetition with Vector Quantization

The classifiers in Chap. 1 can be applied to simple images (the MNIST exercises at the end of the chapter, for example), but they will annoy you if you try to apply them as described to more complicated signals. All the methods described apply to feature vectors of fixed length. But typical of signals like speech, images, video, or accelerometer outputs is that different versions of the same thing have different lengths. For example, pictures appear at different resolutions, and it seems clumsy to insist that every image be 28×28 before it can be classified. As another example, some speakers are slow, and others are fast, but it's hard to see much future for a speech understanding system that insisted that everyone speak at the same speed so the classifier could operate. We need a construction that will take a signal and produce a useful feature vector of fixed length. This section shows one of the most useful such constructions (but be aware, this is an enormous topic).

Repetition is an important feature of many interesting signals. For example, images contain *textures*, which are orderly patterns that look like large numbers of small structures that are repeated. Examples include the spots of animals such as leopards or cheetahs; the stripes of animals such as tigers or zebras; the patterns on bark, wood, and skin. Similarly, speech signals contain *phonemes*—characteristic, stylized sounds that people assemble together to produce speech (for example, the "ka" sound followed by the "tuh" sound leading to "cat"). Another example comes from accelerometers. If a subject wears an accelerometer while moving around, the signals record the accelerations during their movements. So, for example, brushing one's teeth involves a lot of repeated twisting movements at the wrist, and walking involves swinging the hand back and forth.

Repetition occurs in subtle forms. The essence is that a small number of local patterns can be used to represent a large number of examples. You see this effect in pictures of scenes. If you collect many pictures of, say, a beach scene, you will expect most to contain some waves, some sky, and some sand. The individual patches of wave, sky, or sand can be surprisingly similar. However, it's fair to model this by saying different images are made by selecting some patches from a vocabulary of patches, then placing them down to form an image. Similarly, pictures of living rooms contain chair patches, TV patches, and carpet patches. Many different living rooms can be made from small vocabularies of patches; but you won't often see wave patches in living rooms, or carpet patches in beach scenes. This suggests that the patches that are used to make an image reveal something about what is in the image. This observation works for speech, for video, and for accelerometer signals too.

An important part of representing signals that repeat is building a vocabulary of patterns that repeat, then describing the signal in terms of those patterns. For many problems, knowing what vocabulary elements appear and how often is much more important than knowing where they appear. For example, if you want to tell the difference between zebras and leopards, you need to know whether stripes or spots are more common, but you don't particularly need to know where they appear. As another example, if you want to tell the difference between brushing teeth and walking using accelerometer signals, knowing that there are lots of (or few) twisting movements is important, but knowing how the movements are linked together in time may not be. As a general rule, one can do quite a good job of classifying video just by knowing what patterns are there (i.e., without knowing where or when the patterns appear). Not all signals are like this. For example, in speech it really matters what sound follows what sound.

8.3.1 Vector Quantization

It is natural to try and find patterns by looking for small pieces of signal of fixed size that appear often. In an image, a piece of signal might be a 10x10 patch, which can be reshaped into a vector. In a sound file, which is likely represented as a vector, it might be a subvector of fixed size. A 3-axis accelerometer signal is usually represented as a $3 \times r$ dimensional array (where r is the number of samples); in this case, a piece might be a 3×10 subarray, which can be reshaped into a vector. But finding patterns that appear often is hard to do, because the signal is continuous—each pattern will be slightly different, so we cannot simply count how many times a particular pattern occurs.

Here is a strategy. We take a training set of signals, and cut each signal into pieces of fixed size and reshape them into d dimensional vectors. We then build a set of clusters out of these pieces. This set of clusters is often thought of as a dictionary, because we expect many or most cluster centers to look like pieces that occur often in the signals and so are repeated.

We can now describe any new piece of signal with the cluster center closest to that piece. This means that a piece of signal is described with a number in the range $[1, \ldots, k]$ (where you get to choose k), and two pieces that are close should be described by the same number. This strategy is known as **vector quantization** (often **VQ**).

This strategy applies to any kind of signal, and is surprisingly robust to details. We could use d dimensional vectors for a sound file; $\sqrt{d} \times \sqrt{d}$ dimensional patches for an image; or $3 \times (d/3)$ dimensional subarrays for an accelerometer signal. In each case, it is easy to compute the distance between two pieces using sum of squared differences. It seems not to matter much if the signals are cut into overlapping or

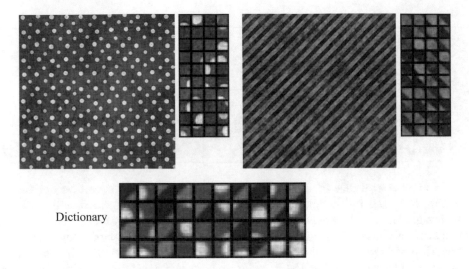

Dictionary

Figure 8.12: **Top:** two images with rather exaggerated repetition, published on flickr.com with a creative commons license by `webtreats`. Next to these images, I have placed zoomed sampled 10×10 patches from those images; although the spots (resp. stripes) aren't necessarily centered in the patches, it's pretty clear which image each patch comes from. **Bottom:** a 40 patch dictionary computed using k-means from 4000 samples from each image. If you look closely, you'll see that some dictionary entries are clearly stripe entries, others clearly spot entries. Stripe images will have patches represented by stripe entries in the dictionary and spot images by spot entries

non-overlapping pieces when forming the dictionary, as long as there are enough pieces.

Procedure: 8.5 *Building a Dictionary for VQ*

Take a training set of signals, and cut each signal into pieces of fixed size. The size of the piece will affect how well your method works, and is usually chosen by experiment. It does not seem to matter much if the pieces overlap. Cluster all the example pieces, and record the k cluster centers. It is usual, but not required, to use k-means clustering.

We can now build features that represent important repeated structure in signals. We take a signal, and cut it up into vectors of length d. These might overlap, or be disjoint. We then take each vector, and compute the number that describes it (i.e., the number of the closest cluster center, as above). We then compute a histogram of the numbers we obtained for all the vectors in the signal. This histogram describes the signal.

Procedure: 8.6 *Representing a Signal Using VQ*

Take your signal, and cut it into pieces of fixed size. The size of the piece will affect how well your method works, and is usually chosen by experiment. It does not seem to matter much if the pieces overlap. For each piece, record the closest cluster center in the dictionary. Represent the signal with a histogram of these numbers, which will be a k dimensional vector.

Notice several nice features to this construction. First, it can be applied to anything that can be thought of in terms of fixed size pieces, so it will work for speech signals, sound signals, accelerometer signals, images, and so on. Another nice feature is the construction can accept signals of different length, and produce a description of fixed length. One accelerometer signal might cover 100 time intervals; another might cover 200; but the description is always a histogram with k buckets, so it's always a vector of length k.

Yet another nice feature is that we don't need to be all that careful how we cut the signal into fixed length vectors. This is because it is hard to hide repetition. This point is easier to make with a figure than in text, so look at Fig. 8.12.

The number of pieces of signal (and so k) might be very big indeed. It is quite reasonable to want to build a dictionary for a million items and use tens to hundreds of thousands of cluster centers. In this case, it is a good idea to use hierarchical k-means, as in Sect. 8.2.3. Hierarchical k-means produces a tree of cluster centers. It is easy to use this tree to vector quantize a query data item. We vector quantize at the first level. Doing so chooses a branch of the tree, and we pass the data item to this branch. It is either a leaf, in which case we report the number of the leaf, or it is a set of clusters, in which case we vector quantize, and pass the data item down. This procedure is efficient both when one clusters and at run time.

Representing a signal as a histogram of cluster centers loses information in two important ways. First, the histogram has little or no information about how the pieces of signal are arranged. So, for example, the representation can tell whether an image has stripy or spotty patches in it, but not where those patches lie. You should not rely on your intuition to tell you whether this lost information is important or not. For many kinds of image classification task, histograms of cluster centers are much better than you might guess, despite not encoding where patches lie (though still better results are now obtained with convolutional neural networks).

Second, replacing a piece of signal with a cluster center must lose some detail, which might be important, and likely results in some classification errors. There is a surprisingly simple construction that can alleviate these problems. Build three (or more) dictionaries, rather than one, using different sets of training pieces. For example, you could cut the same signals into pieces on a different grid. Now use each dictionary to produce a histogram of cluster centers, and classify with those. Finally, use a voting scheme to decide the class of each test signal. In many problems, this approach yields small but useful improvements.

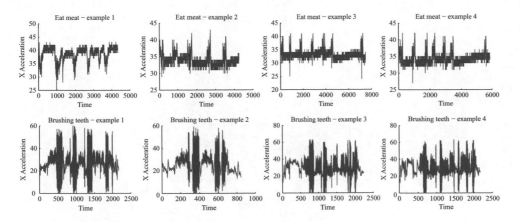

Figure 8.13: Some examples from the accelerometer dataset at https://archive .ics.uci.edu/ml/datasets/Dataset+for+ADL+Recognition+with+Wrist-worn+ Accelerometer. I have labelled each signal by the activity. These show acceleration in the X direction (Y and Z are in the dataset, too). There are four examples for **brushing teeth** and four for **eat meat**. You should notice that the examples don't have the same length in time (some are slower and some faster eaters, etc.), but that there seem to be characteristic features that are shared within a category (brushing teeth seems to involve faster movements than eating meat)

8.3.2 Example: Activity from Accelerometer Data

A complex example dataset appears at https://archive.ics.uci.edu/ml/datasets/ Dataset+for+ADL+Recognition+with+Wrist-worn+Accelerometer. This dataset consists of examples of the signal from a wrist mounted accelerometer, produced as different subjects engaged in different activities of daily life. Activities include: brushing teeth, climbing stairs, combing hair, descending stairs, and so on. Each is performed by 16 volunteers. The accelerometer samples the data at 32 Hz (i.e., this data samples and reports the acceleration 32 times per second). The accelerations are in the x, y, and z-directions. The dataset was collected by Barbara Bruno, Fulvio Mastrogiovanni, and Antonio Sgorbissa. Figure 8.13 shows the x-component of various examples of toothbrushing.

There is an important problem with using data like this. Different subjects take quite different amounts of time to perform these activities. For example, some subjects might be more thorough toothbrushers than other subjects. As another example, people with longer legs walk at somewhat different frequencies than people with shorter legs. This means that the same activity performed by different subjects will produce data vectors *that are of different lengths*. It's not a good idea to deal with this by warping time and resampling the signal. For example, doing so will make a thorough toothbrusher look as though they are moving their hands very fast (or a careless toothbrusher look ludicrously slow: think speeding up or slowing down a movie). So we need a representation that can cope with signals that are a bit longer or shorter than other signals.

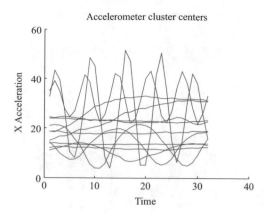

Figure 8.14:   Some cluster centers from the accelerometer dataset. Each cluster center represents a 1-s burst of activity. There are a total of 480 in my model, which I built using hierarchical k-means. Notice there are a couple of centers that appear to represent movement at about 5 Hz; another few that represent movement at about 2 Hz; some that look like 0.5 Hz movement; and some that seem to represent much lower frequency movement. These cluster centers are samples (rather than chosen to have this property)

Another important property of these signals is that all examples of a particular activity should contain repeated patterns. For example, brushing teeth should show fast accelerations up and down; walking should show a strong signal at somewhere around 2 Hz; and so on. These two points should suggest vector quantization to you. Representing the signal in terms of stylized, repeated structures is probably a good idea because the signals probably contain these structures. And if we represent the signal in terms of the relative frequency with which these structures occur, the representation will have a fixed length, even if the signal doesn't. To do so, we need to consider (a) over what time scale we will see these repeated structures and (b) how to ensure we segment the signal into pieces so that we see these structures.

Generally, repetition in activity signals is so obvious that we don't need to be smart about segment boundaries. I broke these signals into 32 sample segments, one following the other. Each segment represents 1 s of activity. This is long enough for the body to do something interesting, but not so long that our representation will suffer if we put the segment boundaries in the wrong place. This resulted in about 40,000 segments. I then used hierarchical k-means to cluster these segments. I used two levels, with 40 cluster centers at the first level, and 12 at the second. Figure 8.14 shows some cluster centers at the second level.

I then computed histogram representations for different example signals (Fig. 8.15). You should notice that when the activity label is different, the histogram looks different, too.

Another useful way to check this representation is to compare the average within class chi-squared distance with the average between class chi-squared

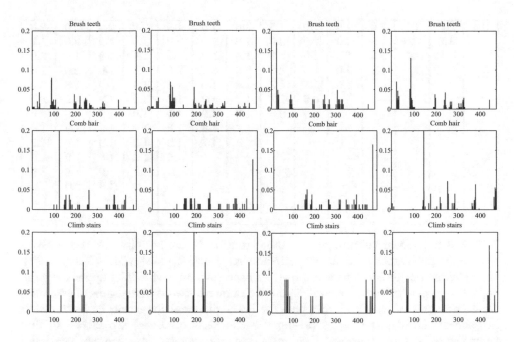

Figure 8.15:   Histograms of cluster centers for the accelerometer dataset, for different activities. You should notice that (a) these histograms look somewhat similar for different actors performing the same activity and (b) these histograms look somewhat different for different activities

distance. I computed the histogram for each example. Then, for each pair of examples, I computed the chi-squared distance between the pair. Finally, for each pair of *activity labels*, I computed the average distance between pairs of examples where one example has one of the activity labels and the other example has the other activity label. In the ideal case, all the examples with the same label would be very close to one another, and all examples with different labels would be rather different. Table 8.1 shows what happens with the real data. You should notice that for some pairs of activity label, the mean distance between examples is smaller than one would hope for (perhaps some pairs of examples are quite close?). But generally, examples of activities with different labels tend to be further apart than examples of activities with the same label.

Yet another way to check the representation is to try classification with nearest neighbors, using the chi-squared distance to compute distances. I split the dataset into 80 test pairs and 360 training pairs; using 1-nearest neighbors, I was able to get a held-out error rate of 0.79. This suggests that the representation is fairly good at exposing what is important.

0.9	2.0	1.9	2.0	2.0	2.0	1.9	2.0	1.9	1.9	2.0	2.0	2.0	2.0
	1.6	2.0	1.8	2.0	2.0	2.0	1.9	1.9	2.0	1.9	1.9	2.0	1.7
		1.5	2.0	1.9	1.9	1.9	1.9	1.9	1.9	1.9	1.9	1.9	2.0
			1.4	2.0	2.0	2.0	2.0	2.0	2.0	2.0	2.0	2.0	1.8
				1.5	1.8	1.7	1.9	1.9	1.8	1.9	1.9	1.8	2.0
					0.9	1.7	1.9	1.9	1.8	1.9	1.9	1.9	2.0
						0.3	1.9	1.9	1.5	1.9	1.9	1.9	2.0
							1.8	1.8	1.9	1.9	1.9	1.9	1.9
								1.7	1.9	1.9	1.9	1.9	1.9
									1.6	1.9	1.9	1.9	2.0
										1.8	1.9	1.9	1.9
											1.8	2.0	1.9
												1.5	2.0
													1.5

TABLE 8.1: Each column of the table represents an activity for the activity dataset https://archive.ics.uci.edu/ml/datasets/Dataset+for+ADL+Recognition+ with+Wrist-worn+Accelerometer, as does each row. In each of the upper diagonal cells, I have placed the average chi-squared distance between histograms of examples from that pair of classes (I dropped the lower diagonal for clarity). Notice that in general the diagonal terms (average within class distance) are rather smaller than the off diagonal terms. This quite strongly suggests that we can use these histograms to classify examples successfully

8.4 You Should

8.4.1 Remember These Terms

8.4.2 Remember These Facts

8.4.3 Remember These Procedures

Programming Exercises

8.1. You can find a dataset dealing with European employment in 1979 at http://dasl.datadesk.com/data/view/47. This dataset gives the percentage of people employed in each of a set of areas in 1979 for each of a set of European countries.

    (a) Use an agglomerative clusterer to cluster this data. Produce a dendrogram of this data for each of single link, complete link, and group average clustering. You should label the countries on the axis. What structure in the data does each method expose? It's fine to look for code, rather than writing your own. **Hint:** I made plots I liked a lot using R's `hclust` clustering function, and then turning the result into a phylogenetic tree and using a fan plot, a trick I found on the web; try `plot(as.phylo(hclustresult), type=''fan'')`. You should see dendrograms that "make sense" (at least if you remember some European history), and have interesting differences.

    (b) Using k-means, cluster this dataset. What is a good choice of k for this data and why?

8.2. Obtain the liver disorder dataset from UC Irvine machine learning website (http://archive.ics.uci.edu/ml/datasets/Liver%20Disorders; data provided by Richard S. Forsyth). The first five values in each row represent various biological measurements, and the sixth is a measure of the amount of alcohol consumed daily. We will analyze this data following the recipe of Sect. 8.2.5. Divide the data into four blocks using the amount of alcohol, where each block is a quantile (so data representing the lowest quarter of consumption rates go into the first block, etc.). Now cluster the data using the first five values and k-means. For each block, compute a histogram of the cluster centers (as in Fig. 8.11). Plot these histograms. What do you conclude?

8.3. Obtain the Portuguese student math grade dataset from UC Irvine machine learning website (https://archive.ics.uci.edu/ml/datasets/student+performance; data provided by Paulo Cortez). There are two datasets at the URL—you are looking for the one that relates to math grades. Each row contains some numerical attributes (columns 3, 7, 8, 13, 14, 15, 24, 25, 26, 27, 28, 29, 30) and some other attributes. We will consider only the numerical attributes. Column 33 contains the grade at the end of the year in numerical form. We will analyze this data following the recipe of Sect. 8.2.5. Divide the data into four blocks using the final grade, where each block is a quantile (so data representing the lowest quarter of grades go into the first block, etc.). Now cluster the data using the numerical values and k-means. For each block, compute a histogram of the cluster centers (as in Fig. 8.11). Plot these histograms. What do you conclude?

8.4. Obtain the activities of daily life dataset from the UC Irvine machine learning website (https://archive.ics.uci.edu/ml/datasets/Dataset+for+ADL+Recognition+with+Wrist-worn+Accelerometer; data provided by Barbara Bruno, Fulvio Mastrogiovanni, and Antonio Sgorbissa).

    (a) Build a classifier that classifies sequences into one of the 14 activities provided. To make features, you should vector quantize, then use a histogram of cluster centers (as described in the subsection; this gives a pretty explicit set of steps to follow). You will find it helpful to use hierarchical k-means to vector quantize. You may use whatever multiclass classifier you wish, though I'd start with R's decision forest, because it's easy to use and effective. You should report (a) the total error rate and (b) the class-confusion matrix of your classifier.

(b) Now see if you can improve your classifier by (a) modifying the number of cluster centers in your hierarchical k-means and (b) modifying the size of the fixed length samples that you use.

8.5. This is a fairly ambitious exercise. It will demonstrate how to use vector quantization to handle extremely sparse data. The 20 newsgroups dataset is a famous text dataset. It consists of posts collected from 20 different newsgroups. There are a variety of tricky data issues that this presents (for example, what aspects of the header should one ignore? should one reduce words to their stems, so "winning" goes to "win," "hugely" to "huge," and so on?). We will ignore these issues, and deal with a cleaned up version of the dataset. This consists of three items each for train and test: a document-word matrix, a set of labels, and a map. You can find this cleaned up version of the dataset at http://qwone.com/~jason/20Newsgroups/. You should look for the cleaned up version, identified as 20news-bydate-matlab.tgz on that page. The usual task is to label a test article with which newsgroup it came from. Instead, we will assume you have a set of test articles, all from the same newsgroup, and you need to identify the newsgroup. The document-word matrix is a table of counts of how many times a particular word appears in a particular document. The collection of words is very large (53,975 distinct words), and most words do not appear in most documents, so most entries of this matrix are zero. The file train.data contains this matrix for a collection of training data; each row represents a distinct document (there are 11,269), and each column represents a distinct word.

(a) Cluster the rows of this matrix to get a set of cluster centers using k-means. You should have about one center for every 10 documents. Use k-means, and you should find an efficient package rather than using your own implementation. In particular, implementations of k-means differ in important ways from my rather high-level description of the algorithm; you should look for a package that uses the Lloyd–Hartigan method. **Hint:** Clustering all these points is a bit of a performance; check your code on small subsets of the data first, because the size of this dataset means that clustering the whole thing will be slow.

(b) You can now think of each cluster center as a document "type." For each newsgroup, plot a histogram of the "types" of document that appear in the training data for that newsgroup. You'll need to use the file train.label, which will tell you what newsgroup a particular item comes from.

(c) Now train a classifier that accepts a small set of documents (10–100) from a single newsgroup, and predicts which of 20 newsgroups it comes from. You should use the histogram of types from the previous subexercise as a feature vector. Compute the performance of this classifier on the test data (test.data and test.label).

8.6. *This is a substantial exercise.* The MNIST dataset is a dataset of 60,000 training and 10,000 test examples of handwritten digits, originally constructed by Yann Lecun, Corinna Cortes, and Christopher J.C. Burges. It is very widely used to check simple methods. There are 10 classes in total ("0" to "9"). This dataset has been extensively studied, and there is a history of methods and feature constructions at https://en.wikipedia.org/wiki/MNIST_database and at http://yann.lecun.com/exdb/mnist/. You should notice that the best methods perform extremely well. The original dataset is at http://yann.lecun.com/exdb/mnist/. It is stored in an unusual format, described in detail on that website. Writing your own reader is pretty simple, but web search yields readers for standard packages. There is reader code in Matlab available (at least) at http://ufldl.stanford.edu/wiki/index.php/Using_the_MNIST_Dataset. There is reader code for R available (at least) at https://stackoverflow.com/questions/21521571/how-to-read-mnist-database-in-r.

The dataset consists of 28×28 images. These were originally binary images, but appear to be grey level images as a result of some anti-aliasing. I will ignore mid grey pixels (there aren't many of them) and call dark pixels "ink pixels," and light pixels "paper pixels." The digit has been centered in the image by centering the center of gravity of the image pixels. For this exercise, we will use raw pixels in untouched images.

(a) We will use hierarchical k-means to build a dictionary of image patches. For untouched images, construct a collection of 10×10 image patches. You should extract these patches from the training images on an overlapping 4×4 grid, meaning that each training image produces 16 overlapping patches (so you could have 960,000 training patches!). For each training image, choose one of these patches uniformly and at random. Now subsample this dataset of 60,000 patches uniformly and at random to produce a 6000 element dataset. Cluster this dataset to 50 centers. Now build 50 datasets, one per cluster center. Do this by taking each element of the 60,000 patch dataset, finding which of the cluster centers is closest to it, and putting the patch in that center's dataset. Now cluster each of these datasets to 50 centers.

(b) You now have a dictionary of 2500 entries. For each query image, construct a set of 10×10 patches on an overlapping 4×4 grid. Now for each of the centers, you should extract 9 patches. Assume the center is at (x, y); obtain 9 patches by extracting a patch centered at $(x - 1, y - 1), (x, y - 1), \ldots, (x + 1, y + 1)$. This means each test image will have 144 associated patches. Now use your dictionary to find the closest center to each patch, and construct a histogram of patches for each test image.

(c) Train a classifier (I'd use a decision forest) using this histogram of patches representation. Evaluate this classifier on the test data.

(d) Can you improve this classifier by modifying how you extract patches, the size of the patches, or the number of centers?

(e) At this point, you're likely tired of MNIST, but very well informed. Compare your methods to the table of methods at http://yann.lecun.com/exdb/mnist/.

8.7. CIFAR-10 is a dataset of 32x32 images in 10 categories, collected by Alex Krizhevsky, Vinod Nair, and Geoffrey Hinton. It is often used to evaluate machine learning algorithms. You can download this dataset from https://www.cs.toronto.edu/~kriz/cifar.html. There are 10 classes, 50,000 training images, and 10,000 test images.

(a) We will use hierarchical k-means to build a dictionary of image patches. For untouched images, construct a collection of $10 \times 10 \times 3$ image patches. You should extract these patches from the training images at random locations (you don't know where the good stuff in the image is), and you should extract two patches per training image. Now subsample this dataset of 100,000 patches uniformly and at random to produce a 10,000 element dataset. Cluster this dataset to 50 centers. Now build 50 datasets, one per cluster center. Do this by taking each element of the 100,000 patch dataset, finding which of the cluster centers is closest to it, and putting the patch in that center's dataset. Now cluster each of these datasets to 50 centers.

(b) You now have a dictionary of 2500 entries. For each query image, construct a set of 10×10 patches. You should extract patches using centers that are on a grid spaced two pixels apart horizontally and vertically, so there will be a lot of overlap between the patches. Now use your dictionary to find the closest center to each patch, and construct a histogram of patches for each test image.

(c) Train a classifier (I'd use a decision forest) using this histogram of patches representation. Evaluate this classifier on the test data.

(d) Can you improve this classifier by modifying how you extract patches, the size of the patches, or the number of centers?

(e) At this point, you're likely tired of CIFAR-10 but very well informed. Compare your methods to Roderigo Benenson's table of methods at http://rodrigob.github.io/are_we_there_yet/build/classification_datasets_results.html.

CHAPTER 9

Clustering Using Probability Models

Clustering objects requires some notion of how similar they are. We have seen how to cluster using distance in feature space, which is a natural way of thinking about similarity. Another way to think about similarity is to ask whether two objects have high probability under the same probability model. This can be a convenient way of looking at things when it is easier to build probability models than it is to measure distances. It turns out to be a natural way of obtaining soft clustering weights (which emerge from the probability model). And it provides a framework for our first encounter with an extremely powerful and general algorithm, which you should see as a very aggressive generalization of k-means.

9.1 Mixture Models and Clustering

It is natural to think of clustering in the following way. The data was created by a collection of distinct probability models (one per cluster). For each data item, something (nature?) chose which model was to produce a point, and then an IID sample of that model produces the point. We see the points: we'd like to know what the models were, but (and this is crucial) we don't know which model produced which point. If we knew the models, it would be easy to decide which model produced which point. Similarly, if we knew which point went to which model, we could determine what the models were. One encounters this situation—or problems that can be mapped to this situation—again and again. It is very deeply embedded in clustering problems.

You should notice a resonance with k-means here. In k-means, if we knew the centers, which point belongs to which center would be easy; if we knew which point belongs to which center, the centers would be easy. We dealt with this situation quite effectively by repeatedly fixing one, then estimating the other. It is pretty clear that a natural algorithm for dealing with the probability models is to iterate between estimating which model gets which point, and the model parameters. This is the key to a standard, and very important, algorithm for estimation here, called **EM** (or **expectation maximization**, if you want the long version). I will develop this algorithm in two simple cases, and we will see it in a more general form later.

Notation: This topic lends itself to a glorious festival of indices, limits of sums and products, etc. I will do one example in quite gory detail; the other follows the same form, and for that we'll proceed more expeditiously. Writing the limits of sums or products explicitly is usually even more confusing than adopting a compact notation. When I write $\sum_i$ or $\prod_i$, I mean a sum (or product) over all values of i. When I write $\sum_{i,\hat{j}}$ or $\prod_{i,\hat{j}}$, I mean a sum (or product) over all values of i *except* for the jth item. I will write vectors, as usual, as $\mathbf{x}$; the ith such vector

© Springer Nature Switzerland AG 2019
D. Forsyth, *Applied Machine Learning*,
https://doi.org/10.1007/978-3-030-18114-7_9

in a collection is $\mathbf{x}_i$, and the kth component of the ith vector in a collection is x_{ik}. In what follows, I will construct a vector δ_i corresponding to the ith data item $\mathbf{x}_i$ (it will tell us what cluster that item belongs to). I will write δ to mean all the δ_i (one for each data item). The jth component of δ_i is δ_{ij}. When I write $\sum_{\delta_u}$, I mean a sum over all values that δ_u can take. When I write $\sum_{\delta}$, I mean a sum over all values that each δ can take. When I write $\sum_{\delta,\delta_v}$, I mean a sum over all values that all δ can take, *omitting* all cases for the vth vector δ_v.

9.1.1 A Finite Mixture of Blobs

A blob of data points is quite easily modelled with a single normal distribution. Obtaining the parameters is straightforward (estimate the mean and covariance matrix with the usual expressions). Now imagine I have t blobs of data, and I know t. A normal distribution is likely a poor model, but I could think of the data as being produced by t normal distributions. I will assume that each normal distribution has a fixed, *known* covariance matrix Σ, but the mean of each is unknown. Because the covariance matrix is fixed, and *known*, we can compute a factorization $\Sigma = \mathcal{A}\mathcal{A}^T$. The factors must have full rank, because the covariance matrix must be positive definite. This means that we can apply $\mathcal{A}^{-1}$ to all the data, so that each blob covariance matrix (and so each normal distribution) is the identity.

Write μ_j for the mean of the jth normal distribution. We can model a distribution that consists of t distinct blobs by forming a weighted sum of the blobs, where the jth blob gets weight π_j. We ensure that $\sum_j \pi_j = 1$, so that we can think of the overall model as a probability distribution. We can then model the data as samples from the probability distribution

$$p(\mathbf{x}|\mu_1,\ldots,\mu_k,\pi_1,\ldots,\pi_k) = \sum_j \pi_j \left[\frac{1}{\sqrt{(2\pi)^d}} \exp\left(-\frac{1}{2}(\mathbf{x} - \mu_j)^T(\mathbf{x} - \mu_j) \right) \right].$$

The way to think about this probability distribution is that a point is generated by first choosing one of the normal distributions (the jth is chosen with probability π_j), then generating a point from that distribution. This is a pretty natural model of clustered data. Each mean is the center of a blob. Blobs with many points in them have a high value of π_j, and blobs with a few points have a low value of π_j. We must now use the data points to estimate the values of π_j and μ_j (again, I am assuming that the blobs—and the normal distribution modelling each—have the identity as a covariance matrix). A distribution of this form is known as a **mixture of normal distributions**, and the π_j terms are usually called **mixing weights**.

Writing out the likelihood will reveal a problem: we have a product of many sums. The usual trick of taking the log will not work, because then you have a sum of logs of sums, which is hard to differentiate and hard to work with. A much more productive approach is to think about a set of hidden variables which tell us which blob each data item comes from. For the ith data item, we construct a vector δ_i. The jth component of this vector is δ_{ij}, where $\delta_{ij} = 1$ if $\mathbf{x}_i$ comes from blob (equivalently, normal distribution) j and zero otherwise. Notice there is exactly one 1 in δ_i, because each data item comes from one blob. I will write δ to mean all the δ_i (one for each data item). Assume we know the values of these terms. I will write $\theta = (\mu_1, \ldots, \mu_k, \pi_1, \ldots, \pi_k)$ for the unknown parameters. Then we can write

$$p(\mathbf{x}_i | \delta_i, \theta) = \prod_j \left[\frac{1}{\sqrt{(2\pi)^d}} \exp\left(-\frac{1}{2}(\mathbf{x}_i - \mu_j)^T (\mathbf{x}_i - \mu_j) \right) \right]^{\delta_{ij}}$$

(because $\delta_{ij} = 1$ means that $\mathbf{x}_i$ comes from blob j, so the terms in the product are a collection of 1's and the probability we want). We also have

$$p(\delta_{ij} = 1 | \theta) = \pi_j$$

allowing us to write

$$p(\delta_i | \theta) = \prod_j [\pi_j]^{\delta_{ij}}$$

(because this is the probability that we select blob j to produce a data item; again, the terms in the product are a collection of 1's and the probability we want). This means that

$$p(\mathbf{x}_i, \delta_i | \theta) = \prod_j \left\{ \left[\frac{1}{\sqrt{(2\pi)^d}} \exp\left(-\frac{1}{2}(\mathbf{x}_i - \mu_j)^T (\mathbf{x}_i - \mu_j) \right) \right] \pi_j \right\}^{\delta_{ij}}$$

and we can write a log-likelihood. The data are the observed values of $\mathbf{x}$ and δ (remember, we pretend we know these; I'll fix this in a moment), and the parameters are the unknown values of $\mu_1, \ldots, \mu_k$ and $\pi_1, \ldots, \pi_k$. We have

$$
\begin{aligned}
\mathcal{L}(\mu_1, \ldots, \mu_k, \pi_1, \ldots, \pi_k; \mathbf{x}, \delta) &= \mathcal{L}(\theta; \mathbf{x}, \delta) \\
&= \sum_{ij} \left\{ \left[\left(-\frac{1}{2}(\mathbf{x}_i - \mu_j)^T (\mathbf{x}_i - \mu_j) \right) \right] + \log \pi_j \right\} \delta_{ij} \\
&\quad + K,
\end{aligned}
$$

where K is a constant that absorbs the normalizing constants for the normal distributions. You should check this expression. I have used the δ_{ij} as a "switch"—for one term, $\delta_{ij} = 1$ and the term in curly brackets is "on," and for all others that term is multiplied by zero. The problem with all this is that we don't know δ. I will deal with this when we have another example.

9.1.2 Topics and Topic Models

We have already seen that word counts expose similarities between documents (Sect. 6.3). We now assume that documents with similar word counts will come from the same **topic** (mostly, a term of art for cluster used in the natural language processing community). A really useful model is to assume that words are conditionally independent, conditioned on the topic. This means that, once you know the topic, words are IID samples of a multinomial distribution that is given by the topic (the **word probabilities** for that topic). If it helps, you can think of the topic as multi-sided die with a different word on each face. Each document has one topic. If you know the topic, you make a document by rolling this die—which is likely not a fair die—some number of times.

This model of documents has problems. Word order doesn't matter in this model nor does where a word appears in a document or what words are near in the document and what others are far away. We've already seen that ignoring word order, word position, and neighbors can still produce useful representations (Sect. 6.3). Despite its problems, this model clusters documents rather well, is easy to work with, and is the basis for more complex models.

A single document is a set of word counts that is obtained by (a) selecting a topic then (b) drawing words as IID samples from that topic. We now have a collection of documents, and we want to know (a) what topic each document came from and (b) the word probabilities for each topic. Now imagine we know which document comes from which topic. Then we could estimate the word probabilities using the documents in each topic by simply counting. In turn, imagine we know the word probabilities for each topic. Then we could tell (at least in principle) which topic a document comes from by looking at the probability each topic generates the document, and choosing the topic with the highest probability. This procedure should strike you as being very like k-means, though the details have changed.

To construct a probabilistic model more formally, we will assume that a document is generated in two steps. We will have t topics. First, we choose a topic, choosing the jth topic with probability π_j. Then we will obtain a set of words by repeatedly drawing IID samples from that topic, and record the count of each word in a count vector. Each topic is a multinomial probability distribution. The vocabulary is d-dimensional. Write $\mathbf{p}_j$ for the d-dimensional vector of word probabilities for the jth topic. Now write $\mathbf{x}_i$ for the ith vector of word counts (there are N vectors in the collection). We assume that words are generated independently, conditioned on the topic. Write x_{ik} for the kth component of $\mathbf{x}_i$, and so on. Notice that $\mathbf{x}_i^T\mathbf{1}$ is the sum of entries in $\mathbf{x}_i$, and so the number of words in document i. Then the probability of observing the counts in $\mathbf{x}_i$ when the document was generated by topic j is

$$p(\mathbf{x}_i|\mathbf{p}_j) = \left(\frac{(\mathbf{x}_i^T\mathbf{1})!}{\prod_v x_{iv}!}\right)\prod_u p_{ju}^{x_{iu}}.$$

We can now write the probability of observing a document. Again, we write $\theta = (\mathbf{p}_1,\ldots,\mathbf{p}_t,\pi_1,\ldots,\pi_t)$ for the vector of unknown parameters. We have

$$p(\mathbf{x}_i|\theta) = \sum_l p(\mathbf{x}_i|\text{topic is }l)p(\text{topic is }l|\theta)$$

$$= \sum_l \left[\left(\frac{(\mathbf{x}_i^T\mathbf{1})!}{\prod_v x_{iv}!}\right)\prod_u p_{lu}^{x_{iu}}\right]\pi_l.$$

This model is widely called a **topic model**; be aware that there are many kinds of topic model, and this is a simple one. The expression should look unpromising, in a familiar way. If you write out a likelihood, you will see a product of sums; and if you write out a log-likelihood, you will see a sum of logs of sums. Neither is enticing. We could use the same trick we used for a mixture of normals. Write

$\delta_{ij} = 1$ if $\mathbf{x}_i$ comes from topic j, and $\delta_{ij} = 0$ otherwise. Then we have

$$p(\mathbf{x}_i|\delta_{ij} = 1, \theta) = \left[\left(\frac{(\mathbf{x}_i^T \mathbf{1})!}{\prod_v x_{iv}!} \right) \prod_u p_{ju}^{x_{iu}} \right]$$

(because $\delta_{ij} = 1$ means that $\mathbf{x}_i$ comes from topic j). This means we can write

$$p(\mathbf{x}_i|\delta_i, \theta) = \prod_j \left\{ \left[\left(\frac{(\mathbf{x}_i^T \mathbf{1})!}{\prod_v x_{iv}!} \right) \prod_u p_{ju}^{x_{iu}} \right] \right\}^{\delta_{ij}}$$

(because $\delta_{ij} = 1$ means that $\mathbf{x}_i$ comes from topic j, so the terms in the product are a collection of 1's and the probability we want). We also have

$$p(\delta_{ij} = 1|\theta) = \pi_j$$

(because this is the probability that we select topic j to produce a data item), allowing us to write

$$p(\delta_i|\theta) = \prod_j [\pi_j]^{\delta_{ij}}$$

(again, the terms in the product are a collection of 1's and the probability we want). This means that

$$p(\mathbf{x}_i, \delta_i|\theta) = \prod_j \left[\left(\frac{(\mathbf{x}_i^T \mathbf{1})!}{\prod_v x_{iv}!} \right) \prod_u \left(p_{ju}^{x_{iu}} \right) \pi_j \right]^{\delta_{ij}}$$

and we can write a log-likelihood. The data are the observed values of $\mathbf{x}$ and δ (remember, we pretend we know these for the moment), and the parameters are the unknown values collected in θ. We have

$$\mathcal{L}(\theta; \mathbf{x}, \delta) = \sum_i \left\{ \sum_j \left[\sum_u x_{iu} \log p_{ju} + \log \pi_j \right] \delta_{ij} \right\} + K,$$

where K is a term that contains all the

$$\log \left(\frac{(\mathbf{x}_i^T \mathbf{1})!}{\prod_v x_{iv}!} \right)$$

terms. This is of no interest to us, because it doesn't depend on any of our parameters. It takes a fixed value for each dataset. You should check this expression, noticing that, again, I have used the δ_{ij} as a "switch"—for one term, $\delta_{ij} = 1$ and the term in curly brackets is "on," and for all others that term is multiplied by zero. The problem with all this, as before, is that we don't know δ_{ij}. But there is a recipe.

9.2 The EM Algorithm

There is a straightforward, natural, and very powerful recipe for estimating θ for both models. In essence, we will average out the things we don't know. But this average will depend on our estimate of the parameters, so we will average, then re-estimate parameters, then re-average, and so on. If you lose track of what's going on here, think of the example of k-means with soft weights (Sect. 8.2.2; this is close to what the equations for the case of a mixture of normals will boil down to). In this analogy, the δ tell us which cluster center a data item came from. Because we don't know the values of the δ, we assume we have a set of cluster centers; these allow us to make an estimate of the δ; then we use this estimate to re-estimate the centers; and so on.

This is an instance of a general recipe. Recall we wrote θ for a vector of parameters. In the mixture of normals case, θ contained the means and the mixing weights; in the topic model case, it contained the topic distributions and the mixing weights. Assume we have an estimate of the value of this vector, say $\theta^{(n)}$. We could then compute $p(\delta|\theta^{(n)}, \mathbf{x})$. In the mixture of normals case, this is a guide to which example goes to which cluster. In the topic case, it is a guide to which example goes to which topic.

We could use this to compute the expected value of the likelihood with respect to δ. We compute

$$\mathcal{Q}(\theta; \theta^{(n)}) = \sum_\delta \mathcal{L}(\theta; \mathbf{x}, \delta) p(\delta|\theta^{(n)}, \mathbf{x}) = \mathbb{E}_{p(\delta|\theta^{(n)}, \mathbf{x})}[\mathcal{L}(\theta; \mathbf{x}, \delta)]$$

(where the sum is over all values of δ). Notice that $Q(\theta; \theta^{(n)})$ is a *function* of θ (because $\mathcal{L}$ was), but now does not have any unknown δ terms in it. This $Q(\theta; \theta^{(n)})$ encodes what we know about δ.

For example, assume that $p(\delta|\theta^{(n)}, \mathbf{x})$ has a single, narrow peak in it, at (say) $\delta = \delta^0$. In the mixture of normals case, this would mean that there is one allocation of points to clusters that is significantly better than all others, given $\theta^{(n)}$. For this example, $Q(\theta; \theta^{(n)})$ will be approximately $\mathcal{L}(\theta; \mathbf{x}, \delta^0)$.

Now assume that $p(\delta|\theta^{(n)}, \mathbf{x})$ is about uniform. In the mixture of normals case, this would mean that any particular allocation of points to clusters is about as good as any other. For this example, $Q(\theta; \theta^{(n)})$ will average $\mathcal{L}$ over all possible δ values with about the same weight for each.

We obtain the next estimate of θ by computing

$$\theta^{(n+1)} = \begin{array}{c} \text{argmax} \\ \theta \end{array} Q\left(\theta; \theta^{(n)}\right)$$

and iterate this procedure until it converges (which it does, though I shall not prove that). The algorithm I have described is extremely general and powerful, and is known as **expectation maximization** or (more usually) **EM**. The step where we compute $Q(\theta; \theta^{(n)})$ is called the **E-step**; the step where we compute the new estimate of θ is known as the **M-step**.

One trick to be aware of: it is quite usual to ignore additive constants in the log-likelihood, because they have no effect. When you do the E-step, taking the expectation of a constant gets you a constant; in the M-step, the constant can't change the outcome. As a result, additive constants may disappear without notice (they do so regularly in the research literature). In the mixture of normals example, below, I've tried to keep track of them; for the mixture of multinomials, I've been looser.

9.2.1 Example: Mixture of Normals: The E-step

Now let us do the actual calculations for a mixture of normal distributions. The E-step requires a little work. We have

$$\mathcal{Q}\left(\theta; \theta^{(n)}\right) = \sum_{\delta} \mathcal{L}(\theta; \mathbf{x}, \delta) p\left(\delta | \theta^{(n)}, \mathbf{x}\right).$$

If you look at this expression, it should strike you as deeply worrying. There are a very large number of different possible values of δ. In this case, there are t^N cases (there is one δ_i for each data item, and each of these can have a one in each of t locations). It isn't obvious how we could compute this average.

But notice

$$p(\delta | \theta^{(n)}, \mathbf{x}) = \frac{p(\delta, \mathbf{x} | \theta^{(n)})}{p(\mathbf{x} | \theta^{(n)})}$$

and let us deal with numerator and denominator separately. For the numerator, notice that the $\mathbf{x}_i$ and the δ_i are independent, identically distributed samples, so that

$$p(\delta, \mathbf{x} | \theta^{(n)}) = \prod_i p(\delta_i, \mathbf{x}_i | \theta^{(n)}).$$

The denominator is slightly more work. We have

$$
\begin{aligned}
p\left(\mathbf{x} | \theta^{(n)}\right) &= \sum_{\delta} p\left(\delta, \mathbf{x} | \theta^{(n)}\right) \\
&= \sum_{\delta} \left[\prod_i p\left(\delta_i, \mathbf{x}_i | \theta^{(n)}\right) \right] \\
&= \prod_i \left[\sum_{\delta_i} p\left(\delta_i, \mathbf{x}_i | \theta^{(n)}\right) \right].
\end{aligned}
$$

You should check the last step; one natural thing to do is check with $N = 2$ and

$t = 2$. This means that we can write

$$
\begin{aligned}
p(\delta|\theta^{(n)}, \mathbf{x}) &= \frac{p\left(\delta, \mathbf{x}|\theta^{(n)}\right)}{p\left(\mathbf{x}|\theta^{(n)}\right)} \\
&= \frac{\prod_i p\left(\delta_i, \mathbf{x}_i|\theta^{(n)}\right)}{\prod_i \left[\sum_{\delta_i} p\left(\delta_i, \mathbf{x}_i|\theta^{(n)}\right)\right]} \\
&= \prod_i \frac{p\left(\delta_i, \mathbf{x}_i|\theta^{(n)}\right)}{\sum_{\delta_i} p\left(\delta_i, \mathbf{x}_i|\theta^{(n)}\right)} \\
&= \prod_i p\left(\delta_i|\mathbf{x}_i, \theta^{(n)}\right).
\end{aligned}
$$

Now we need to look at the log-likelihood. We have

$$
\mathcal{L}(\theta; \mathbf{x}, \delta) = \sum_{ij} \left\{\left[\left(-\frac{1}{2}(\mathbf{x}_i - \mu_j)^T(\mathbf{x}_i - \mu_j)\right)\right] + \log \pi_j\right\} \delta_{ij} + K.
$$

The K term is of no interest—it will result in a constant—but we will try to keep track of it. To simplify the equations we need to write, I will construct a t dimensional vector $\mathbf{c}_i$ for the ith data point. The jth component of this vector will be

$$
\left\{\left[\left(-\frac{1}{2}(\mathbf{x}_i - \mu_j)^T(\mathbf{x}_i - \mu_j)\right)\right] + \log \pi_j\right\}
$$

so we can write

$$
\mathcal{L}(\theta; \mathbf{x}, \delta) = \sum_i \mathbf{c}_i^T \delta_i + K.
$$

Now all this means that

$$
\begin{aligned}
\mathcal{Q}(\theta; \theta^{(n)}) &= \sum_\delta \mathcal{L}(\theta; \mathbf{x}, \delta) p(\delta|\theta^{(n)}, \mathbf{x}) \\
&= \sum_\delta \left(\sum_i \mathbf{c}_i^T \delta_i + K\right) p(\delta|\theta^{(n)}, \mathbf{x}) \\
&= \sum_\delta \left(\sum_i \mathbf{c}_i^T \delta_i + K\right) \prod_u p(\delta_u|\theta^{(n)}, \mathbf{x}) \\
&= \sum_\delta \left(\mathbf{c}_1^T \delta_1 \prod_u p(\delta_u|\theta^{(n)}, \mathbf{x}) + \ldots \mathbf{c}_N^T \delta_N \prod_u p(\delta_u|\theta^{(n)}, \mathbf{x})\right).
\end{aligned}
$$

We can simplify further. We have that $\sum_{\delta_i} p(\delta_i|\mathbf{x}_i, \theta^{(n)}) = 1$, because this is a probability distribution. Notice that, for any index v

$$
\begin{aligned}
\sum_\delta \left(\mathbf{c}_v^T \delta_v \prod_u p(\delta_u|\theta^{(n)}, \mathbf{x})\right) &= \sum_{\delta_v} \left(\mathbf{c}_v^T \delta_v p(\delta_v|\theta^{(n)}, \mathbf{x})\right) \left[\sum_{\delta, \hat{\delta}_v} \prod_{u, \hat{v}} p(\delta_u|\theta^{(n)}, \mathbf{x})\right] \\
&= \sum_{\delta_v} \left(\mathbf{c}_v^T \delta_v p(\delta_v|\theta^{(n)}, \mathbf{x})\right).
\end{aligned}
$$

So we can write

$$
\begin{aligned}
\mathcal{Q}(\theta;\theta^{(n)}) &= \sum_{\delta} \mathcal{L}(\theta;\mathbf{x},\delta)p(\delta|\theta^{(n)},\mathbf{x}) \\
&= \sum_{i}\left[\sum_{\delta_i}\mathbf{c}_i^T\delta_i p(\delta_i|\theta^{(n)},\mathbf{x})\right]+K \\
&= \sum_{i}\left[\left(\sum_{j}\left\{\left[\left(-\frac{1}{2}(\mathbf{x}_i-\mu_j)^T(\mathbf{x}_i-\mu_j)\right)+\log\pi_j\right]w_{ij}\right\}\right)\right]+K,
\end{aligned}
$$

where

$$
\begin{aligned}
w_{ij} &= 1p(\delta_{ij}=1|\theta^{(n)},\mathbf{x})+0p(\delta_{ij}=0|\theta^{(n)},\mathbf{x}) \\
&= p(\delta_{ij}=1|\theta^{(n)},\mathbf{x}).
\end{aligned}
$$

Now

$$
\begin{aligned}
p(\delta_{ij}=1|\theta^{(n)},\mathbf{x}) &= \frac{p(\mathbf{x},\delta_{ij}=1|\theta^{(n)})}{p(\mathbf{x}|\theta^{(n)})} \\
&= \frac{p(\mathbf{x},\delta_{ij}=1|\theta^{(n)})}{\sum_l p(\mathbf{x},\delta_{il}=1|\theta^{(n)})} \\
&= \frac{p(\mathbf{x}_i,\delta_{ij}=1|\theta^{(n)})\prod_{u,\hat{i}}p(\mathbf{x}_u,\delta_u|\theta)}{\left(\sum_l p(\mathbf{x},\delta_{il}=1|\theta^{(n)})\right)\prod_{u,\hat{i}}p(\mathbf{x}_u,\delta_u|\theta)} \\
&= \frac{p(\mathbf{x}_i,\delta_{ij}=1|\theta^{(n)})}{\sum_l p(\mathbf{x},\delta_{il}=1|\theta^{(n)})}.
\end{aligned}
$$

If the last couple of steps puzzle you, remember we obtained $p(\mathbf{x},\delta|\theta)=\prod_i p(\mathbf{x}_i,\delta_i|\theta)$. Also, look closely at the denominator; it expresses the fact that the data must have come from somewhere. So the main question is to obtain $p(\mathbf{x}_i,\delta_{ij}=1|\theta^{(n)})$. But

$$
\begin{aligned}
p(\mathbf{x}_i,\delta_{ij}=1|\theta^{(n)}) &= p(\mathbf{x}_i|\delta_{ij}=1,\theta^{(n)})p(\delta_{ij}=1|\theta^{(n)}) \\
&= \left[\frac{1}{\sqrt{(2\pi)^d}}\exp\left(-\frac{1}{2}(\mathbf{x}_i-\mu_j)^T(\mathbf{x}_i-\mu_j)\right)\right]\pi_j.
\end{aligned}
$$

Substituting yields

$$
p(\delta_{ij}=1|\theta^{(n)},\mathbf{x})=\frac{\left[\exp\left(-\frac{1}{2}(\mathbf{x}_i-\mu_j)^T(\mathbf{x}_i-\mu_j)\right)\right]\pi_j}{\sum_k\left[\exp\left(-\frac{1}{2}(\mathbf{x}_i-\mu_k)^T(\mathbf{x}_i-\mu_k)\right)\right]\pi_k}=w_{ij}.
$$

9.2.2 Example: Mixture of Normals: The M-step

The M-step is more straightforward. Recall

$$
\mathcal{Q}(\theta;\theta^{(n)})=\left(\sum_{ij}\left\{\left[\left(-\frac{1}{2}(\mathbf{x}_i-\mu_j)^T(\mathbf{x}_i-\mu_j)\right)\right]+\log\pi_j\right\}w_{ij}+K\right)
$$

and we have to maximize this with respect to μ and π, and the terms w_{ij} are known. This maximization is easy. We compute

$$\mu_j^{(n+1)} = \frac{\sum_i \mathbf{x}_i w_{ij}}{\sum_i w_{ij}}$$

and

$$\pi_j^{(n+1)} = \frac{\sum_i w_{ij}}{N}.$$

You should check these expressions. When you do so, remember that, because π is a probability distribution, $\sum_j \pi_j = 1$ (otherwise you'll get the wrong answer). You need to either use a Lagrange multiplier or set one probability to $(1 - \text{all others})$.

9.2.3 Example: Topic Model: The E-step

We need to work out two steps. The E-step requires a little calculation. We have

$$
\begin{aligned}
\mathcal{Q}(\theta; \theta^{(n)}) &= \sum_\delta \mathcal{L}(\theta; \mathbf{x}, \delta) p(\delta | \theta^{(n)}, \mathbf{x}) \\
&= \sum_\delta \left(\sum_{ij} \left\{ \left[\sum_u x_{iu} \log p_{ju} \right] + \log \pi_j \right\} \delta_{ij} \right) p(\delta | \theta^{(n)}, \mathbf{x}) \\
&= \left(\sum_{ij} \left\{ \left[\sum_k x_{i,k} \log p_{j,k} \right] + \log \pi_j \right\} w_{ij} \right).
\end{aligned}
$$

Here the last two steps follow from the same considerations as in the mixture of normals. The $\mathbf{x}_i$ and δ_i are IID samples, and so the expectation simplifies as in that case. If you're uncertain, rewrite the steps of Sect. 9.2.1. The form of this Q function is the same as that (a sum of $\mathbf{c}_i^T \delta_i$ terms, but using a different expression for $\mathbf{c}_i$). In this case, as above,

$$
\begin{aligned}
w_{ij} &= 1 p(\delta_{ij} = 1 | \theta^{(n)}, \mathbf{x}) + 0 p(\delta_{ij} = 0 | \theta^{(n)}, \mathbf{x}) \\
&= p(\delta_{ij} = 1 | \theta^{(n)}, \mathbf{x}).
\end{aligned}
$$

Again, we have

$$
\begin{aligned}
p(\delta_{ij} = 1 | \theta^{(n)}, \mathbf{x}) &= \frac{p(\mathbf{x}_i, \delta_{ij} = 1 | \theta^{(n)})}{p(\mathbf{x}_i | \theta^{(n)})} \\
&= \frac{p(\mathbf{x}_i, \delta_{ij} = 1 | \theta^{(n)})}{\sum_l p(\mathbf{x}_i, \delta_{il} = 1 | \theta^{(n)})}
\end{aligned}
$$

and so the main question is to obtain $p(\mathbf{x}_i, \delta_{ij} = 1 | \theta^{(n)})$. But

$$
\begin{aligned}
p(\mathbf{x}_i, \delta_{ij} = 1 | \theta^{(n)}) &= p(\mathbf{x}_i | \delta_{ij} = 1, \theta^{(n)}) p(\delta_{ij} = 1 | \theta^{(n)}) \\
&= = \left[\prod_k p_{j,k}^{x_k} \right] \pi_j.
\end{aligned}
$$

Substituting yields

$$
p(\delta_{ij} = 1 | \theta^{(n)}, \mathbf{x}) = \frac{\left[\prod_k p_{j,k}^{x_k} \right] \pi_j}{\sum_l \left[\prod_k p_{l,k}^{x_k} \right] \pi_l}.
$$

9.2.4 Example: Topic Model: The M-step

The M-step is more straightforward. Recall

$$
\mathcal{Q}(\theta; \theta^{(n)}) = \left(\sum_{ij} \left\{ \left[\sum_k x_{i,k} \log p_{j,k} \right] + \log \pi_j \right\} w_{ij} \right)
$$

and we have to maximize this with respect to μ and π, and the terms w_{ij} are known. This maximization is easy, but remember that the probabilities sum to one, so you need to either use a Lagrange multiplier or set one probability to $(1 - \text{all others})$. You should get

$$
\mathbf{p}_j^{(n+1)} = \frac{\sum_i \mathbf{x}_i w_{ij}}{\sum_i \mathbf{x}_i^T \mathbf{1} w_{ij}}
$$

and

$$
\pi_j^{(n+1)} = \frac{\sum_i w_{ij}}{N}.
$$

You should check these expressions by differentiating and setting to zero.

9.2.5 EM in Practice

The algorithm we have seen is amazingly powerful; I will use it again, ideally with less notation. One could reasonably ask whether it produces a "good" answer. Slightly surprisingly, the answer is yes. The algorithm produces a local maximum of $p(\mathbf{x} | \theta)$, the likelihood of the data conditioned on parameters. This is rather surprising because we engaged in all the activity with δ to avoid directly dealing with this likelihood (which in our cases was an unattractive product of sums). I did not prove this, but it's true anyway. I have summarized the general algorithm, and the two instances we studied, in boxes below for reference. There are some practical issues.

Procedure: 9.1 *EM*

Given a model with parameters θ, data $\mathbf{x}$, and missing data δ, which gives rise to a log-likelihood $\mathcal{L}(\theta; \mathbf{x}, \delta) = \log P(\mathbf{x}, \delta | \theta)$ and some initial estimate of parameters $\theta^{(1)}$, iterate

- **The E-step:** Obtain

$$\mathcal{Q}(\theta; \theta^{(n)}) = \mathbb{E}_{p(\delta | \theta^{(n)}, \mathbf{x})}[\mathcal{L}(\theta; \mathbf{x}, \delta)].$$

- **The M-step:** Compute

$$\theta^{(n+1)} = \underset{\theta}{\text{argmax}} \; Q(\theta; \theta^{(n)}).$$

Diagnose convergence by testing the size of the update to θ.

Procedure: 9.2 *EM for Mixtures of Normals: E-step*

Assume $\theta^{(n)} = (\mu_1, \ldots, \mu_t, \pi_1, \ldots, \pi_t)$ is known. Compute weights w_{ij} linking the ith data item to the jth cluster center, using

$$w_{ij}^{(n)} = \frac{\left[\exp\left(-\frac{1}{2}\left(\mathbf{x}_i - \mu_j^{(n)}\right)^T \left(\mathbf{x}_i - \mu_j^{(n)}\right)\right)\right]\pi_j^{(n)}}{\sum_k \left[\exp\left(-\frac{1}{2}\left(\mathbf{x}_i - \mu_k^{(n)}\right)^T \left(\mathbf{x}_i - \mu_k^{(n)}\right)\right)\right]\pi_k^{(n)}}.$$

Procedure: 9.3 *EM for Mixtures of Normals: M-step*

Assume $\theta^{(n)} = (\mu_1, \ldots, \mu_t, \pi_1, \ldots, \pi_t)$ and weights w_{ij} linking the ith data item to the jth cluster center are known. Then estimate

$$\mu_j^{(n+1)} = \frac{\sum_i \mathbf{x}_i w_{ij}^{(n)}}{\sum_i w_{ij}^{(n)}}$$

and

$$\pi_j^{(n+1)} = \frac{\sum_i w_{ij}^{(n)}}{N}.$$

Procedure: 9.4   *EM for Topic Models: E-step*

Assume $\theta^{(n)} = (\mathbf{p}_1, \ldots, \mathbf{p}_t, \pi_1, \ldots, \pi_t)$ is known. Compute weights $w_{ij}^{(n)}$ linking the ith data item to the jth cluster center, using

$$w_{ij}^{(n)} = \frac{\left[\prod_k \left(p_{j,k}^{(n)} \right)^{x_k} \right] \pi_j^{(n)}}{\sum_l \left[\prod_k \left(p_{j,k}^{(n)} \right)^{x_k} \right] \pi_l^{(n)}}.$$

Procedure: 9.5   *EM for Topic Models: M-step*

Assume $\theta^{(n)} = (\mathbf{p}_1, \ldots, \mathbf{p}_t, \pi_1, \ldots, \pi_t)$ and weights $w_{ij}^{(n)}$ linking the ith data item to the jth cluster center are known. Then estimate

$$\mathbf{p}_j^{(n+1)} = \frac{\sum_i \mathbf{x}_i w_{ij}^{(n)}}{\sum_i \mathbf{x}_i^T \mathbf{1} w_{ij}^{(n)}}$$

and

$$\pi_j^{(n+1)} = \frac{\sum_i w_{ij}^{(n)}}{N}.$$

First, how many cluster centers should there be? Mostly, the answer is a practical one. We are usually clustering data for a reason (vector quantization is a really good reason), and then we search for a k that yields the best results. Second, how should one start the iteration? This depends on the problem you want to solve, but for the two cases I have described, a rough clustering using k-means usually provides an excellent start. In the mixture of normals problem, you can take the cluster centers as initial values for the means, and the fraction of points in each cluster as initial values for the mixture weights. In the topic model problem, you can cluster the count vectors with k-means, use the overall counts within a cluster to get an initial estimate of the multinomial model probabilities, and use the fraction of documents within a cluster to get mixture weights. You need to be careful here, though. You really don't want to initialize a topic probability with a zero value for any word (otherwise no document containing that word can ever go into the cluster, which is a bit extreme). For our purposes, it will be enough to allocate a small value to each zero count, then adjust all the word probabilities to be sure they sum to one. More complicated approaches are possible.

Third, we need to avoid numerical problems in the implementation. Notice that you will be evaluating terms that look like

$$\frac{\pi_k e^{-(\mathbf{x}_i - \mu_k)^T (\mathbf{x}_i - \mu_k)/2}}{\sum_u \pi_u e^{-(\mathbf{x}_i - \mu_u)^T (\mathbf{x}_i - \mu_u)/2}}.$$

Imagine you have a point that is far from all cluster means. If you just blithely exponentiate the negative distances, you could find yourself dividing zero by zero, or a tiny number by a tiny number. This can lead to trouble. There's an easy alternative. Find the center the point is closest to. Now subtract the square of this distance ($d^2_{\min}$ for concreteness) from all the distances. Then evaluate

$$\frac{\pi_k e^{-\left[(\mathbf{x}_i-\mu_k)^T(\mathbf{x}_i-\mu_k)-d^2_{\min}\right]/2}}{\sum_u \pi_u e^{-\left[(\mathbf{x}_i-\mu_u)^T(\mathbf{x}_i-\mu_u)-d^2_{\min}\right]/2}}$$

which is a better way of estimating the same number (notice the $e^{-d^2_{\min}/2}$ terms cancel top and bottom).

The last problem is more substantial. EM will get to a local minimum of $p(\mathbf{x}|\theta)$, but there might be more than one local minimum. For clustering problems, the usual case is there are lots of them. One doesn't really expect a clustering problem to have a single best solution, as opposed to a lot of quite good solutions. Points that are far from all clusters are a particular source of local minima; placing these points in different clusters yields somewhat different sets of cluster centers, each about as good as the other. It's not usual to worry much about this point. A natural strategy is to start the method in a variety of different places (use k-means with different start points), and choose the one that has the best value of Q when it has converged.

Remember This: *You should use the same approach to choosing the number of cluster centers with EM as you use with k-means (try a few different values, and see which yields the most useful clustering). You should initialize an EM clusterer with k-means, but be careful of initial probabilities that are zero when initializing a topic model. You should be careful when computing weights, as it is easy to have numerical problems. Finally, it's a good idea to start EM clustering at multiple start points.*

However, EM isn't magic. There are problems where computing the expectation is hard, typically because you have to sum over a large number of cases which don't have the nice independence structure that helped in the examples I showed. There are strategies for dealing with this problem—essentially, you can get away with an approximate expectation—but they're beyond our reach at present.

There is an important, rather embarrassing, secret about EM. In practice, it isn't usually that much better as a clustering algorithm than k-means. You can only really expect improvements in performance if it is really important that many points can make a contribution to multiple cluster centers, and this doesn't happen very often. For a dataset where this does apply, the data itself may not really be an IID draw from a mixture of normal distributions, so the weights you compute are only approximate. Usually, it is smart to start EM with k-means. Nonetheless, EM is an algorithm you should know, because it is very widely applied in other situations, and because it can cluster data in situations where it isn't obvious how you compute distances.

> **Remember This:** *EM clusterers aren't much better than k-means clusterers, but EM is very general. It is a procedure for estimating the parameters of a probability model in the presence of missing data; this is a scenario that occurs in many applications. In clustering, the missing data was which data item belonged to which cluster.*

9.3   You Should

9.3.1   Remember These Terms

9.3.2   Remember These Facts

9.3.3   Remember These Procedures

9.3.4   Be Able to

- Use EM to cluster points using a mixture of normals model.
- Cluster documents using EM and a topic model.

Problems

9.1. You will derive the expressions for the M-step for mixture of normal clustering. Recall

$$Q(\theta;\theta^{(n)}) = \left(\sum_{ij} \left\{ \left[\left(-\frac{1}{2}(\mathbf{x}_i - \mu_j)^T (\mathbf{x}_i - \mu_j) \right) \right] + \log \pi_j \right\} w_{ij} + K \right)$$

and we have to maximize this with respect to μ and π, and the terms w_{ij} are known. Show that

$$\mu_j^{(n+1)} = \frac{\sum_i \mathbf{x}_i w_{ij}}{\sum_i w_{ij}}$$

and

$$\pi_j^{(n+1)} = \frac{\sum_i w_{ij}}{N}$$

maximize Q. When you do so, remember that, because π is a probability distribution, $\sum_j \pi_j = 1$ (otherwise you'll get the wrong answer). You need to either use a Lagrange multiplier or set one probability to $(1 -$ all others).

9.2. You will derive the expressions for the M-step for topic models. Recall

$$Q(\theta;\theta^{(n)}) = \left(\sum_{ij} \left\{ \left[\sum_k x_{i,k} \log p_{j,k} \right] + \log \pi_j \right\} w_{ij} \right)$$

and we have to maximize this with respect to μ and π, and the terms w_{ij} are known. Show that

$$\mathbf{p}_j^{(n+1)} = \frac{\sum_i \mathbf{x}_i w_{ij}}{\sum_i \mathbf{x}_i^T \mathbf{1} w_{ij}}$$

and

$$\pi_j^{(n+1)} = \frac{\sum_i w_{ij}}{N}.$$

When you do so, remember that, because π is a probability distribution, $\sum_j \pi_j = 1$ (otherwise you'll get the wrong answer). Furthermore, the $\mathbf{p}_j$ are all probability distributions. You need to either use Lagrange multipliers or set one probability to $(1 -$ all others).

Programming Exercises

9.3. Image segmentation is an important application of clustering. One breaks an image into k segments, determined by color, texture, etc. These segments are obtained by clustering image pixels by some representation of the image around the pixel (color, texture, etc.) into k clusters. Then each pixel is assigned to the segment corresponding to its cluster center.

(a) Obtain a color image represented as three arrays (red, green, and blue). You should look for an image where there are long scale color gradients (a sunset is a good choice). Ensure that this image is represented so the darkest pixel takes the value $(0,0,0)$ and the lightest pixel takes the value $(1,1,1)$. Now assume the pixel values have covariance the identity matrix. Cluster its pixels into 10, 20, and 50 clusters, modelling the pixel values as a mixture of normal distributions and using EM. Display the image obtained by replacing each pixel with the mean of its cluster center. What do you see?

(b) The weights linking an image to a cluster center can be visualized as an image. For the case of 10 cluster centers, construct a figure showing the weights linking each pixel to each cluster center (all 10 images). You should notice that the weights linking a given pixel to each cluster center do not vary very much. Why?

(c) Now repeat the previous two subexercises, but now using $0.1 \times \mathcal{I}$ as the covariance matrix. Show the new set of weight maps. What has changed, and why?

(d) Now estimate the covariance of pixel values by assuming that pixels are normally distributed (this is somewhat in tension with assuming they're distributed as a mixture of normals, but it works). Again, cluster the image's pixels into 10, 20, and 50 clusters, modelling the pixel values as a mixture of normal distributions and using EM, but now assuming that each normal distribution has the covariance from your estimate. Display the image obtained by replacing each pixel with the mean of its cluster center. Compare this result from the result of the first exercise. What do you see?

9.4. If you have a careful eye, or you chose a picture fortunately, you will have noticed that the previous exercise can produce image segments that have many connected components. For some applications, this is fine, but for others, we want segments that are compact clumps of pixels. One way to achieve this is to represent each pixel with 5D vector, consisting of its RG and B values *and* its x and y coordinates. You then cluster these 5D vectors.

(a) Obtain a color image represented as three arrays (red, green, and blue). You should look for an image where there are many distinct colored objects (for example, a bowl of fruit). Ensure that this image is represented so the darkest pixel takes the value $(0,0,0)$ and the lightest pixel takes the value $(1,1,1)$. Represent the x and y coordinates of each pixel using the range 0 to 1 as well. Now assume the pixel RGB values have covariance 0.1 times the identity matrix, there is zero covariance between position and color, and the coordinates have covariance σ times the identity matrix where σ is a parameter we will modify. Cluster your image's pixels into 20, 50, and 100 clusters, with $\sigma = (0.01, 0.1, 1)$ (so 9 cases). Again, model the pixel values as a mixture of normal distributions and using EM. For each case, display the image obtained by replacing each pixel with the mean of its cluster center. What do you see?

9.5. EM has applications that don't look like clustering at first glance. Here is one. We will use EM to reject points that don't fit a line well (if you haven't seen least squares line fitting, this exercise isn't for you).

(a) Construct a dataset of 10 2D points which are IID samples from the following mixture distribution. Draw the x coordinate from the uniform distribution on the range $[0, 10]$. With probability 0.8, draw ξ a normal random variable with mean 0 and standard deviation 0.001 and form the y coordinate as $y = x + \xi$. With probability 0.2, draw the y coordinate from the uniform distribution on the range $[0, 10]$. Plot this dataset—you should see about eight points on a line with about two scattered points.

(b) Fit a least squares line to your dataset, and plot the result. It should be bad, because the scattered points may have a significant effect on the line. If you were unlucky, and drew a sample where there were no scattered points or where this line fits well, keep drawing datasets until you get one where the fit is poor.

(c) We will now use EM to fit a good line. Write $N(\mu,\sigma)$ for a normal distribution with mean μ and standard deviation σ, and $U(0,10)$ for the uniform distribution on the range $0-10$. Model the y coordinate using the mixture model $P(y|a,b,\pi,x) = \pi N(ax+b, 0.001) + (1-\pi)U(0,10)$. Now associate a variable δ_i with the ith data point, where $\delta_i = 1$ if the data point comes from the line model and $\delta_i = 0$ otherwise. Write an expression for $P(y_i, \delta_i|a,b,\pi,x)$.

(d) Assume that $a^{(n)}$, $b^{(n)}$, and $\pi^{(n)}$ are known. Show that

$$Q\left(a,b,\pi; a^{(n)}, b^{(n)}, \pi^{(n)}\right) = -\sum_i w_i \frac{(ax_i + b - y_i)^2}{20.001^2} + (1-w_i)(1/10) + K$$

(where K is a constant). Here

$$w_i = \mathbb{E}_{P(\delta_i|a^{(n)}, b^{(n)}, \pi^{(n)}, x)}[\delta_i].$$

(e) Show that

$$w_i = P(\delta_i|a^{(n)}, b^{(n)}, \pi^{(n)}, x) = \frac{\pi^{(n)} e^{-\frac{(a^{(n)}x_i + b^{(n)} - y_i)^2}{20.001^2}}}{\pi^{(n)} e^{-\frac{(a^{(n)}x_i + b^{(n)} - y_i)^2}{20.001^2}} + (1 - \pi^{(n)})\frac{1}{10}}.$$

(f) Now implement an EM algorithm using this information, and estimate the line for your data. You should try multiple start points. Do you get a better line fit? Why?

9.6. This is a fairly ambitious exercise. We will use the document clustering method of Sect. 9.1.2 to identify clusters of documents, which we will associate with topics. The 20 newsgroups dataset is a famous text dataset. It consists of posts collected from 20 different newsgroups. There are a variety of tricky data issues that this presents (for example, what aspects of the header should one ignore? should one reduce words to their stems, so "winning" goes to "win," "hugely" to "huge," and so on?). We will ignore these issues, and deal with a cleaned up version of the dataset. This consists of three items each for train and test: a document-word matrix, a set of labels, and a map. You can find this cleaned up version of the dataset at http://qwone.com/~jason/20Newsgroups/. You should look for the cleaned up version, identified as 20news-bydate-matlab.tgz on that page. The usual task is to label a test article with which newsgroup it came from. The document-word matrix is a table of counts of how many times a particular word appears in a particular document. The collection of words is very large (53,975 distinct words), and most words do not appear in most documents, so most entries of this matrix are zero. The file train.data contains this matrix for a collection of training data; each row represents a distinct document (there are 11,269), and each column represents a distinct word.

(a) Cluster the rows of this matrix, using the method of Sect. 9.1.2, to get a set of cluster centers which we will identify as topics. **Hint:** Clustering all these points is a bit of a performance; check your code on small subsets of the data first, because the size of this dataset means that clustering the whole thing will be slow.

(b) You can now think of each cluster center as a document "type." Assume you have k clusters (topics). Represent each document by a k-dimensional vector. Each entry of the vector should be the negative log probability of the document under that cluster model. Now use this information to build a classifier that identifies the newsgroup using the vector. You'll need to use the file `train.label`, which will tell you what newsgroup a particular item comes from. I advise you to use a randomized decision forest, but other choices are plausible. Evaluate your classifier using the test data (`test.data` and `test.label`).

PART FOUR

Regression

C H A P T E R 10

Regression

Classification tries to predict a class from a data item. **Regression** tries to predict a value. For example, we know the zip code of a house, the square footage of its lot, the number of rooms, and the square footage of the house, and we wish to predict its likely sale price. As another example, we know the cost and condition of a trading card for sale, and we wish to predict a likely profit in buying it and then reselling it. As yet another example, we have a picture with some missing pixels— perhaps there was text covering them, and we want to replace it—and we want to fill in the missing values. As a final example, you can think of classification as a special case of regression, where we want to predict either $+1$ or -1; this isn't usually the best way to proceed, however. Predicting values is very useful, and so there are many examples like this.

10.1 Overview

We want to build a model that predicts some number y from a feature vector $\mathbf{x}$. An appropriate choice of $\mathbf{x}$ (details below) will mean that the predictions made by this model will lie on a straight line. Figure 10.1 shows two example regressions that work rather well. The data are plotted with a scatterplot, and the line gives the prediction of the model for each value on the horizontal axis. Figure 10.2 shows two example regressions that work fairly badly. These examples capture some important points.

For most data, y isn't really a *function* of $\mathbf{x}$—you might have two examples where the same $\mathbf{x}$ gives different y values. This occurs in each of the figures. Usually, we think of the data as samples from a distribution $P(X, Y)$, and the regression estimates the mean of $P(Y \mid \{X = \mathbf{x}\})$. Thinking about the problem this way should make it clear that we're not relying on any exact, physical, or causal relationship between Y and X. It's enough that their joint probability makes useful predictions possible, something we will test by experiment. This means that you can build regressions that work in somewhat surprising circumstances. For example, regressing children's reading ability against their foot size can be quite successful. This isn't because having big feet somehow helps you read; it's because on the whole, older children read better, and also have bigger feet.

The model might predict well (the weight of the fish), but it's unlikely to make perfect predictions. Even when the y associated with a given value of $\mathbf{x}$ changes quite a lot, the predictions can be useful (the crickets—you really can make a fair guess at temperature from frequency). Sometimes regressions work rather badly (there is very little relationship between heart rate and temperature).

In this chapter, I show how to fit a model; how to tell whether the model is good; and some things we can do to make the model better. Mostly, we will use regression for prediction, but there are other applications. It should be fairly clear

© Springer Nature Switzerland AG 2019
D. Forsyth, *Applied Machine Learning*,
https://doi.org/10.1007/978-3-030-18114-7_10

that if test data isn't "like" training data, the regression might not work. Formal guarantees require that both test and training data be IID samples from the same distribution. It isn't usually possible to tell whether this occurs with practical data, so that we ignore this point—we'll simply try to build the best regressions we can with the data we have.

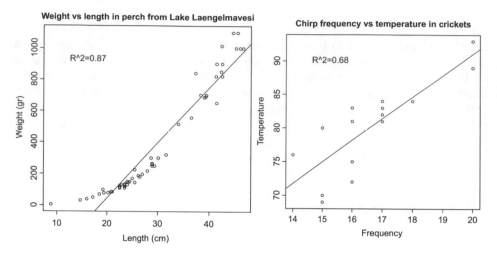

Figure 10.1: On the **left**, a regression of weight against length for perch from a Finnish lake (you can find this dataset, and the back story at http://www.amstat. org/publications/jse/jse_data_archive.htm; look for "fishcatch" on that page). Notice that the linear regression fits the data fairly well, meaning that you should be able to predict the weight of a perch from its length fairly well. On the **right**, a regression of air temperature against chirp frequency for crickets. The data is fairly close to the line, meaning that you should be able to tell the temperature from the pitch of cricket's chirp fairly well. This data is from http://mste.illinois.edu/patel/ amar430/keyprob1.html. The R^2 you see on each figure is a measure of the goodness of fit of the regression (Sect. 10.2.4)

10.1.1 Regression to Spot Trends

Regression isn't only used to predict values. Another reason to build a regression model is to compare trends in data. Doing so can make it clear what is really happening. Here is an example from Efron ("Computer-Intensive methods in statistical regression," B. Efron, SIAM Review, 1988). Table 10.1 in the Appendix shows some data from medical devices, which sit in the body and release a hormone. The data shows the amount of hormone currently in a device after it has spent some time in service, and the time the device spent in service. The data describes devices from three production lots (A, B, and C). Each device, from each lot, is supposed to have

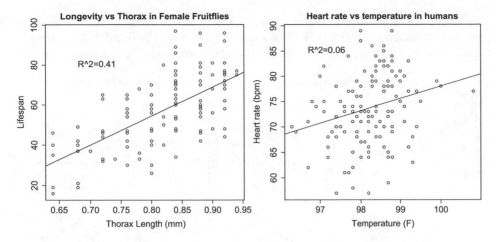

Figure 10.2:   Regressions do not necessarily yield good predictions or good model fits. On the **left**, a regression of the lifespan of female fruitflies against the length of their torso as adults (apparently, this doesn't change as a fruitfly ages; you can find this dataset, and the back story at http://www.amstat.org/publications/jse/ jse_data_archive.htm; look for "fruitfly" on that page). The figure suggests you can make some prediction of how long your fruitfly will last by measuring its torso, but not a particularly accurate one. On the **right**, a regression of heart rate against body temperature for adults. You can find the data at http://www.amstat.org/ publications/jse/jse_data_archive.htm as well; look for "temperature" on that page. Notice that predicting heart rate from body temperature isn't going to work that well, either

the same behavior. The important question is: Are the lots the same? The amount of hormone changes over time, so we can't just compare the amounts currently in each device. Instead, we need to determine the relationship between time in service and hormone, and see if this relationship is different between batches. We can do so by regressing hormone against time.

Figure 10.3 shows how a regression can help. In this case, we have modelled the amount of hormone in the device as

$$a \times (\text{time in service}) + b$$

for a, b chosen to get the best fit (much more on this point later!). This means we can plot each data point on a scatterplot, together with the best fitting line. This plot allows us to ask whether any particular batch behaves differently from the overall model in any interesting way.

However, it is hard to evaluate the distances between data points and the best fitting line by eye. A sensible alternative is to subtract the amount of hormone predicted by the model from the amount that was measured. Doing so yields a

residual—the difference between a measurement and a prediction. We can then plot those residuals (Fig. 10.3). In this case, the plot suggests that lot A is special— all devices from this lot contain less hormone than our model predicts.

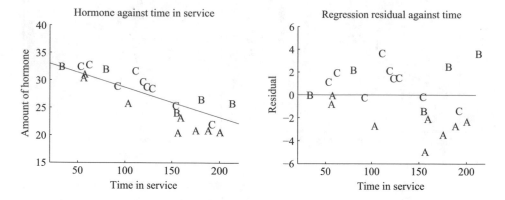

Figure 10.3: On the **left**, a scatterplot of hormone against time for devices from Table 10.1. Notice that there is a pretty clear relationship between time and amount of hormone (the longer the device has been in service, the less hormone there is). The issue now is to understand that relationship so that we can tell whether lots A, B, and C are the same or different. The best fit line to all the data is shown as well, fitted using the methods of Sect. 10.2. On the **right**, a scatterplot of residual—the distance between each data point and the best fit line—against time for the devices from Table 10.1. Now you should notice a clear difference; some devices from lots B and C have positive and some negative residuals, but all lot A devices have negative residuals. This means that, when we account for loss of hormone over time, lot A devices still have less hormone in them. This is a pretty good evidence that there is a problem with this lot

---

Useful Fact: 10.1 *Definition: Regression*

Regression accepts a feature vector and produces a prediction, which is usually a number, but can sometimes have other forms. You can use these predictions as predictions, or to study trends in data. It is possible, but not usually particularly helpful, to see classification as a form of regression.

10.2   Linear Regression and Least Squares

Assume we have a dataset consisting of a set of N pairs $(\mathbf{x}_i, y_i)$. We think of y_i as the value of some function evaluated at $\mathbf{x}_i$, with some random component added. This means there might be two data items where the $\mathbf{x}_i$ are the same, and the y_i are different. We refer to the $\mathbf{x}_i$ as **explanatory variables** and the y_i as a **dependent variable**. We want to use the examples we have—the **training examples**—to build a model of the dependence between y and $\mathbf{x}$. This model will be used to predict values of y for new values of $\mathbf{x}$, which are usually called **test examples**. It can also be used to understand the relationships between the $\mathbf{x}$. The model needs to have some probabilistic component; we do not expect that y is a function of $\mathbf{x}$, and there is likely some error in evaluating y anyhow.

10.2.1 Linear Regression

We cannot expect that our model makes perfect predictions. Furthermore, y may not be a function of $\mathbf{x}$—it is quite possible that the same value of $\mathbf{x}$ could lead to different y's. One way that this could occur is that y is a measurement (and so subject to some measurement noise). Another is that there is some randomness in y. For example, we expect that two houses with the same set of features (the $\mathbf{x}$) might still sell for different prices (the y's).

A good, simple model is to assume that the dependent variable (i.e., y) is obtained by evaluating a linear function of the explanatory variables (i.e., $\mathbf{x}$), then adding a zero mean normal random variable. We can write this model as

$$y = \mathbf{x}^T \beta + \xi,$$

where ξ represents random (or at least, unmodelled) effects. We will always assume that ξ has zero mean. In this expression, β is a vector of weights, which we must estimate. When we use this model to predict a value of y for a particular set of explanatory variables $\mathbf{x}^*$, we cannot predict the value that ξ will take. Our best available prediction is the mean value (which is zero). Notice that if $\mathbf{x} = 0$, the model predicts $y = 0$. This may seem like a problem to you—you might be concerned that we can fit only lines through the origin—but remember that $\mathbf{x}$ contains explanatory variables, and we can choose what appears in $\mathbf{x}$. The two examples show how a sensible choice of $\mathbf{x}$ allows us to fit a line with an arbitrary y-intercept.

Useful Fact: 10.2 *Definition: Linear Regression*

A linear regression takes the feature vector $\mathbf{x}$ and predicts $\mathbf{x}^T \beta$, for some vector of coefficients β. The coefficients are adjusted, using data, to produce the best predictions.

> **Example: 10.1** *A Linear Model Fitted to a Single Explanatory Variable*
>
> Assume we fit a linear model to a single explanatory variable. Then the model has the form $y = x\beta + \xi$, where ξ is a zero mean random variable. For any value x^* of the explanatory variable, our best estimate of y is βx^*. In particular, if $x^* = 0$, the model predicts $y = 0$, which is unfortunate. We can draw the model by drawing a line through the origin with slope β in the x, y plane. The y-intercept of this line must be zero.

> **Example: 10.2** *A Linear Model with a Non-zero y-Intercept*
>
> Assume we have a single explanatory variable, which we write u. We can then create a *vector* $\mathbf{x} = [u, 1]^T$ from the explanatory variable. We now fit a linear model to this vector. Then the model has the form $y = \mathbf{x}^T\beta + \xi$, where ξ is a zero mean random variable. For any value $\mathbf{x}^* = [u^*, 1]^T$ of the explanatory variable, our best estimate of y is $(\mathbf{x}^*)^T\beta$, which can be written as $y = \beta_1 u^* + \beta_2$. If $x^* = 0$, the model predicts $y = \beta_2$. We can draw the model by drawing a line through the origin with slope β_1 and y-intercept β_2 in the x, y plane.

10.2.2 Choosing β

We must determine β. We can proceed in two ways. I show both because different people find different lines of reasoning more compelling. Each will get us to the same solution. One is probabilistic, the other isn't. Generally, I'll proceed as if they're interchangeable, although at least in principle they're different.

Probabilistic Approach: We could assume that ξ is a zero mean normal random variable with unknown variance. Then $P(y|\mathbf{x}, \beta)$ is normal, with mean $\mathbf{x}^T\beta$, and so we can write out the log-likelihood of the data. Write σ^2 for the variance of ξ, which we don't know, but will not worry about right now. We have that

$$
\begin{aligned}
\log \mathcal{L}(\beta) &= \sum_i \log P(y_i|\mathbf{x}_i, \beta) \\
&= -\frac{1}{2\sigma^2} \sum_i (y_i - \mathbf{x}_i^T\beta)^2 + \text{term not depending on } \beta.
\end{aligned}
$$

Maximizing the log-likelihood of the data is equivalent to minimizing the negative log-likelihood of the data. Furthermore, the term $\frac{1}{2\sigma^2}$ does not affect the location of the minimum, so we must have that the β we want minimizes $\sum_i (y_i - \mathbf{x}_i^T \beta)^2$, or anything proportional to it. It is helpful to minimize an expression that is an average of squared errors, because (hopefully) this doesn't grow much when we add data. We therefore minimize

$$\left(\frac{1}{N} \right) \left(\sum_i (y_i - \mathbf{x}_i^T \beta)^2 \right).$$

Direct Approach: Notice that, if we have an estimate of β, we have an estimate of the values of the unmodelled effects ξ_i for each example. We just take $\xi_i = y_i - \mathbf{x}_i^T \beta$. It is quite natural to make the unmodelled effects "small." A good measure of size is the mean of the squared values, which means we want to minimize

$$\left(\frac{1}{N} \right) \left(\sum_i (y_i - \mathbf{x}_i^T \beta)^2 \right).$$

We can write all this more conveniently using vectors and matrices. Write $\mathbf{y}$ for the vector

$$\begin{pmatrix} y_1 \\ y_2 \\ \ldots \\ y_n \end{pmatrix}$$

and $\mathcal{X}$ for the matrix

$$\begin{pmatrix} \mathbf{x}_1^T \\ \mathbf{x}_2^T \\ \ldots \\ \mathbf{x}_n^T \end{pmatrix}.$$

Then we want to minimize

$$\left(\frac{1}{N} \right) (\mathbf{y} - \mathcal{X}\beta)^T (\mathbf{y} - \mathcal{X}\beta)$$

which means that we must have

$$\mathcal{X}^T \mathcal{X} \beta - \mathcal{X}^T \mathbf{y} = 0.$$

For reasonable choices of features, we could expect that $\mathcal{X}^T \mathcal{X}$—which should strike you as being a lot like a covariance matrix—has full rank. If it does, which is the usual case, this equation is easy to solve. If it does not, there is more to do, which we will do in Sect. 10.4.2.

Remember This: *The vector of coefficients β for a linear regression is usually estimated using a least-squares procedure.*

10.2.3 Residuals

Assume we have produced a regression by solving

$$\mathcal{X}^T \mathcal{X} \hat{\beta} - \mathcal{X}^T \mathbf{y} = 0$$

for the value of $\hat{\beta}$. I write $\hat{\beta}$ because this is an *estimate*; we likely don't have the true value of the β that generated the data (the model might be wrong, etc.). We cannot expect that $\mathcal{X}\hat{\beta}$ is the same as $\mathbf{y}$. Instead, there is likely to be some error. The **residual** is the vector

$$\mathbf{e} = \mathbf{y} - \mathcal{X}\hat{\beta}$$

which gives the difference between the true value and the model's prediction at each point. Each component of the residual is an estimate of the unmodelled effects for that data point. The **mean-squared error** is

$$m = \frac{\mathbf{e}^T \mathbf{e}}{N}$$

and this gives the average of the squared error of prediction on the training examples.

Notice that the mean-squared error is not a great measure of how good the regression is. This is because the value depends on the units in which the dependent variable is measured. So, for example, if you measure y in meters you will get a different mean-squared error than if you measure y in kilometers.

10.2.4 *R*-squared

There is an important quantitative measure of how good a regression is which doesn't depend on units. Unless the dependent variable is a constant (which would make prediction easy), it has some variance. If our model is of any use, it should explain some aspects of the value of the dependent variable. This means that the variance of the residual should be smaller than the variance of the dependent variable. If the model made perfect predictions, then the variance of the residual should be zero.

We can formalize all this in a relatively straightforward way. We will ensure that $\mathcal{X}$ always has a column of ones in it, so that the regression can have a non-zero y-intercept. We now fit a model

$$\mathbf{y} = \mathcal{X}\beta + \mathbf{e}$$

(where $\mathbf{e}$ is the vector of residual values) by choosing the value $\hat{\beta}$ of β such that $\mathbf{e}^T \mathbf{e}$ is minimized. Then we get some useful technical results.

> **Useful Facts: 10.3** *Regression*
>
> We write $\mathbf{y} = \mathcal{X}\hat{\beta} + \mathbf{e}$, where $\mathbf{e}$ is the residual. Assume $\mathcal{X}$ has a column of ones, and $\hat{\beta}$ is chosen to minimize $\mathbf{e}^T\mathbf{e}$. Then we have
>
> 1. $\mathbf{e}^T\mathcal{X} = \mathbf{0}$, i.e., $\mathbf{e}$ is orthogonal to any column of $\mathcal{X}$. This is because, if $\mathbf{e}$ is not orthogonal to some column of $\mathbf{e}$, we can increase or decrease the $\hat{\beta}$ term corresponding to that column to make the error smaller. Another way to see this is to notice that $\hat{\beta}$ is chosen to minimize $\frac{1}{N}\mathbf{e}^T\mathbf{e}$, which is $\frac{1}{N}(\mathbf{y} - \mathcal{X}\hat{\beta})^T(\mathbf{y} - \mathcal{X}\hat{\beta})$. Now because this is a minimum, the gradient with respect to $\hat{\beta}$ is zero, so $(\mathbf{y} - \mathcal{X}\hat{\beta})^T(-\mathcal{X}) = -\mathbf{e}^T\mathcal{X} = 0$.
> 2. $\mathbf{e}^T\mathbf{1} = 0$ (recall that $\mathcal{X}$ has a column of all ones, and apply the previous result).
> 3. $\mathbf{1}^T(\mathbf{y} - \mathcal{X}\hat{\beta}) = 0$ (same as previous result).
> 4. $\mathbf{e}^T\mathcal{X}\hat{\beta} = 0$ (first result means that this is true).

Now $\mathbf{y}$ is a one-dimensional dataset arranged into a vector, so we can compute $\mathsf{mean}\,(\{y\})$ and $\mathsf{var}[y]$. Similarly, $\mathcal{X}\hat{\beta}$ is a one-dimensional dataset arranged into a vector (its elements are $\mathbf{x}_i^T\hat{\beta}$), as is $\mathbf{e}$, so we know the meaning of mean and variance for each. We have a particularly important result:

$$\mathsf{var}[y] = \mathsf{var}\left[\mathcal{X}\hat{\beta}\right] + \mathsf{var}[e].$$

This is quite easy to show, with a little more notation. Write $\overline{\mathbf{y}} = (1/N)(\mathbf{1}^T\mathbf{y})\mathbf{1}$ for the vector whose entries are all $\mathsf{mean}\,(\{y\})$; similarly for $\overline{\mathbf{e}}$ and for $\mathcal{X}\hat{\beta}$. We have

$$\mathsf{var}[y] = (1/N)(\mathbf{y} - \overline{\mathbf{y}})^T(\mathbf{y} - \overline{\mathbf{y}})$$

and so on for $\mathsf{var}[e_i]$, etc. Notice from the facts that $\overline{\mathbf{y}} = \overline{\mathcal{X}\hat{\beta}}$. Now

$$
\begin{aligned}
\mathsf{var}[y] &= (1/N)\left(\left[\mathcal{X}\hat{\beta} - \overline{\mathcal{X}\hat{\beta}}\right] + [\mathbf{e} - \overline{\mathbf{e}}]\right)^T\left(\left[\mathcal{X}\hat{\beta} - \overline{\mathcal{X}\hat{\beta}}\right] + [\mathbf{e} - \overline{\mathbf{e}}]\right) \\
&= (1/N)\left(\left[\mathcal{X}\hat{\beta} - \overline{\mathcal{X}\hat{\beta}}\right]^T\left[\mathcal{X}\hat{\beta} - \overline{\mathcal{X}\hat{\beta}}\right] + 2\,[\mathbf{e} - \overline{\mathbf{e}}]^T\left[\mathcal{X}\hat{\beta} - \overline{\mathcal{X}\hat{\beta}}\right]\right. \\
&\qquad \left. + [\mathbf{e} - \overline{\mathbf{e}}]^T[\mathbf{e} - \overline{\mathbf{e}}]\right) \\
&= (1/N)\left(\left[\mathcal{X}\hat{\beta} - \overline{\mathcal{X}\hat{\beta}}\right]^T\left[\mathcal{X}\hat{\beta} - \overline{\mathcal{X}\hat{\beta}}\right] + [\mathbf{e} - \overline{\mathbf{e}}]^T[\mathbf{e} - \overline{\mathbf{e}}]\right) \\
&\qquad \text{because } \overline{\mathbf{e}} = 0 \text{ and } \mathbf{e}^T\mathcal{X}\hat{\beta} = 0 \text{ and } \mathbf{e}^T\mathbf{1} = 0 \\
&= \mathsf{var}\left[\mathcal{X}\hat{\beta}\right] + \mathsf{var}[e].
\end{aligned}
$$

This is extremely important, because it allows us to think about a regression as explaining variance in $\mathbf{y}$. As we are better at explaining $\mathbf{y}$, $\mathsf{var}[e]$ goes down. In turn, a natural measure of the goodness of a regression is what percentage of the variance of **y** it explains. This is known as R^2 (the r-squared measure). We have

$$R^2 = \frac{\text{var}\left[\mathcal{X}\hat{\beta}\right]}{\text{var}[\mathbf{y}]}$$

which gives some sense of how well the regression explains the training data. Notice that the value of R^2 is not affected by the units of **y** (exercises).

Good predictions result in high values of R^2, and a perfect model will have $R^2 = 1$ (which doesn't usually happen). For example, the regression of Fig. 10.3 has an R^2 value of 0.87. Figures 10.1 and 10.2 show the R^2 values for the regressions plotted there; notice how better models yield larger values of R^2. Notice that if you look at the summary that R provides for a linear regression, it will offer you *two* estimates of the value for R^2. These estimates are obtained in ways that try to account for (a) the amount of data in the regression, and (b) the number of variables in the regression. For our purposes, the differences between these numbers and the R^2 I defined are not significant. For the figures, I computed R^2 as I described in the text above, but if you substitute one of R's numbers nothing terrible will happen.

> **Remember This:** *The quality of predictions made by a regression can be evaluated by looking at the fraction of the variance in the dependent variable that is explained by the regression. This number is called R^2, and lies between zero and one; regressions with larger values make better predictions.*

10.2.5 Transforming Variables

Sometimes the data isn't in a form that leads to a good linear regression. In this case, transforming explanatory variables, the dependent variable, or both can lead to big improvements. Figure 10.4 shows one example, based on the idea of word frequencies. Some words are used very often in text; most are used seldom. The dataset for this figure consists of counts of the number of time a word occurred for the 100 most common words in Shakespeare's printed works. It was originally collected from a concordance, and has been used to attack a variety of interesting questions, including an attempt to assess how many words Shakespeare knew. This is hard, because he likely knew many words that he didn't use in his works, so one can't just count. If you look at the plot of Fig. 10.4, you can see that a linear regression of count (the number of times a word is used) against rank (how common a word is, 1–100) is not really useful. The most common words are used very often, and the number of times a word is used falls off very sharply as one looks at less common words. You can see this effect in the scatterplot of residual against dependent variable in Fig. 10.4—the residual depends rather strongly on the dependent variable. This is an extreme example that illustrates how poor linear regressions can be.

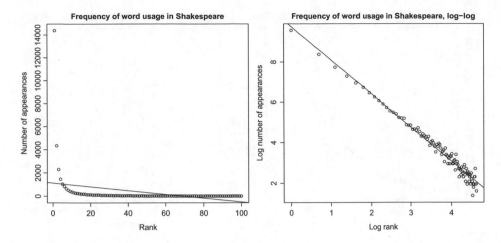

Figure 10.4: On the **left**, word count plotted against rank for the 100 most common words in Shakespeare, using a dataset that comes with R (called "bard," and quite likely originating in an unpublished report by J. Gani and I. Saunders). I show a regression line too. This is a poor fit by eye, and the R^2 is poor, too ($R^2 = 0.1$). On the **right**, log word count plotted against log rank for the 100 most common words in Shakespeare, from that dataset. The regression line is very close to the data

However, if we regress log-count against log rank, we get a very good fit indeed. This suggests that Shakespeare's word usage (at least for the 100 most common words) is consistent with **Zipf's law**. This gives the relation between frequency f and rank r for a word as

$$f \propto \frac{1}{r}^s,$$

where s is a constant characterizing the distribution. Our linear regression suggests that s is approximately 1.67 for this data.

In some cases, the natural logic of the problem will suggest variable transformations that improve regression performance. For example, one could argue that humans have approximately the same density, and so that weight should scale as the cube of height; in turn, this suggests that one regress weight against the cube root of height. Generally, shorter people tend not to be scaled versions of taller people, so the cube root might be too aggressive, and so one thinks of the square root.

Remember This: *The performance of a regression can be improved by transforming variables. Transformations can follow from looking at plots, or thinking about the logic of the problem.*

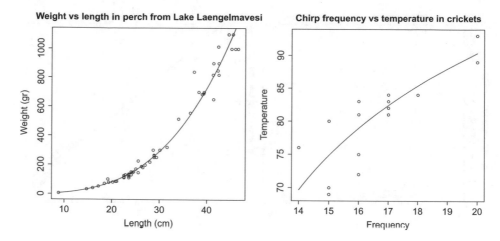

Figure 10.5: The Box-Cox transformation suggests a value of $\lambda = 0.303$ for the regression of weight against height for the perch data of Fig. 10.1. You can find this dataset, and the back story at http://www.amstat.org/publications/jse/jse_data_archive.htm; look for "fishcatch" on that page. On the **left**, a plot of the resulting curve overlaid on the data. For the cricket temperature data of that figure (from http://mste.illinois.edu/patel/amar430/keyprob1.html), the transformation suggests a value of $\lambda = 4.75$. On the **right**, a plot of the resulting curve overlaid on the data

The **Box-Cox transformation** is a method that can search for a transformation of the dependent variable that improves the regression. The method uses a one-parameter family of transformations, with parameter λ, then searches for the best value of this parameter using maximum likelihood. A clever choice of transformation means that this search is relatively straightforward. We define the Box-Cox transformation of the dependent variable to be

$$y_i^{(bc)} = \begin{cases} \frac{y_i^\lambda - 1}{\lambda} & \text{if } \lambda \neq 0 \\ \log y_i & \text{if } \lambda = 0 \end{cases}.$$

It turns out to be straightforward to estimate a good value of λ using maximum likelihood. One searches for a value of λ that makes residuals look most like a normal distribution. Statistical software will do it for you. This transformation can produce significant improvements in a regression. For example, the transformation suggests a value of $\lambda = 0.303$ for the fish example of Fig. 10.1. It isn't natural to plot weight$^{0.303}$ against height, because we don't really want to predict weight$^{0.303}$. Instead, we plot the predictions of weight that come from this model, which will lie on a curve with the form $(ax + b)^{\frac{1}{0.303}}$, rather than on a straight line. Similarly, the transformation suggests a value of $\lambda = 0.475$ for the cricket data. Figure 10.5 shows the result of these transforms.

> **Remember This:** *The Box-Cox transformation seeks a power of the dependent variable that is better predicted by a linear function of the independent variables. The process works by constructing a likelihood in the power, then finding the maximum likelihood; statistical software will do the details. The transformation applies only to datasets.*

10.2.6 Can You Trust Your Regression?

Linear regression is useful, but it isn't magic. Some regressions make poor predictions (recall the regressions of Fig. 10.2). As another example, regressing the first digit of someone's telephone number against the length of their foot won't work.

We have some straightforward tests to tell whether a regression is working. You can **look at a plot** for a dataset with one explanatory variable and one dependent variable. You plot the data on a scatterplot, then plot the model as a line on that scatterplot. Just looking at the picture can be informative (compare Fig. 10.1 and Fig. 10.2).

You can check if the regression **predicts a constant**. This is usually a bad sign. You can check this by looking at the predictions for each of the training data items. If the variance of these predictions is small compared to the variance of the independent variable, the regression isn't working well. If you have only one explanatory variable, then you can plot the regression line. If the line is horizontal, or close, then the value of the explanatory variable makes very little contribution to the prediction. This suggests that there is no particular relationship between the explanatory variable and the independent variable.

You can also check, by eye, if **the residual isn't random**. If $y - \mathbf{x}^T\beta$ is a zero mean normal random variable, then the value of the residual vector should not depend on the corresponding y-value. Similarly, if $y - \mathbf{x}^T\beta$ is just a zero mean collection of unmodelled effects, we want the value of the residual vector to not depend on the corresponding y-value either. If it does, that means there is some phenomenon we are not modelling. Looking at a scatterplot of $\mathbf{e}$ against $\mathbf{y}$ will often reveal trouble in a regression (Fig. 10.7). In the case of Fig. 10.7, the trouble is caused by a few data points that are very different from the others severely affecting the regression. We will discuss how to identify and deal with such points in Sect. 10.3. Once they have been removed, the regression improves markedly (Fig. 10.8).

> **Remember This:** *Linear regressions can make bad predictions. You can check for trouble by: evaluating R^2; looking at a plot; looking to see if the regression makes a constant prediction; or checking whether the residual is random.*

Procedure: 10.1 *Linear Regression Using Least Squares*

We have a dataset containing N pairs $(\mathbf{x}_i, y_i)$. Each x_i is a d-dimensional explanatory vector, and each y_i is a single dependent variable. We assume that each data point conforms to the model

$$y_i = \mathbf{x}_i^T \beta + \xi_i,$$

where ξ_i represents unmodelled effects. We assume that ξ_i are samples of a random variable with 0 mean and unknown variance. Sometimes, we assume the random variable is normal. Write

$$\mathbf{y} = \begin{pmatrix} y_1 \\ y_2 \\ \ldots \\ y_n \end{pmatrix}$$

and

$$\mathcal{X} = \begin{pmatrix} \mathbf{x}_1^T \\ \mathbf{x}_2^T \\ \ldots \\ \mathbf{x}_n^T \end{pmatrix}.$$

We estimate $\hat{\beta}$ (the value of β) by solving the linear system

$$\mathcal{X}^T \mathcal{X} \hat{\beta} - \mathcal{X}^T \mathbf{y} = 0.$$

For a data point $\mathbf{x}$, our model predicts $\mathbf{x}^T \hat{\beta}$. The residuals are

$$\mathbf{e} = \mathbf{y} - \mathcal{X} \hat{\beta}.$$

We have that $\mathbf{e}^T \mathbf{1} = 0$. The mean-squared error is given by

$$m = \frac{\mathbf{e}^T \mathbf{e}}{N}.$$

The R^2 is given by

$$R^2 = \frac{\operatorname{var}\left[\mathcal{X}\hat{\beta}\right]}{\operatorname{var}[\mathbf{y}]}.$$

Values of R^2 range from 0 to 1; a larger value means the regression is better at explaining the data.

10.3 Visualizing Regressions to Find Problems

I have described regressions on a single explanatory variable, because it is easy to plot the line in this case. You can find most problems by looking at the line and the data points. But a single explanatory variable isn't the most common or useful case. If we have many explanatory variables, it can be hard to plot the regression in a way that exposes problems. Data points that are significantly different from others are the main source of problems. This is most easily seen in plots with one explanatory variable, but the effect applies in other regressions, too. This section mainly describes methods to identify and solve difficulties when you can't simply plot the regression. Mostly, we will focus on problem data points.

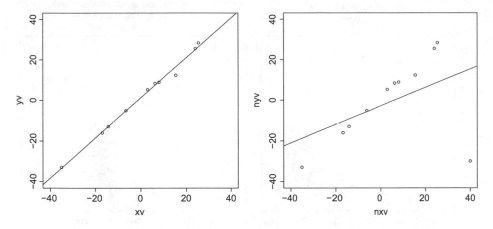

Figure 10.6: On the **left**, a synthetic dataset with one independent and one explanatory variable, with the regression line plotted. Notice the line is close to the data points, and its predictions seem likely to be reliable. On the **right**, the result of adding a single outlying data point to that dataset. The regression line has changed significantly, because the regression line tries to minimize the sum of squared vertical distances between the data points and the line. Because the outlying data point is far from the line, the squared vertical distance to this point is enormous. The line has moved to reduce this distance, at the cost of making the other points further from the line

10.3.1 Problem Data Points Have Significant Impact

When we construct a regression, we are solving for the β that minimizes $\sum_i (y_i - \mathbf{x}_i^T \beta)^2$, equivalently for the β that produces the smallest value of $\sum_i e_i^2$. This means that residuals with large value can have a very strong influence on the outcome—we are squaring that large value, resulting in an enormous value. Generally, many residuals of medium size will have a smaller cost than one large residual and the rest tiny. This means that a data point that lies far from the others can swing the regression line very significantly. Figure 10.6 illustrates this effect, which occurs commonly in real datasets and is an important nuisance.

Data points like the one in Fig. 10.6 are often known as **outliers**. This isn't a term one can use with much precision. Such data points can come from a variety of sources. Failures of equipment, transcription errors, someone guessing a value to replace lost data, and so on are some methods that might produce outliers. In the distant past, academics blamed secretaries for outliers. There could be some important but relatively rare effect that produces a couple of odd data points. Major scientific discoveries have resulted from investigators taking outliers seriously, and trying to find out what caused them (though you shouldn't see a Nobel Prize lurking behind every outlier).

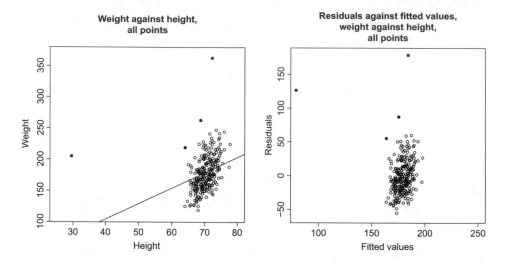

Figure 10.7:   On the **left**, weight regressed against height for the bodyfat dataset. The line doesn't describe the data particularly well, because it has been strongly affected by a few data points (filled-in markers). On the **right**, a scatterplot of the residual against the value predicted by the regression. This doesn't look like noise, which is a sign of trouble

Outlying data points can significantly weaken the usefulness of a regression. For some regression problems, we can identify data points that might be a problem, and then resolve how to deal with them. One possibility is that they are true outliers—someone recorded a data item wrong, or they represent an effect that just doesn't occur all that often. Then removing these data points is justified. Another is that they are important data, and our linear model may not be good enough. In this case, removing the points might not be justified. But it might be acceptable to remove outliers, as the result could be a linear model that is well behaved for most cases, but doesn't model some rare effect and so occasionally makes wild errors.

Figure 10.7 shows a regression of human weight against height using a dataset that has four outliers. This data is taken from a dataset published by Dr. John Rasp on human body measurements. I found it at http://www2.stetson.edu/~jrasp/ data.htm (look for bodyfat.xls). I will use this dataset several times in this chapter to produce regressions predicting weight from height, and from a variety of other measurements. Notice how the regression line doesn't fit the main blob of data particularly well, and there are some odd residuals. One fitted value is very different from the others, and there are some residuals that are about the same size as the predicted values. Figure 10.8 shows what happens when the four oddest points are removed. The line is much closer to the blob, and the residuals are much less erratic.

There are two problems here. The first is a visualization problem. It's relatively straightforward to spot outliers in regressions you can plot, but finding outliers in regressions where there are many independent variables requires new tools. The second is a modelling problem. We could construct a model that fits all data, but is poor; we could seek some transformation so that the model works better on all data (which might be hard to obtain); or we could throw out or discount the outliers and settle for a model that works very well on all data, but ignores some effects.

The simplest strategy, which is powerful but dangerous, is to identify outliers and remove them. The danger is you might find that each time you remove a few problematic data points, some more data points look strange to you. Following this line of thought too far can lead to models that fit some data very well, but are essentially pointless because they don't handle most effects. An alternative strategy is to build methods that can discount the effects of outliers. I describe some such methods, which can be technically complex, in the following chapter.

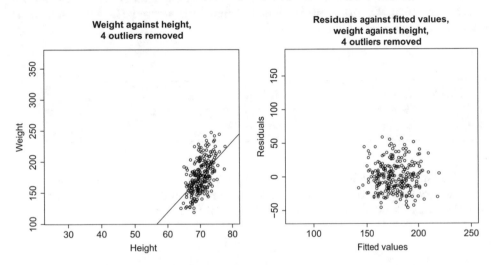

Figure 10.8: On the **left**, weight regressed against height for the bodyfat dataset. I have now removed the four suspicious looking data points, identified in Fig. 10.7 with filled-in markers; these seemed the most likely to be outliers. On the **right**, a scatterplot of the residual against the value predicted by the regression. Notice that the residual looks like noise. The residual seems to be uncorrelated to the predicted value; the mean of the residual seems to be zero; and the variance of the residual doesn't depend on the predicted value. All these are good signs, consistent with our model, and suggest the regression will yield good predictions

> **Remember This:** *Outliers can affect linear regressions significantly. Usually, if you can plot the regression, you can look for outliers by eyeballing the plot. Other methods exist, but are beyond the scope of this text.*

10.3.2 The Hat Matrix and Leverage

Write $\hat{\beta}$ for the estimated value of β, and $\mathbf{y}^{(p)} = \mathcal{X}\hat{\beta}$ for the predicted y values. Then we have

$$\hat{\beta} = \left(\mathcal{X}^T \mathcal{X}\right)^{-1}\left(\mathcal{X}^T \mathbf{y}\right)$$

so that

$$\mathbf{y}^{(p)} = (\mathcal{X}\left(\mathcal{X}^T \mathcal{X}\right)^{-1}\mathcal{X}^T)\mathbf{y}.$$

What this means is that the values the model predicts at training points are a linear function of the true values at the training points. The matrix $(\mathcal{X}\left(\mathcal{X}^T \mathcal{X}\right)^{-1}\mathcal{X}^T)$ is sometimes called the **hat matrix**. The hat matrix is written $\mathcal{H}$, and I shall write the i, j'th component of the hat matrix h_{ij}.

> **Remember This:** *The predictions of a linear regression at training points are a linear function of the y-values at the training points. The linear function is given by the hat matrix.*

The hat matrix has a variety of important properties. I won't prove any here, but the proofs are in the exercises. It is a symmetric matrix. The eigenvalues can be only 1 or 0. And the row sums have the important property that

$$\sum_j h_{ij}^2 \leq 1.$$

This is important, because it can be used to find data points that have values that are hard to predict. The **leverage** of the i'th training point is the i'th diagonal element, h_{ii}, of the hat matrix $\mathcal{H}$. Now we can write the prediction at the i'th training point $y_{p,i} = h_{ii}y_i + \sum_{j \neq i} h_{ij}y_j$. But if h_{ii} has large absolute value, then all the other entries in that row of the hat matrix must have small absolute value. This means that, if a data point has high leverage, the model's value at that point is predicted almost entirely by the observed value at that point. Alternatively, it's hard to use the other training data to predict a value at that point.

Here is another way to see this importance of h_{ii}. Imagine we change the value of y_i by adding Δ; then $y_i^{(p)}$ becomes $y_i^{(p)} + h_{ii}\Delta$. In turn, a large value of h_{ii} means that the predictions at the i'th point are very sensitive to the value of y_i.

> **Remember This:** *Ideally, the value predicted for a particular data point depends on many other data points. Leverage measures the importance of a data point in producing a prediction at that data point. If the leverage of a point is high, other points are not contributing much to the prediction for that point, and it may well be an outlier.*

10.3.3 Cook's Distance

Another way to find points that may be creating problems is to look at the effect of omitting the point from the regression. We could compute $\mathbf{y}^{(p)}$ using the whole dataset. We then omit the i'th point from the dataset, and compute the regression coefficients from the remaining data (which I will write $\hat{\beta}_{\hat{i}}$). Now write $\mathbf{y}_{\hat{i}}^{(p)} = \mathcal{X}\hat{\beta}_{\hat{i}}$. This vector is the predictions that the regression makes at all points when it is trained with the i'th point removed from the training data. Now one can compare $\mathbf{y}^{(p)}$ to $\mathbf{y}_{\hat{i}}^{(p)}$. Points with large values of Cook's distance are suspect, because omitting such a point strongly changes the predictions of the regression. The score for the comparison is called **Cook's distance**. If a point has a large value of Cook's distance, then it has a strong influence on the regression and might well be an outlier. Typically, one computes Cook's distance for each point, and takes a closer look at any point with a large value. This procedure is described in more detail in the box below. Notice the rough similarity to cross-validation (omit some data and recompute). But in this case, we are using the procedure to identify points we might not trust, rather than to get an unbiased estimate of the error.

> **Procedure: 10.2** *Computing Cook's Distance*
>
> We have a dataset containing N pairs $(\mathbf{x}_i, y_i)$. Each x_i is a d-dimensional explanatory vector, and each y_i is a single dependent variable. Write $\hat{\beta}$ for the coefficients of a linear regression (see Procedure 10.1), and $\hat{\beta}_{\hat{i}}$ for the coefficients of the linear regression computed by omitting the i'th data point, $\mathbf{y}^{(p)}$ for $\mathcal{X}\hat{\beta}$, m for the mean-squared error, and
>
> $$\mathbf{y}_{\hat{i}}^{(p)} = \mathcal{X}\hat{\beta}_{\hat{i}}.$$
>
> The Cook's distance of the i'th data point is
>
> $$\frac{(\mathbf{y}^{(p)} - \mathbf{y}_{\hat{i}}^{(p)})^T(\mathbf{y}^{(p)} - \mathbf{y}_{\hat{i}}^{(p)})}{dm}.$$
>
> Large values of this distance suggest that a point may present problems. Statistical software will compute and plot this distance for you.

> **Remember This:**   *The Cook's distance of a training data point measures the effect on predictions of leaving that point out of the regression. A large value of Cook's distance suggests other points are poor at predicting the value at a given point, so a point with a large value of Cook's distance may be an outlier.*

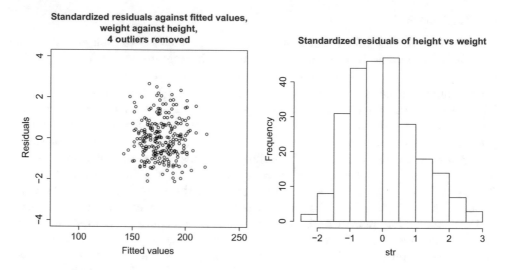

Figure 10.9:   On the **left**, standardized residuals plotted against predicted value for weight regressed against height for the bodyfat dataset. I removed the four suspicious looking data points, identified in Fig. 10.7 with filled-in markers. These seemed the most likely to be outliers. You should compare this plot with the residuals in Fig. 10.8, which are not standardized. Notice that relatively few residuals are more than two standard deviations away from the mean, as one expects. On the **right**, a histogram of the residual values. Notice this looks rather like a histogram of a standard normal random variable, though there are slightly more large positive residuals than one would like. This suggests that the regression is working tolerably

10.3.4   Standardized Residuals

The hat matrix has another use. It can be used to tell how "large" a residual is. The residuals that we measure depend on the units in which y was expressed, meaning we have no idea what a "large" residual is. For example, if we were to express y in kilograms, then we might want to think of 0.1 as a small residual. Using exactly the same dataset, but now with y expressed in grams, that residual value becomes 100—is it really "large" because we changed units?

Now recall that we assumed, in Sect. 10.2.1, that $y - \mathbf{x}^T \beta$ was a zero mean normal random variable, but we didn't know its variance. It can be shown that, under our assumption, the i'th residual value, e_i, is a sample of a normal random variable whose variance is

$$\left(\frac{(\mathbf{e}^T \mathbf{e})}{N} \right) (1 - h_{ii}).$$

This means we can tell whether a residual is large by **standardizing** it—that is, dividing by its standard deviation. Write s_i for the standard residual at the i'th training point. Then we have that

$$s_i = \frac{e_i}{\sqrt{\left(\frac{(\mathbf{e}^T \mathbf{e})}{N} \right) (1 - h_{ii})}}.$$

When the regression is behaving, this standardized residual should look like a sample of a standard normal random variable (Fig. 10.9). Three simple properties of standard normal random variables that you should commit to memory (if you haven't already) appear in the box below. Large (or odd) values of the standard residuals are a sign of trouble.

> **Remember This:** *About 66% of the sampled values of a standard normal random variable are in the range $[-1, 1]$. About 95% of the sampled values of a standard normal random variable are in the range $[-2, 2]$. About 99% of the sampled values of a standard normal random variable are in the range $[-3, 3]$.*

R produces a nice diagnostic plot that can be used to look for problem data points. The plot is a scatterplot of the standardized residuals against leverage, with level curves of Cook's distance superimposed. Figure 10.10 shows an example. Some bad points that are likely to present problems are identified with a number (you can control how many, and the number, with arguments to `plot`). Problem points will have high leverage and/or high Cook's distance and/or high residual. The figure shows this plot for three different versions of the dataset (original; two problem points removed; and two further problem points removed).

10.4 Many Explanatory Variables

In earlier sections, I implied you could put anything into the explanatory variables. This is correct, and makes it easy to do the math for the general case. However, I have plotted only cases where there was one explanatory variable (together with a constant, which hardly counts). In some cases (Sect. 10.4.1), we can add explanatory variables and still have an easy plot. Adding explanatory variables can cause the matrix $\mathcal{X}^T \mathcal{X}$ to have poor condition number; there's an easy strategy to deal with this (Sect. 10.4.2).

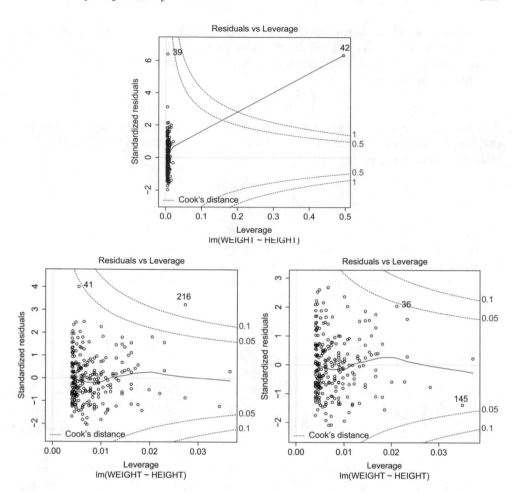

Figure 10.10: A diagnostic plot, produced by R, of a linear regression of weight against height for the bodyfat dataset. **Top:** the whole dataset. Notice that two points are very odd indeed. The point labelled 42 has gigantic Cook's distance and quite extraordinary leverage. The point labelled 39 has a standardized residual that is six standard deviations away from the mean (i.e., never happens for a normal random variable). **Bottom left:** The diagnostic plot with the two most extreme points in the top figure removed. Notice that the points labelled 39 and 216 have very large standardized residual and large Cook's distance. **Bottom right:** The diagnostic plot with these two further points removed. At this point, the regression seems well behaved (or, at least, removing more points seems more dangerous than stopping)

Most cases are hard to plot successfully, and one needs better ways to visualize the regression than just plotting. The value of R^2 is still a useful guide to the goodness of the regression, but the way to get more insight is to use the tools of the previous section.

10.4.1 Functions of One Explanatory Variable

Imagine we have only one measurement to form explanatory variables. For example, in the perch data of Fig. 10.1, we have only the length of the fish. If we evaluate functions of that measurement, and insert them into the vector of explanatory variables, the resulting regression is still easy to plot. It may also offer better predictions. The fitted line of Fig. 10.1 looks quite good, but the data points look as though they might be willing to follow a curve. We can get a curve quite easily. Our current model gives the weight as a linear function of the length with a noise term (which we wrote $y_i = \beta_1 x_i + \beta_0 + \xi_i$). But we could expand this model to incorporate other functions of the length. In fact, it's quite surprising that the weight of a fish should be predicted by its length. If the fish doubled in each direction, say, its weight should go up by a factor of eight. The success of our regression suggests that fish do not just scale in each direction as they grow. But we might try the model $y_i = \beta_2 x_i^2 + \beta_1 x_i + \beta_0 + \xi_i$. This is easy to do. The i'th row of the matrix $\mathcal{X}$ currently looks like $[x_i, 1]$. We build a new matrix $\mathcal{X}^{(b)}$, where the i'th row is $[x_i^2, x_i, 1]$, and proceed as before. This gets us a new model. The nice thing about this model is that it is easy to plot—our predicted weight is still a function of the length, it's just not a linear function of the length. Several such models are plotted in Fig. 10.11.

You should notice that it can be quite easy to add a lot of functions like this (in the case of the fish, I tried x_i^3 as well). However, it's hard to decide whether the regression has actually gotten better. The least-squares error *on the training data* will *never* go up when you add new explanatory variables, so the R^2 will *never* get worse. This is easy to see, because you could always use a coefficient of zero with the new variables and get back the previous regression. However, the models that you choose are likely to produce worse and worse predictions as you add explanatory variables. Knowing when to stop can be tough (Sect. 11.1), though it's sometimes obvious that the model is untrustworthy (Fig. 10.11).

> **Remember This:** *If you have only one measurement, you can construct a high dimensional* **x** *by using functions of that measurement. This produces a regression that has many explanatory variables, but is still easy to plot. Knowing when to stop is hard, but insight into the underlying problem can help.*

10.4.2 Regularizing Linear Regressions

Our current regression strategy requires solving $\mathcal{X}^T \mathcal{X} \hat{\beta} = \mathcal{X}^T \mathbf{y}$. If $\mathcal{X}^T \mathcal{X}$ has small (or zero) eigenvalues, we may make serious errors. A matrix with small eigenvalues can turn large vectors into small ones, so if $\mathcal{X}^T \mathcal{X}$ has small eigenvalues, then there is some large $\mathbf{w}$ so that $\mathcal{X}^T \mathcal{X} (\hat{\beta} + \mathbf{w})$ is not much different from $\mathcal{X}^T \mathcal{X} \hat{\beta}$. This means that a small change in $\mathcal{X}^T \mathbf{y}$ can lead to a large change in the estimate of $\hat{\beta}$.

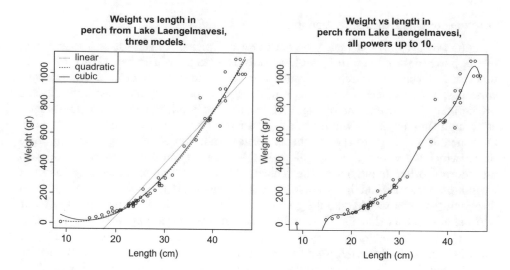

Figure 10.11: On the **left**, several different models predicting fish weight from length. The line uses the explanatory variables 1 and x_i; and the curves use other monomials in x_i as well, as shown by the legend. This allows the models to predict curves that lie closer to the data. It is important to understand that, while you can make a curve go closer to the data by inserting monomials that doesn't mean you necessarily have a better model. On the **right**, I have used monomials up to x_i^{10}. This curve lies very much closer to the data points than any on the other side, at the cost of some very odd looking wiggles in between data points (look at small lengths; the model goes quite strongly negative there, but I can't bring myself to change the axes and show predictions that are obvious nonsense). I can't think of any reason that these structures would come from true properties of fish, and it would be hard to trust predictions from this model

This is a problem, because we can expect that different samples from the same data will have somewhat different values of $\mathcal{X}^T\mathbf{y}$. For example, imagine the person recording fish measurements in Lake Laengelmavesi recorded a different set of fish; we expect changes in $\mathcal{X}$ and $\mathbf{y}$. But, if $\mathcal{X}^T\mathcal{X}$ has small eigenvalues, these changes could produce large changes in our model.

Small (or zero) eigenvalues are quite common in practical problems. They arise from correlations between explanatory variables. Correlations between explanatory variables mean that we can predict, quite accurately, the value of one explanatory variable using the values of the other variables. In turn, there must be a vector $\mathbf{w}$ so that $\mathcal{X}\mathbf{w}$ is small. The exercises give the construction in detail, but the idea is simple. If (say) you can predict the first explanatory variable very well as a linear function of the others, then a $\mathbf{w}$ that looks like

$$[-1, \text{coefficients of linear function}]$$

should result in a small $\mathcal{X}\mathbf{w}$. But if $\mathcal{X}\mathbf{w}$ is small, $\mathbf{w}^T\mathcal{X}^T\mathcal{X}\mathbf{w}$ must be small, so that $\mathcal{X}^T\mathcal{X}$ has at least one small eigenvalue.

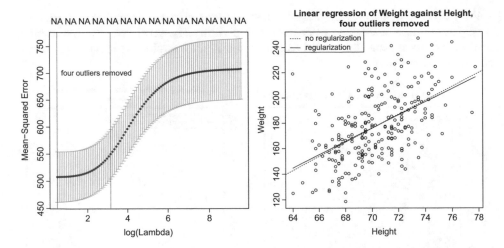

Figure 10.12:   On the **left**, cross-validated error estimated for different choices of regularization constant for a linear regression of weight against height for the bodyfat dataset, with four outliers removed. The horizontal axis is log regression constant; the vertical is cross-validated error. The mean of the error is shown as a spot, with vertical error bars. The vertical lines show a range of reasonable choices of regularization constant (**left** yields the lowest observed error, **right** the error whose mean is within one standard error of the minimum). On the **right**, two regression lines on a scatterplot of this dataset; one is the line computed without regularization, the other is obtained using the regularization parameter that yields the lowest observed error. In this case, the regularizer doesn't change the line much, but may produce improved values on new data. Notice how the cross-validated error is fairly flat with low values of the regularization constant—there is a range of values that works quite well

The problem is relatively easy to control. When there are small eigenvalues in $\mathcal{X}^T\mathcal{X}$, we expect that $\hat{\beta}$ will be large (because we can add components in the direction of at least one **w** without changing anything much), and so the largest components in $\hat{\beta}$ might be very inaccurately estimated. If we are trying to predict new y values, we expect that large components in $\hat{\beta}$ turn into large errors in prediction (exercises).

An important and useful way to suppress these errors is to try to find a $\hat{\beta}$ that isn't large, and also gives a low error. We can do this by regularizing, using the same trick we saw in the case of classification. Instead of choosing the value of β that minimizes

$$\left(\frac{1}{N}\right)(\mathbf{y} - \mathcal{X}\beta)^T(\mathbf{y} - \mathcal{X}\beta)$$

we minimize

$$\left(\frac{1}{N}\right)(\mathbf{y} - \mathcal{X}\beta)^T(\mathbf{y} - \mathcal{X}\beta) \quad + \quad \lambda\beta^T\beta$$

$$\text{Error} \quad + \quad \text{Regularizer.}$$

Here $\lambda > 0$ is a constant that weights the two requirements of small error and small $\hat{\beta}$ relative to one another. Notice also that dividing the total error by the number of data points means that our choice of λ shouldn't be affected by changes in the size of the dataset.

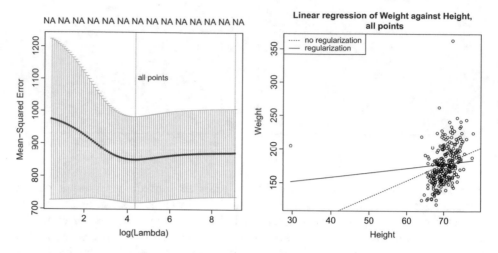

Figure 10.13: Regularization doesn't make outliers go away. On the **left**, cross-validated error estimated for different choices of regularization constant for a linear regression of weight against height for the bodyfat dataset, with all points. The horizontal axis is log regression constant; the vertical is cross-validated error. The mean of the error is shown as a spot, with vertical error bars. The vertical lines show a range of reasonable choices of regularization constant (**left** yields the lowest observed error, **right** the error whose mean is within one standard error of the minimum). On the **right**, two regression lines on a scatterplot of this dataset; one is the line computed without regularization, the other is obtained using the regularization parameter that yields the lowest observed error. In this case, the regularizer doesn't change the line much, but may produce improved values on new data. Notice how the cross-validated error is fairly flat with low values of the regularization constant—the precise value of the regularization constant doesn't matter much hear. Notice also how the error in this regression is very much greater than that in the regression of Fig. 10.12. Outliers in the training part of a cross-validation split cause the model to be poor, *and* outliers in the test part result in very poor predictions, too

Regularization helps to deal with the small eigenvalue, because to solve for β we must solve the equation

$$\left[\left(\frac{1}{N}\right)\mathcal{X}^T\mathcal{X} + \lambda\mathcal{I}\right]\hat{\beta} = \left(\frac{1}{N}\right)\mathcal{X}^T\mathbf{y}$$

(obtained by differentiating with respect to β and setting to zero) and the smallest eigenvalue of the matrix $((\frac{1}{N})(\mathcal{X}^T\mathcal{X}+\lambda\mathcal{I}))$ will be at least λ (exercises). Penalizing a regression with the size of β in this way is sometimes known as **ridge regression**.

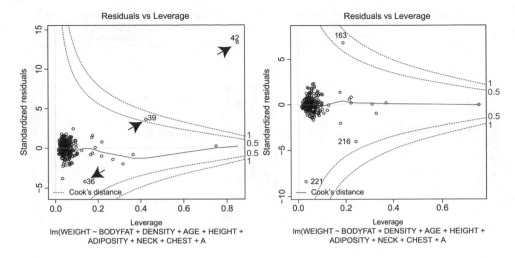

Figure 10.14:   On the **left**, residuals plotted against leverage for a regression of weight against all other measurements for the bodyfat dataset. I did *not* remove the outliers. The contours on the plot are contours of Cook's distance; I have overlaid arrows showing points with suspiciously large Cook's distance. Notice also that several points have high leverage, without having a large residual value. These points may or may not present problems. On the **right**, the same plot for this dataset with points 36, 39, 41, and 42 removed (these are the points I have been removing for each such plot). Notice that another point now has high Cook's distance, but mostly the residual is much smaller

We choose λ in the same way we used for classification; split the training set into a training piece and a validation piece, train for different values of λ, and test the resulting regressions on the validation piece. The error is a random variable, random because of the random split. It is a fair model of the error that would occur on a randomly chosen test example, assuming that the training set is "like" the test set, in a way that I do not wish to make precise. We could use multiple splits, and average over the splits. Doing so yields both an average error for a value of λ and an estimate of the standard deviation of error.

Statistical software will do all the work for you. I used the `glmnet` package in R (see exercises for details). Figure 10.12 shows an example, for weight regressed against height. Notice the regularization doesn't change the model (plotted in the figure) all that much. For each value of λ (horizontal axis), the method has computed the mean error and standard deviation of error using cross-validation splits, and displays these with error bars. Notice that $\lambda = 0$ yields poorer predictions than a larger value; large $\hat{\beta}$ really are unreliable. Notice that there is now no λ that yields the smallest validation error, because the value of error depends on the random splits used in cross-validation. A reasonable choice of λ lies between the one that yields the smallest error encountered (one vertical line in the plot) and the largest value whose mean error is within one standard deviation of the minimum (the other vertical line in the plot) (Fig. 10.13).

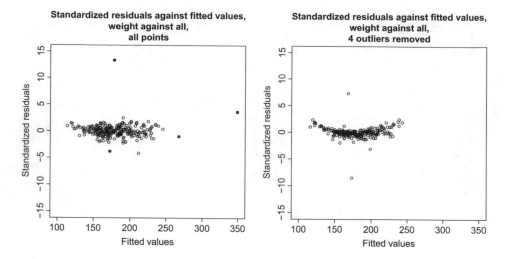

Figure 10.15:   On the **left**, standardized residuals plotted against predicted value for weight regressed against all variables for the bodyfat dataset. Four data points appear suspicious, and I have marked these with a filled-in marker. On the **right**, standardized residuals plotted against predicted value for weight regressed against all variables for the bodyfat dataset, but with the four suspicious looking data points removed. Notice two other points stick out markedly

All this is quite similar to regularizing a classification problem. We started with a cost function that evaluated the errors caused by a choice of β, then added a term that penalized β for being "large." This term is the squared length of β, as a vector. It is sometimes known as the L_2 **norm** of the vector. In Sect. 11.4, I describe the consequences of using other norms.

> **Remember This:** *The performance of a regression can be improved by regularizing, particularly if some explanatory variables are correlated. The procedure is similar to that used for classification.*

10.4.3   Example: Weight Against Body Measurements

We can now look at regressing weight against all body measurements for the bodyfat dataset. We can't plot this regression (too many independent variables), but we can approach the problem in a series of steps.

Finding Suspect Points: Figure 10.14 shows the R diagnostic plots for a regression of weight against all body measurements for the bodyfat dataset. We've already seen there are outliers, so the odd structure of this plot should be no particular surprise. There are several really worrying points here. As the figure shows,

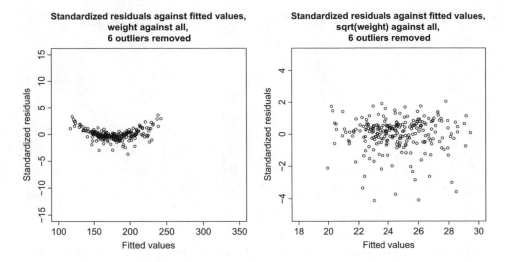

Figure 10.16: On the **left**, standardized residuals plotted against predicted value for weight regressed against all variables for the bodyfat dataset. I removed the four suspicious data points of Fig. 10.15, and the two others identified in that figure. Notice a suspicious "banana" shape—the residuals are distinctly larger for small and for large predicted values. This suggests that a non-linear transformation of something might be helpful. I used a Box-Cox transformation, which suggested a value of 0.5 (i.e., regress $2(\sqrt{\text{weight}} - 1)$) against all variables. On the **right**, the standardized residuals for this regression. Notice that the "banana" has gone, though there is a suspicious tendency for the residuals to be smaller rather than larger. Notice also the plots are on different axes. It's fair to compare these plots by eye; but it's not fair to compare details, because the residual of a predicted square root means something different than the residual of a predicted value

removing the four points identified in the caption, based on their very high standardized residuals, high leverage, and high Cook's distance, yields improvements. We can get some insight by plotting standardized residuals against predicted value (Fig. 10.9). There is clearly a problem here; the residual seems to depend quite strongly on the predicted value. Removing the four outliers we have already identified leads to a much improved plot, also shown in Fig. 10.15. This is banana-shaped, which is suspicious. There are two points that seem to come from some other model (one above the center of the banana, one below). Removing these points gives the residual plot shown in Fig. 10.16.

Transforming Variables: The banana shape of the plot of standardized residuals against value means that large predictions and small predictions are each a bit too big, but predictions that are close to the mean are a bit too small. This is a suggestion that some non-linearity somewhere would improve the regression. One option is a non-linear transformation of the independent variables. Finding the right one might require some work, so it's natural to try a Box-Cox transformation first. This gives the best value of the parameter as 0.5 (i.e., the dependent variable should be $\sqrt{\text{weight}}$), which makes the residuals look much better (Fig. 10.16).

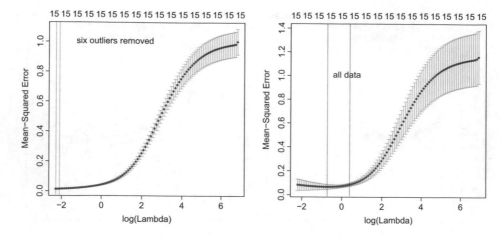

Figure 10.17:   Plots of mean-squared error as a function of log regularization parameter (i.e., $\log \lambda$) for a regression of weight$^{1/2}$ against all variables for the bodyfat dataset. These plots show that mean-squared error averaged over cross-validation folds with a vertical one standard deviation bar. On the **left**, the plot for the dataset with the six outliers identified in Fig. 10.15 removed. On the **right**, the plot for the whole dataset. Notice how the outliers increase the variability of the error, and the best error

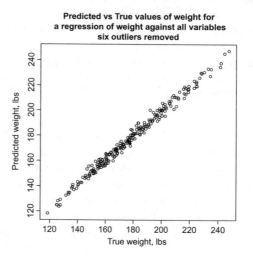

Figure 10.18:   A scatterplot of the predicted weight against the true weight for the bodyfat dataset. The prediction is made with all variables, but the six outliers identified above are omitted. I used a Box-Cox transformation with parameter 1/2, and the regularization parameter that yielded the smallest mean-squared error in Fig. 10.17

Choosing a Regularizing Value: Figure 10.17 shows the `glmnet` plot of cross-validated error as a function of regularizer weight. A sensible choice of value here seems to be a bit smaller than -2 (between the value that yields the smallest error encountered—one vertical line in the plot—and the largest value whose mean error is within one standard deviation of the minimum—the other vertical line in the plot). I chose -2.2.

How Good Are the Resulting Predictions Likely to Be: the standardized residuals don't seem to depend on the predicted values, but how good are the predictions? We already have some information on this point. Figure 10.17 shows cross-validation errors for regressions of weight$^{1/2}$ against height for different regularization weights, but some will find this slightly indirect. We want to predict weight, not weight$^{1/2}$. I chose the regularization weight that yielded the lowest mean-squared error for the model of Fig. 10.17, omitting the six outliers previously mentioned. I then computed the predicted weight for each data point using that model (which predicts weight$^{1/2}$, remember; but squaring takes care of that). Figure 10.18 shows the predicted values plotted against the true values. You should not regard this plot as a safe way to estimate generalization (the points were used in training the model; Fig. 10.17 is better for that), but it helps to visualize the errors. This regression looks as though it is quite good at predicting bodyweight from other measurements.

10.5 You Should

10.5.1 Remember These Terms

10.5.2 Remember These Facts

10.5.3 Remember These Procedures

10.5.4 Be Able to

- Construct a linear regression, using ridge regularization as required.
- Use leverage, Cook's distance, and standardized residuals to identify possible outliers.
- Apply simple variable transformations to improve a linear regression.
- Evaluate a linear regression.

Appendix: Data

Batch A			Batch B			Batch C	
Amount of hormone	Time in service		Amount of hormone	Time in service		Amount of hormone	Time in service
25.8	99		16.3	376		28.8	119
20.5	152		11.6	385		22.0	188
14.3	293		11.8	402		29.7	115
23.2	155		32.5	29		28.9	88
20.6	196		32.0	76		32.8	58
31.1	53		18.0	296		32.5	49
20.9	184		24.1	151		25.4	150
20.9	171		26.5	177		31.7	107
30.4	52		25.8	209		28.5	125

TABLE 10.1: A table showing the amount of hormone remaining and the time in service for devices from lot A, lot B, and lot C. The numbering is arbitrary (i.e., there's no relationship between device 3 in lot A and device 3 in lot B). We expect that the amount of hormone goes down as the device spends more time in service, so cannot compare batches just by comparing numbers

Problems

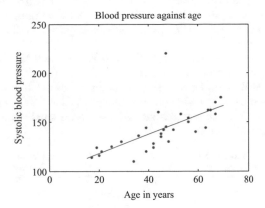

Figure 10.19: A regression of blood pressure against age, for 30 data points

10.1. Figure 10.19 shows a linear regression of systolic blood pressure against age. There are 30 data points.
   (a) Write $e_i = y_i - \mathbf{x}_i^T \beta$ for the residual. What is the $\mathsf{mean}\,(\{e\})$ for this regression?
   (b) For this regression, $\mathsf{var}\,(\{y\}) = 509$ and the R^2 is 0.4324. What is $\mathsf{var}\,(\{e\})$ for this regression?
   (c) How well does the regression explain the data?
   (d) What could you do to produce better predictions of blood pressure (without actually measuring blood pressure)?

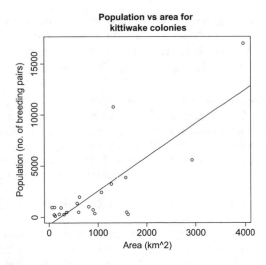

Figure 10.20: A regression of the number of breeding pairs of kittiwakes against the area of an island, for 22 data points

10.2. At http://www.statsci.org/data/general/kittiwak.html, you can find a dataset collected by D.K. Cairns in 1988 measuring the area available for a seabird (black-legged kittiwake) colony and the number of breeding pairs for a variety of different colonies. Figure 10.20 shows a linear regression of the number of breeding pairs against the area. There are 22 data points.

(a) Write $e_i = y_i - \mathbf{x}_i^T \beta$ for the residual. What is the mean $(\{e\})$ for this regression?

(b) For this regression, var $(\{y\}) = 16{,}491{,}357$ and the R^2 is 0.62. What is var $(\{e\})$ for this regression?

(c) How well does the regression explain the data? If you had a large island, to what extent would you trust the prediction for the number of kittiwakes produced by this regression? If you had a small island, would you trust the answer more?

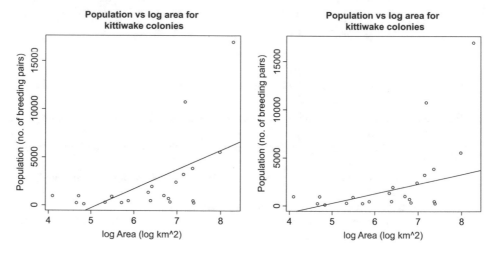

Figure 10.21:   **Left:** A regression of the number of breeding pairs of kittiwakes against the log of area of an island, for 22 data points. **Right:** A regression of the number of breeding pairs of kittiwakes against the log of area of an island, for 22 data points, using a method that ignores two likely outliers

10.3. At http://www.statsci.org/data/general/kittiwak.html, you can find a dataset collected by D.K. Cairns in 1988 measuring the area available for a seabird (black-legged kittiwake) colony and the number of breeding pairs for a variety of different colonies. Figure 10.21 shows a linear regression of the number of breeding pairs against the log of area. There are 22 data points.

(a) Write $e_i = y_i - \mathbf{x}_i^T \beta$ for the residual. What is the mean $(\{e\})$ for this regression?

(b) For this regression, var $(\{y\}) = 16{,}491{,}357$ and the R^2 is 0.31. What is var $(\{e\})$ for this regression?

(c) How well does the regression explain the data? If you had a large island, to what extent would you trust the prediction for the number of kittiwakes produced by this regression? If you had a small island, would you trust the answer more? Why?

(d) Figure 10.21 shows the result of a linear regression that ignores two likely outliers. Would you trust the predictions of this regression more? Why?

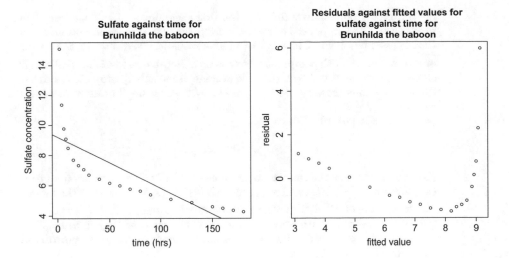

Figure 10.22: **Left:** A regression of the concentration of sulfate in the blood of Brunhilda the baboon against time. **Right:** For this regression, a plot of residual against fitted value

10.4. At http://www.statsci.org/data/general/brunhild.html, you will find a dataset that measures the concentration of a sulfate in the blood of a baboon named Brunhilda as a function of time. Figure 10.22 plots this data, with a linear regression of the concentration against time. I have shown the data, and also a plot of the residual against the predicted value. The regression appears to be unsuccessful.

    (a) What suggests the regression has problems?

    (b) What is the cause of the problem, and why?

    (c) What could you do to improve the problems?

10.5. Assume we have a dataset where $\mathbf{Y} = \mathcal{X}\beta + \xi$, for some unknown β and ξ. The term ξ is a normal random variable with zero mean, and covariance $\sigma^2 \mathcal{I}$ (i.e., this data really does follow our model).

    (a) Write $\hat{\beta}$ for the estimate of β recovered by least squares, and $\hat{\mathbf{Y}}$ for the values predicted by our model for the training data points. Show that

$$\hat{\mathbf{Y}} = \mathcal{X}\left(\mathcal{X}^T \mathcal{X}\right)^{-1} \mathcal{X}^T . \mathbf{Y}$$

    (b) Show that

$$\mathbb{E}[\hat{y}_i - y_i] = 0$$

for each training data point y_i, where the expectation is over the probability distribution of ξ.

    (c) Show that

$$\mathbb{E}\left[(\hat{\beta} - \beta)\right] = 0,$$

where the expectation is over the probability distribution of ξ.

10.6. This exercise investigates the effect of correlation on a regression. Assume we have N data items $(\mathbf{x}_i, y_i)$. We will investigate what happens when the data have the property that the first component is relatively accurately predicted by the other components. Write x_{i1} for the first component of $\mathbf{x}_i$, and $\mathbf{x}_{i,\hat{1}}$

for the vector obtained by deleting the first component of $\mathbf{x}_i$. Choose $\mathbf{u}$ to predict the first component of the data from the rest *with minimum error*, so that $x_{i1} = \mathbf{x}_{i\hat{1}}^T \mathbf{u} + w_i$. The error of prediction is w_i. Write $\mathbf{w}$ for the vector of errors (i.e., the i'th component of $\mathbf{w}$ is w_i). Because $\mathbf{w}^T \mathbf{w}$ is minimized by choice of $\mathbf{u}$, we have $\mathbf{w}^T \mathbf{1} = 0$ (i.e., the average of the w_i's is zero). Assume that these predictions are very good, so that there is some small positive number ϵ so that $\mathbf{w}^T \mathbf{w} \leq \epsilon$.

(a) Write $\mathbf{a} = [-1, \mathbf{u}]^T$. Show that

$$\mathbf{a}^T \mathcal{X}^T \mathcal{X} \mathbf{a} \leq \epsilon.$$

(b) Now show that the smallest eigenvalue of $\mathcal{X}^T \mathcal{X}$ is less than or equal to ϵ.

(c) Write $s_k = \sum_u x_{uk}^2$, and $s_{\max}$ for $\max(s_1, \ldots, s_d)$. Show that the largest eigenvalue of $\mathcal{X}^T \mathcal{X}$ is greater than or equal to $s_{\max}$.

(d) The **condition number** of a matrix is the ratio of largest to smallest eigenvalue of a matrix. Use the information above to bound the **condition number** of $\mathcal{X}^T \mathcal{X}$.

(e) Assume that $\hat{\beta}$ is the solution to $\mathcal{X}^T \mathcal{X} \hat{\beta} = \mathcal{X}^T \mathbf{Y}$. Show that the

$$(\mathcal{X}^T \mathbf{Y} - \mathcal{X}^T \mathcal{X}(\hat{\beta} + \mathbf{a}))^T (\mathcal{X}^T \mathbf{Y} - \mathcal{X}^T \mathcal{X}(\hat{\beta} + \mathbf{a}))$$

(for $\mathbf{a}$ as above) is bounded above by

$$\epsilon^2 (1 + \mathbf{u}^T \mathbf{u}).$$

(f) Use the last subexercises to explain why correlated data will lead to a poor estimate of $\hat{\beta}$.

10.7. This exercise explores the effect of regularization on a regression. Assume we have N data items $(\mathbf{x}_i, y_i)$. We will investigate what happens when the data have the property that the first component is relatively accurately predicted by the other components. Write x_{i1} for the first component of $\mathbf{x}_i$, and $\mathbf{x}_{i,\hat{1}}$ for the vector obtained by deleting the first component of $\mathbf{x}_i$. Choose $\mathbf{u}$ to predict the first component of the data from the rest *with minimum error*, so that $x_{i1} = \mathbf{x}_{i\hat{1}}^T \mathbf{u} + w_i$. The error of prediction is w_i. Write $\mathbf{w}$ for the vector of errors (i.e., the i'th component of $\mathbf{w}$ is w_i). Because $\mathbf{w}^T \mathbf{w}$ is minimized by choice of $\mathbf{u}$, we have $\mathbf{w}^T \mathbf{1} = 0$ (i.e., the average of the w_i's is zero). Assume that these predictions are very good, so that there is some small positive number ϵ so that $\mathbf{w}^T \mathbf{w} \leq \epsilon$.

(a) Show that, *for any vector* $\mathbf{v}$,

$$\mathbf{v}^T \left(\mathcal{X}^T \mathcal{X} + \lambda \mathcal{I} \right) \mathbf{v} \geq \lambda \mathbf{v}^T \mathbf{v}$$

and use this to argue that the smallest eigenvalue of $\left(\mathcal{X}^T \mathcal{X} + \lambda \mathcal{I} \right)$ is greater than λ.

(b) Write $\mathbf{b}$ for an eigenvector of $\mathcal{X}^T \mathcal{X}$ with eigenvalue $\lambda_{\mathbf{b}}$. Show that $\mathbf{b}$ is an eigenvector of $\left(\mathcal{X}^T \mathcal{X} + \lambda \mathcal{I} \right)$ with eigenvalue $\lambda_{\mathbf{b}} + \lambda$.

(c) Recall $\mathcal{X}^T \mathcal{X}$ is a $d \times d$ matrix which is symmetric, and so has d orthonormal eigenvectors. Write $\mathbf{b}_i$ for the i'th such vector, and $\lambda_{\mathbf{b}_i}$ for the corresponding eigenvalue. Show that

$$\mathcal{X}^T \mathcal{X} \beta - \mathcal{X}^T \mathbf{Y} = 0$$

is solved by

$$\beta = \sum_{i=1}^{d} \frac{\mathbf{Y}^T \mathcal{X} \mathbf{b}_i}{\lambda_{\mathbf{b}_i}}.$$

(d) Using the notation of the previous sub exercise, show that

$$(\mathcal{X}^T \mathcal{X} + \lambda \mathcal{I})\beta - \mathcal{X}^T \mathbf{Y} = 0$$

is solved by

$$\beta = \sum_{i=1}^{d} \frac{\mathbf{Y}^T \mathcal{X} \mathbf{b}_i}{\lambda_{\mathbf{b}_i} + \lambda}.$$

Use this expression to explain why a regularized regression may produce better results on test data than an unregularized regression.

10.8. We will study the hat matrix, $\mathcal{H} = \mathcal{X}\left(\mathcal{X}^T \mathcal{X}\right)^{-1} \mathcal{X}^T$. We assume that $\left(\mathcal{X}^T \mathcal{X}\right)^{-1}$ exists, so that (at least) $N \geq d$.

(a) Show that $\mathcal{H}\mathcal{H} = \mathcal{H}$. This property is sometimes referred to as **idempotence**.

(b) Now an SVD yields $\mathcal{X} = \mathcal{U}\Sigma\mathcal{V}^T$. Show that $\mathcal{H} = \mathcal{U}\mathcal{U}^T$.

(c) Use the result of the previous exercise to show that each eigenvalue of $\mathcal{H}$ is either 0 or 1.

(d) Show that $\sum_j h_{ij}^2 \leq 1$.

Programming Exercises

10.9. At http://www.statsci.org/data/general/brunhild.html, you will find a dataset that measures the concentration of a sulfate in the blood of a baboon named Brunhilda as a function of time. Build a linear regression of the log of the concentration against the log of time.

(a) Prepare a plot showing (a) the data points and (b) the regression line in log–log coordinates.

(b) Prepare a plot showing (a) the data points and (b) the regression curve in the original coordinates.

(c) Plot the residual against the fitted values in log–log and in original coordinates.

(d) Use your plots to explain whether your regression is good or bad and why.

10.10. At http://www.statsci.org/data/oz/physical.html, you will find a dataset of measurements by M. Larner, made in 1996. These measurements include body mass, and various diameters. Build a linear regression of predicting the body mass from these diameters.

(a) Plot the residual against the fitted values for your regression.

(b) Now regress the cube root of mass against these diameters. Plot the residual against the fitted values in both these cube root coordinates and in the original coordinates.

(c) Use your plots to explain which regression is better.

10.11. At https://archive.ics.uci.edu/ml/datasets/Abalone, you will find a dataset of measurements by W. J. Nash, T. L. Sellers, S. R. Talbot, A. J. Cawthorn, and W. B. Ford, made in 1992. These are a variety of measurements of blacklip abalone (*Haliotis rubra*; delicious by repute) of various ages and genders.

(a) Build a linear regression predicting the age from the measurements, ignoring gender. Plot the residual against the fitted values.

(b) Build a linear regression predicting the age from the measurements, including gender. There are three levels for gender; I'm not sure whether this has to do with abalone biology or difficulty in determining gender. You can represent gender numerically by choosing 1 for one level, 0 for another, and −1 for the third. Plot the residual against the fitted values.

(c) Now build a linear regression predicting the log of age from the measurements, ignoring gender. Plot the residual against the fitted values.

(d) Now build a linear regression predicting the log age from the measurements, including gender, represented as above. Plot the residual against the fitted values.

(e) It turns out that determining the age of an abalone is possible, but difficult (you section the shell, and count rings). Use your plots to explain which regression you would use to replace this procedure, and why.

(f) Can you improve these regressions by using a regularizer? Use glmnet to obtain plots of the cross-validated prediction error.

10.12. At https://archive.ics.uci.edu/ml/machine-learning-databases/housing/housing.data, you will find the famous Boston Housing dataset. This consists of 506 data items. Each is 13 measurements, and a house price. The data was collected by Harrison, D. and Rubinfeld, D.L in the 1970s (a date which explains the very low house prices). The dataset has been widely used in regression exercises, but seems to be waning in popularity. At least one of the independent variables measures the fraction of population nearby that is "Black" (their word, not mine). This variable appears to have had a significant effect on house prices then (and, sadly, may still now).

(a) Regress house price (variable 14) against all others, and use leverage, Cook's distance, and standardized residuals to find possible outliers. Produce a diagnostic plot that allows you to identify possible outliers.

(b) Remove all points you suspect as outliers, and compute a new regression. Produce a diagnostic plot that allows you to identify possible outliers.

(c) Apply a Box-Cox transformation to the dependent variable—what is the best value of the parameter?

(d) Now transform the dependent variable, build a linear regression, and check the standardized residuals. If they look acceptable, produce a plot of fitted house price against true house price.

(e) Assume you remove a total of six outliers. There are a variety of reasonable choices. How strongly does the Box-Cox variable depend on your choice of points to remove? Does the Box-Cox variable change if you remove no outliers?

CHAPTER 11

Regression: Choosing and Managing Models

This chapter generalizes our understanding of regression in a number of ways. The previous chapter showed we could at least reduce training error, and quite likely improve predictions, by inserting new independent variables into a regression. The difficulty was knowing when to stop. In Sect. 11.1, I will describe some methods to search a family of models (equivalently, a set of subsets of independent variables) to find a good model. In the previous chapter, we saw how to find outlying points and remove them. In Sect. 11.2, I will describe methods to compute a regression that is largely unaffected by outliers. The resulting methods are powerful, but fairly intricate.

To date, we have used regression to predict a number. With a linear model, it is difficult to predict a probability—because linear models can predict negative numbers or numbers bigger than one—or a count—because linear models can predict non-integers. A very clever trick (Sect. 11.3) uses regression to predict the parameter of a carefully chosen probability distribution, and so probabilities, counts, and so on.

Finally, Sect. 11.4 describes methods to force regression models to choose a small set of predictors from a large set, and so produce sparse models. These methods allow us to fit regressions to data where we have more predictors than examples, and often result in significantly improved predictions. Most of the methods in this chapter can be used together to build sophisticated and accurate regressions in quite surprising circumstances.

11.1 Model Selection: Which Model Is Best?

It is usually quite easy to have many explanatory variables in a regression problem. Even if you have only one measurement, you could always compute a variety of non-linear functions of that measurement. As we have seen, inserting variables into a model will reduce the fitting cost, but that doesn't mean that better predictions will result (Sect. 10.4.1). We need to choose which explanatory variables we will use. A linear model with few explanatory variables may make poor predictions because the model itself is incapable of representing the independent variable accurately. A linear model with many explanatory variables may make poor predictions because we can't estimate the coefficients well. Choosing which explanatory variables we will use (and so which model we will use) requires that we balance these effects.

© Springer Nature Switzerland AG 2019
D. Forsyth, *Applied Machine Learning*,
https://doi.org/10.1007/978-3-030-18114-7_11

11.1.1 Bias and Variance

We now look at the process of finding a model in a fairly abstract way. Doing so makes plain three distinct and important effects that cause models to make predictions that are wrong. One is **irreducible error**. Even a perfect choice of model can make mistaken predictions, because more than one prediction could be correct for the same $\mathbf{x}$. Another way to think about this is that there could be many future data items, all of which have the same $\mathbf{x}$, but each of which has a different y. In this case some of our predictions must be wrong, and the effect is unavoidable.

A second effect is **bias**. We must use some collection of models. Even the best model in the collection may not be capable of predicting all the effects that occur in the data. Errors that are caused by the best model still not being able to predict the data accurately are attributed to bias.

The third effect is **variance**. We must choose our model from the collection of models. The model we choose is unlikely to be the best model. This might occur, for example, because our estimates of the parameters aren't exact because we have a limited amount of data. Errors that are caused by our choosing a model that is not the best in the family are attributed to variance.

All this can be written out in symbols. We have a vector of predictors $\mathbf{x}$, and a random variable Y. At any given point $\mathbf{x}$, we have

$$Y = f(\mathbf{x}) + \xi$$

where ξ is noise and f is an unknown function. We have

$$\mathbb{E}[\xi] = 0 \text{ and } \mathbb{E}\big[\xi^2\big] = \mathsf{var}\left(\{\xi\}\right) = \sigma_\xi^2.$$

The noise ξ is independent of X. We have some procedure that takes a selection of training data, consisting of pairs $(\mathbf{x}_i, y_i)$, and selects a model $\hat{f}$. We will use this model to predict values for future $\mathbf{x}$. It is highly unlikely that $\hat{f}$ is the same as f; assuming that it is involves assuming that we can perfectly estimate the best model with a finite dataset, which doesn't happen.

We need to understand the error that will occur when we use $\hat{f}$ to predict for some data item that isn't in the training set. This is the error that we will encounter in practice. The error at any point $\mathbf{x}$ is

$$\mathbb{E}\Big[(Y - \hat{f}(\mathbf{x}))^2\Big]$$

where the expectation is taken over $P(Y, \text{training data}|\mathbf{x})$. But the new query point $\mathbf{x}$ does not depend on the training data and the value Y does not depend on the training data either, so the distribution is $P(Y|\mathbf{x}) \times P(\text{training data})$.

The expectation can be written in an extremely useful form. Recall $\mathsf{var}[U] = \mathbb{E}\big[U^2\big] - \mathbb{E}[U]^2$. This means we have

$$
\begin{aligned}
\mathbb{E}\Big[(Y - \hat{f}(\mathbf{x}))^2\Big] &= \mathbb{E}\big[Y^2\big] - 2\mathbb{E}\Big[Y\hat{f}\Big] + \mathbb{E}\Big[\hat{f}^2\Big] \\
&= \mathsf{var}[Y] + \mathbb{E}[Y]^2 - 2\mathbb{E}\Big[Y\hat{f}\Big] + \mathsf{var}\Big[\hat{f}\Big] + \mathbb{E}\Big[\hat{f}\Big]^2.
\end{aligned}
$$

Now $Y = f(X) + \xi$, $\mathbb{E}[\xi] = 0$, and ξ is independent of X so we have $\mathbb{E}[Y] = \mathbb{E}[f]$, $\mathbb{E}\left[Y\hat{f}\right] = \mathbb{E}\left[(f + \xi)\hat{f}\right] = \mathbb{E}\left[f\hat{f}\right]$, and $\text{var}[Y] = \text{var}[\xi] = \sigma_\xi^2$. This yields

$$
\begin{aligned}
\mathbb{E}\left[(Y - \hat{f}(\mathbf{x}))^2\right] &= \text{var}[Y] + \mathbb{E}[f]^2 - 2\mathbb{E}\left[f\hat{f}\right] + \text{var}\left[\hat{f}\right] + \mathbb{E}\left[\hat{f}\right]^2 \\
&= \text{var}[Y] + f^2 - 2f\mathbb{E}\left[\hat{f}\right] + \text{var}\left[\hat{f}\right] + \mathbb{E}\left[\hat{f}\right]^2 (f \text{ isn't random}) \\
&= \sigma_\xi^2 + (f - \mathbb{E}\left[\hat{f}\right])^2 + \text{var}\left[\hat{f}\right]
\end{aligned}
$$

The expected error on all future data is the sum of three terms.

- The irreducible error is σ_ξ^2; even the true model must produce this error, on average. There is nothing we can do about this error.
- The bias is $(f - \mathbb{E}\left[\hat{f}\right])^2$. This term reflects the fact that even the best choice of model ($\mathbb{E}\left[\hat{f}\right]$) may not be the same as the true source of data (f).
- The variance is $\text{var}\left[\hat{f}\right] = \mathbb{E}\left[(\hat{f} - \mathbb{E}\left[\hat{f}\right])^2\right]$. To interpret this term, notice the best model to choose would be $\mathbb{E}\left[\hat{f}\right]$ (remember, the expectation is over choices of training data; this model would be the one that best represented all possible attempts to train). Then the variance represents the fact that the model we chose ($\hat{f}$) is different from the best model ($\mathbb{E}\left[\hat{f}\right]$). The difference arises because our training data is a subset of all data, and our model is chosen to be good on the training data, rather than on every possible training set.

Irreducible error is easily dealt with; nothing we do will improve this error, so there is no need to do anything. But there is an important practical trade-off between bias and variance. Generally, when a model comes from a "small" or "simple" family, we expect that (a) we can estimate the best model in the family reasonably accurately (so the variance will be low) but (b) the model may have real difficulty reproducing the data (meaning the bias is large). Similarly, if the model comes from a "large" or "complex" family, the variance is likely to be high (because it will be hard to estimate the best model in the family accurately) but the bias will be low (because the model can more accurately reproduce the data). All modelling involves managing this trade-off between bias and variance. I am avoiding being precise about the complexity of a model because it can be tricky to do. One reasonable proxy is the number of parameters we have to estimate to determine the model.

You can see a crude version of this trade-off in the perch example of Sect. 10.4.1 and Fig. 10.11. Recall that, as I added monomials to the regression of weight against length, the fitting error went down; but the model that uses length^{10} as an explanatory variable makes very odd predictions away from the training data. When I use low degree monomials, the dominant source of error is bias; and when I use high degree monomials, the dominant source of error is variance. A common mistake is to feel that the major difficulty is bias, and so to use extremely complex models. Usually the result is poor estimates of model parameters, leading to huge errors from variance. Experienced modellers fear variance far more than they fear bias.

The bias–variance discussion suggests it isn't a good idea simply to use all the explanatory variables that you can obtain (or think of). Doing so might lead to a model with serious variance problems. Instead, we must choose a model that uses a subset of the explanatory variables that is small enough to control variance, and large enough that the bias isn't a problem. We need some strategy to choose explanatory variables. The simplest (but by no means the best; we'll see better in this chapter) approach is to search sets of explanatory variables for a good set. The main difficulty is knowing when you have a good set.

> **Remember This:** *There are three kinds of error. Nothing can be done about irreducible error. Bias is the result of a family of models none of which can fit the data exactly. Variance is the result of difficulty estimating which model in the family to use. Generally, there is a payoff between bias and variance—using simpler model families causes more bias and less variance, and so on.*

11.1.2 Choosing a Model Using Penalties: AIC and BIC

We would like to choose one of a set of models. We cannot do so using just the training error, because more complex models will tend to have lower training error, and so the model with the lowest training error will tend to be the most complex model. Training error is a poor guide to test error, because lower training error is evidence of lower bias on the models part; but with lower bias, we expect to see greater variance, and the training error doesn't take that into account.

One strategy is to penalize the model for complexity. We add some penalty, reflecting the complexity of the model, to the training error. We then expect to see the general behavior of Fig. 11.1. The training error goes down, and the penalty goes up as the model gets more complex, so we expect to see a point where the sum is at a minimum.

There are a variety of ways of constructing penalties. **AIC** (short for an information criterion) is a method due originally to H. Akaike, described in "A new look at the statistical model identification," IEEE Transactions on Automatic Control, 1974. Rather than using the training error, AIC uses the maximum value of the log-likelihood of the model. Write $\mathcal{L}$ for this value. Write k for the number of parameters estimated to fit the model. Then the AIC is

$$2k - 2\mathcal{L}$$

and a better model has a smaller value of AIC (remember this by remembering that a larger log-likelihood corresponds to a better model). Estimating AIC is

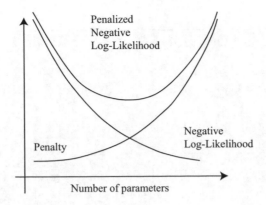

Figure 11.1: This is a standard abstract picture of a family of models. As we add explanatory variables (and so parameters) to produce a more complex model, the value of the negative log-likelihood of the best model can't go up, and usually goes down. This means that we cannot use the value as a guide to how many explanatory variables there should be. Instead, we add a penalty that increases as a function of the complexity of the model, and search for the model that minimizes the sum of negative log-likelihood and penalty. AIC and BIC penalize complexity with a penalty that is linear in the number of parameters, but there are other possible penalties. In this figure, I am following the usual convention of plotting the penalty as a curve rather than a straight line

straightforward for regression models if you assume that the noise is a zero mean normal random variable. You estimate the mean-squared error, which gives the variance of the noise, and so the log-likelihood of the model. You do have to keep track of two points. First, k is the total number of parameters estimated to fit the model. For example, in a linear regression model, where you model y as $\mathbf{x}^T\beta + \xi$, you need to estimate d parameters to estimate $\hat{\beta}$ *and* the variance of ξ (to get the log-likelihood). So in this case $k = d + 1$. Second, log-likelihood is usually only known up to a constant, so that different software implementations often use different constants. This is wildly confusing when you don't know about it (why would `AIC` and `extractAIC` produce different numbers on the same model?) but of no real significance—you're looking for the smallest value of the number, and the actual value doesn't mean anything. Just be careful to compare only numbers computed with the same routine.

An alternative is **BIC** (Bayes' information criterion), given by

$$2k \log N - 2\mathcal{L}$$

(where N is the size of the training dataset). You will often see this written as $2\mathcal{L} - 2k \log N$; I have given the form above so that one always wants the smaller value as with AIC. There is a considerable literature comparing AIC and BIC. AIC has a mild reputation for overestimating the number of parameters required, but is often argued to have firmer theoretical foundations.

> **Worked Example 11.1**   *AIC and BIC*
>
> Write M_d for the model that predicts weight from length for the perch dataset as $\sum_{j=0}^{j=d} \beta_j \text{length}^j$. Choose an appropriate value of $d \in [1, 10]$ using AIC and BIC.
>
> **Solution:** I used the R functions AIC and BIC, and got the table below.
>
> | | 1 | 2 | 3 | 4 | 5 | 6 | 7 | 8 | 9 | 10 |
> |-----|-----|-----|-----|-----|-----|-----|-----|-----|-----|-----|
> | AIC | 677 | 617 | 617 | 613 | 615 | 617 | 617 | 612 | 613 | 614 |
> | BIC | 683 | 625 | 627 | 625 | 629 | 633 | 635 | 633 | 635 | 638 |
>
> The best model by AIC has (rather startlingly!) $d = 8$. One should not take small differences in AIC too seriously, so models with $d = 4$ and $d = 9$ are fairly plausible, too. BIC suggests $d = 2$.

> **Remember This:**   *AIC and BIC are methods for computing a penalty that increases as the complexity of the model increases. We choose a model that gets a low value of penalized negative log-likelihood.*

11.1.3 Choosing a Model Using Cross-Validation

AIC and BIC are estimates of error on future data. An alternative is to measure this error on held-out data, using a cross-validation strategy (as in Sect. 1.1.3). One splits the training data into F **folds**, where each data item lies in exactly one fold. The case $F = N$ is sometimes called "leave-one-out" cross-validation. One then sets aside one fold in turn, fitting the model to the remaining data, and evaluating the model error on the left-out fold. The model error is then averaged. This process gives us an estimate of the performance of a model on held-out data. Numerous variants are available, particularly when lots of computation and lots of data are available. For example, one might not average over all folds; one might use fewer or more folds; and so on.

Worked Example 11.2   *Cross-Validation*

Write M_d for the model that predicts weight from length for the perch dataset as $\sum_{j=0}^{j=d} \beta_j \text{length}^j$. Choose an appropriate value of $d \in [1, 10]$ using leave-one-out cross-validation.

Solution: I used the R functions CVlm, which takes a bit of getting used to. I found:

1	2	3	4	5	6	7	8	9	10
1.9e4	4.0e3	7.2e3	4.5e3	6.0e3	5.6e4	1.2e6	4.0e6	3.9e6	1.9e8

where the best model is $d = 2$.

11.1.4   Greedy Search with Stagewise Regression

Assume we have a set of explanatory variables and we wish to build a model, choosing some of those variables for our model. Our explanatory variables could be many distinct measurements, or they could be different non-linear functions of the same measurement, or a combination of both. We can evaluate models relative to one another fairly easily (AIC, BIC, or cross-validation, your choice). However, choosing which set of explanatory variables to use can be quite difficult, because there are so many sets. The problem is that you cannot predict easily what adding or removing an explanatory variable will do. Instead, when you add (or remove) an explanatory variable, the errors that the model makes change, and so the usefulness of all other variables changes too. This means that (at least in principle) you have to look at every subset of the explanatory variables. Imagine you start with a set of F possible explanatory variables (including the original measurement, and a constant). You don't know how many to use, so you might have to try every different group, of each size, and there are far too many groups to try. There are two useful alternatives.

In **forward stagewise regression**, you start with an empty working set of explanatory variables. You then iterate the following process. For each of the explanatory variables not in the working set, you construct a new model using the working set and that explanatory variable, and compute the model evaluation score. If the best of these models has a better score than the model based on the working set, you insert the appropriate variable into the working set and iterate. If no variable improves the working set, you decide you have the best model and stop. This is fairly obviously a greedy algorithm.

Backward stagewise regression is pretty similar, but you start with a working set containing all the variables, and remove variables one-by-one and greedily. As usual, greedy algorithms are very helpful but not capable of exact optimization. Each of these strategies can produce rather good models, but neither is guaranteed to produce the best model.

> **Remember This:** *Forward and backward stagewise regression are greedy searches for sets of independent variables that predict effectively. In forward stagewise regression, one adds variables to a regression; in backward, one removes variables from the regression. Success can be checked with AIC, BIC, or cross-validation. The search stops when adding (resp. removing) a variable makes the regression worse.*

11.1.5 What Variables Are Important?

Imagine you regress some measure of risk of death against blood pressure, whether someone smokes or not, and the length of their thumb. Because high blood pressure and smoking tend to increase risk of death, you would expect to see "large" coefficients for these explanatory variables. Since changes in the thumb length have no effect, you would expect to see "small" coefficients for these explanatory variables. You might think that this suggests a regression can be used to determine what effects are important in building a model. It can, but doing so correctly involves serious difficulties that I will not deal with in detail. Instead, I will sketch what can go wrong so that you're discouraged from doing this without learning quite a lot more.

One difficulty is the result of variable scale. If you measure thumb length in kilometers, the coefficient is likely small; if you measure thumb length in micrometers, the coefficient is likely large. But this change has nothing to do with how important the variable is for prediction. This means that interpreting the coefficient is tricky.

Another difficulty is the result of sampling variance. Imagine that we have an explanatory variable that has absolutely no relationship to the dependent variable. If we had an arbitrarily large amount of data, and could exactly identify the correct model, we'd find that, in the correct model, the coefficient of that variable was zero. But we don't have an arbitrarily large amount of data. Instead, we have a sample of data. Hopefully, our sample is random so that (with some work) our estimate of the coefficient is the value of a random variable whose expected value is zero, but whose variance isn't. This means we are very unlikely to see a zero, but should see a value which is a small number of standard deviations away from zero. Dealing with this requires a way to tell whether the difference between a coefficient and zero is meaningful, or is just the result of random effects. There is a theory of **statistical significance** for regression coefficients, but we have other things to do.

Yet another difficulty has to do with practical significance, and is rather harder. We could have explanatory variables that are genuinely linked to the independent variable, but might not matter very much. This is a common phenomenon, particularly in medical statistics. It requires considerable care to disentangle some of these issues. Here is an example. Bowel cancer is a nasty disease, which could kill you. Being screened for bowel cancer is at best embarrassing and unpleasant, and involves some startling risks. There is considerable doubt, from reasonable sources, about whether screening has value and if so, how much (as a start point, you could look at Ransohoff DF. "How Much Does Colonoscopy Reduce Colon Cancer Mortality?" which appears in Ann. Intern. Med. 2009). There is some evidence linking eating red or processed meat to incidence of bowel cancer. A good practical question is: should one abstain from eating red or processed meat based on increased bowel cancer risk?

Coming to an answer is tough; the coefficient in any regression is clearly not zero, but it's pretty small. Here are some numbers. The UK population in 2012 was 63.7 million (this is a summary figure from Google, using World Bank data; there's no reason to believe that it's significantly wrong). I obtained the following figures from the UK cancer research institute website, at http://www.cancerresearchuk.org/ health-professional/cancer-statistics/statistics-by-cancer-type/bowel-cancer. There were 41,900 new cases of bowel cancer in the UK in 2012. Of these cases, 43% occurred in people aged 75 or over. Fifty-seven percent of people diagnosed with bowel cancer survive for 10 years or more after diagnosis. Of diagnosed cases, an estimated 21% is linked to eating red or processed meat, and the best current estimate is that the risk of incidence is between 17 and 30% higher per 100 g of red meat eaten per day (i.e., if you eat 100 g of red meat per day, your risk increases by some number between 17 and 30%; 200 g a day gets you twice that number; and—rather roughly—so on). These numbers are enough to confirm that there is a non-zero coefficient linking the amount of red or processed meat in your diet with your risk of bowel cancer (though you'd have a tough time estimating the exact value of that coefficient from the information here). If you eat more red meat, your risk of dying of bowel cancer really will go up. But the numbers I gave above suggest that (a) it won't go up much and (b) you might well die rather late in life, where the chances of dying of something are quite strong. The coefficient linking eating red meat and bowel cancer is clearly pretty small, because the incidence of the disease is about 1 in 1500 per year. Does the effect of this link matter enough to (say) stop eating red or processed meat? you get to choose, and your choice has consequences.

Remember This: *There are serious pitfalls in trying to interpret the coefficients of a regression. A small coefficient might come from a choice of scale for the associated variable. A large coefficient might still be the result of random effects, and assessing whether it requires a model of statistical significance. Worse, a coefficient might be clearly non-zero, but have little practical significance. It's tempting to look at the coefficients and try and come to conclusions, but you should not do this without much more theory.*

11.2 Robust Regression

We have seen that outlying data points can result in a poor model. This is caused by the squared error cost function: squaring a large error yields an enormous number. One way to resolve this problem is to identify and remove outliers before fitting a model. This can be difficult, because it can be hard to specify precisely when a point is an outlier. Worse, in high dimensions most points will look somewhat like outliers, and we may end up removing almost all the data. The alternative solution I offer here is to come up with a cost function that is less susceptible to problems with outliers. The general term for a regression that can ignore some outliers is a **robust regression**.

11.2.1 M-Estimators and Iteratively Reweighted Least Squares

One way to reduce the effect of outliers on a least squares solution would be to weight each point in the cost function. We need some method to estimate an appropriate set of weights. This would use a large weight for errors at points that are "trustworthy," and a low weight for errors at "suspicious" points.

We can obtain such weights using an **M-estimator**, which estimates parameters by replacing the negative log-likelihood with a term that is better behaved. In our examples, the negative log-likelihood has always been squared error. Write β for the parameters of the model being fitted, and $r_i(\mathbf{x}_i, \beta)$ for the residual error of the model on the ith data point. For us, r_i will always be $y_i - \mathbf{x}_i^T \beta$. So rather than minimizing

$$\sum_i (r_i(\mathbf{x}_i, \beta))^2$$

as a function of β, we will minimize an expression of the form

$$\sum_i \rho(r_i(\mathbf{x}_i, \beta); \sigma),$$

for some appropriately chosen function ρ. Clearly, our negative log-likelihood is one such estimator (use $\rho(u; \sigma) = u^2$). The trick to M-estimators is to make $\rho(u; \sigma)$ look like u^2 for smaller values of u, but ensure that it grows more slowly than u^2 for larger values of u (Fig. 11.2).

The **Huber loss** is one important M-estimator. We use

$$\rho(u; \sigma) = \begin{cases} \frac{u^2}{2} & |u| < \sigma \\ \sigma|u| - \frac{\sigma^2}{2} \end{cases}$$

which is the same as u^2 for $-\sigma \le u \le \sigma$, and then switches to $|u|$ for larger (or smaller) σ. The Huber loss is convex (meaning that there will be a unique minimum for our models) and differentiable, but its derivative is not continuous. The choice of the parameter σ (which is known as **scale**) has an effect on the estimate. You should interpret this parameter as the distance that a point can lie from the fitted function while still being seen as an **inlier** (anything that isn't even partially an outlier) (Fig. 11.3).

Generally, M-estimators are discussed in terms of their **influence function**. This is

$$\frac{\partial \rho}{\partial u}.$$

Its importance becomes evidence when we consider algorithms to fit $\hat{\beta}$ using an

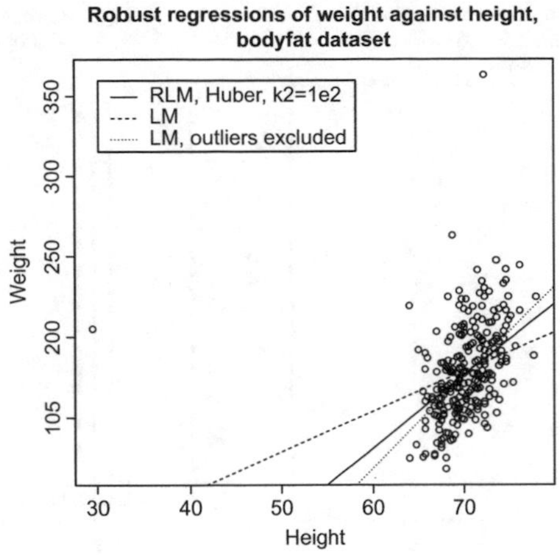

Figure 11.2: Comparing three different linear regression strategies on the bodyfat data, regressing weight against height. Notice that using an M-estimator gives an answer very like that obtained by rejecting outliers by hand. The answer may well be "better" because it isn't certain that each of the four points rejected is an outlier, and the robust method may benefit from some of the information in these points. I tried a range of scales for the Huber loss (the "k2" parameter), but found no difference in the line resulting over scales varying by a factor of 1e4, which is why I plot only one scale

M-estimator. Our minimization criterion is

$$
\nabla_\beta \left(\sum_i \rho(y_i - \mathbf{x}_i^T \beta; \sigma) \right) = \sum_i \left[\frac{\partial \rho}{\partial u} \right] (-\mathbf{x}_i)
$$
$$
= 0.
$$

Here the derivative $\frac{\partial \rho}{\partial u}$ is evaluated at $y_i - \mathbf{x}_i^T \beta$, so it is a function of β. Now write $w_i(\beta)$ for

$$
\frac{\frac{\partial \rho}{\partial u}}{y_i - \mathbf{x}_i^T \beta}
$$

(again, where the derivative is evaluated at $y_i - \mathbf{x}_i^T \beta$, and so w_i is a function of β). We can write the minimization criterion as

$$
\sum_i \left[w_i(\beta) \right] \left[y_i - \mathbf{x}_i^T \beta \right] \left[-\mathbf{x}_i \right] = 0.
$$

Now write $\mathcal{W}(\beta)$ for the diagonal matrix whose i'th diagonal entry is $w_i(\beta)$. Then our fitting criterion is equivalent to

$$
\mathcal{X}^T \left[\mathcal{W}(\beta) \right] \mathbf{y} = \mathcal{X}^T \left[\mathcal{W}(\beta) \right] \mathcal{X} \beta.
$$

The difficulty in solving this is that $w_i(\beta)$ depend on β, so we can't just solve a linear system in β. We could use the following strategy. Find some initial $\hat{\beta}^{(1)}$.

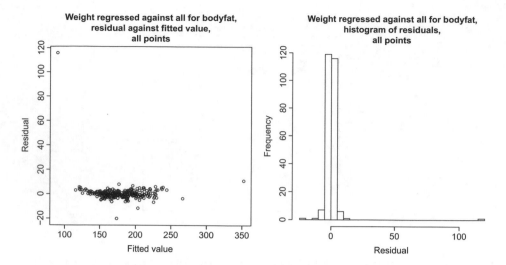

Figure 11.3: A robust linear regression of weight against all variables for the bodyfat dataset, using the Huber loss and all data points. On the **left**, residual plotted against fitted value (the residual is not standardized). Notice that there are some points with very large residual, but most have much smaller residual; this wouldn't happen with a squared error. On the **right**, a histogram of the residual. If one ignores the extreme residual values, this looks normal. The robust process has been able to discount the effect of the outliers, without us needing to identify and reject outliers by hand

Now evaluate $\mathcal{W}$ using that estimate, and re-estimate by solving the linear system. Iterate this until it settles down. This process uses $\mathcal{W}$ to downweight points that are suspiciously inconsistent with our current estimate of β, then update β using those weights. The strategy is known as **iteratively reweighted least squares**, and is very effective.

We assume we have an estimate of the correct parameters $\hat{\beta}^{(n)}$, and consider updating it to $\hat{\beta}^{(n+1)}$. We compute

$$w_i^{(n)} = w_i(\hat{\beta}^{(n)}) = \frac{\frac{\partial \rho}{\partial u}(y_i - \mathbf{x}_i^T \beta^{(n)}; \sigma)}{y_i - \mathbf{x}_i^T \hat{\beta}^{(n)}}.$$

We then estimate $\hat{\beta}^{(n+1)}$ by solving

$$\mathcal{X}^T \mathcal{W}^{(n)} \mathbf{y} = \mathcal{X}^T \mathcal{W}^{(n)} \mathcal{X} \hat{\beta}^{(n+1)}.$$

The key to this algorithm is finding good start points for the iteration. One strategy is randomized search. We select a small subset of points uniformly at random, and fit some $\hat{\beta}$ to these points, then use the result as a start point. If we do this often enough, one of the start points will be an estimate that is not contaminated by outliers.

Procedure: 11.1 *Fitting a Regression with Iteratively Reweighted Least Squares*

Write r_i for the residual at the i'th point, $y_i - \mathbf{x}_i^T \beta$. Choose an M-estimator ρ, likely the Huber loss; write $w_i(\beta)$ for

$$\frac{\frac{\partial \rho}{\partial u}}{y_i - \mathbf{x}_i^T \beta}.$$

We will minimize

$$\sum_i \rho(r_i(\mathbf{x}_i, \beta); \sigma)$$

by repeatedly

- finding some initial $\hat{\beta}^{(1)}$ by selecting a small subset of points uniformly at random and fitting a regression to those points;
- iterating the following procedure until the update is very small
    1. compute $\mathcal{W}^{(n)} = \operatorname{diag}(w_i(\hat{\beta}^{(n)}))$;
    2. solve
       $$\mathcal{X}^T \mathcal{W}^{(n)} \mathbf{y} = \mathcal{X}^T \mathcal{W}^{(n)} \mathcal{X} \hat{\beta}^{(n+1)}$$
       for $\hat{\beta}^{(n+1)}$;
- keep the resulting $\hat{\beta}$ if $\sum_i \rho(r_i(\mathbf{x}_i, \hat{\beta}); \sigma)$ is smaller than any seen so far.

11.2.2 Scale for M-Estimators

The estimators require a sensible estimate of σ, which is often referred to as **scale**. Typically, the scale estimate is supplied at each iteration of the solution method. One reasonable estimate is the **MAD** or **median absolute deviation**, given by

$$\sigma^{(n)} = 1.4826 \operatorname{median}_i |r_i^{(n)}(x_i; \hat{\beta}^{(n-1)})|.$$

Another popular estimate of scale is obtained with **Huber's proposal 2** (that is what everyone calls it!). Choose some constant $k_1 > 0$, and define $\Xi(u) = \min(|u|, k_1)^2$. Now solve the following equation for σ:

$$\sum_i \Xi\left(\frac{r_i^{(n)}(x_i; \hat{\beta}^{(n-1)})}{\sigma}\right) = N k_2$$

where k_2 is another constant, usually chosen so that the estimator gives the right answer for a normal distribution (exercises). This equation needs to be solved with an iterative method; the MAD estimate is the usual start point. R provides `hubers`, which will compute this estimate of scale (and figures out k_2 for itself). The choice of k_1 depends somewhat on how contaminated you expect your data to be. As $k_1 \to \infty$, this estimate becomes more like the standard deviation of the data.

11.3 Generalized Linear Models

We have used a linear regression to predict a value from a feature vector, but implicitly have assumed that this value is a real number. Other cases are important, and some of them can be dealt with using quite simple generalizations of linear regression. When we derived linear regression, I said one way to think about the model was

$$y = \mathbf{x}^T \beta + \xi$$

where ξ was a normal random variable with zero mean and variance σ_ξ^2. Another way to write this is to think of y as the value of a random variable Y. In this case, Y has mean $\mathbf{x}^T \beta$ and variance σ_ξ^2. This can be written as

$$Y \sim N(\mathbf{x}^T \beta, \sigma_\xi^2).$$

This offers a fruitful way to generalize: we replace the normal distribution with some other parametric distribution, and predict the parameter of that distribution using $\mathbf{x}^T \beta$. This is a **generalized linear model** or **GLM**. Three examples are particularly important.

11.3.1 Logistic Regression

Assume the y values can be either 0 or 1. You could think of this as a two-class classification problem, and deal with it using an SVM. There are sometimes advantages to seeing it as a regression problem. One is that we get to see a new classification method that explicitly models class posteriors, which an SVM doesn't do.

We build the model by asserting that the y values represent a draw from a Bernoulli random variable (definition below, for those who have forgotten). The parameter of this random variable is θ, the probability of getting a one. But $0 \le \theta \le 1$, so we can't just model θ as $\mathbf{x}^T \beta$. We will choose some **link function** g so that we can model $g(\theta)$ as $\mathbf{x}^T \beta$. This means that, in this case, g must map the interval between 0 and 1 to the whole line, and must be 1–1. The link function maps θ to $\mathbf{x}^T \beta$; the direction of the map is chosen by convention. We build our model by asserting that $g(\theta) = \mathbf{x}^T \beta$.

Remember This: *A generalized linear model predicts the parameter of a probability distribution from a regression. The link function ensures that the prediction of the regression meets the constraints required by the distribution.*

Useful Fact: 11.1 *Definition: Bernoulli Random Variable*

A Bernoulli random variable with parameter θ takes the value 1 with probability θ and 0 with probability $1 - \theta$. This is a model for a coin toss, among other things.

Notice that, for a Bernoulli random variable, we have that

$$\log\left[\frac{P(y=1|\theta)}{P(y=0|\theta)}\right] = \log\left[\frac{\theta}{1-\theta}\right]$$

and the **logit function** $g(u) = \log\left[\frac{u}{1-u}\right]$ meets our needs for a link function (it maps the interval between 0 and 1 to the whole line, and is 1–1). This means we can build our model by asserting that

$$\log\left[\frac{P(y=1|\mathbf{x})}{P(y=0|\mathbf{x})}\right] = \mathbf{x}^T\beta$$

then solving for the β that maximizes the log-likelihood of the data. Simple manipulation yields

$$P(y=1|\mathbf{x}) = \frac{e^{\mathbf{x}^T\beta}}{1+e^{\mathbf{x}^T\beta}} \text{ and } P(y=0|\mathbf{x}) = \frac{1}{1+e^{\mathbf{x}^T\beta}}.$$

In turn, this means the log-likelihood of a dataset will be

$$\mathcal{L}(\beta) = \sum_i \left[\mathbb{I}_{[y=1]}(y_i)\mathbf{x}_i^T\beta - \log\left(1+e^{\mathbf{x}_i^T\beta}\right)\right].$$

You can obtain β from this log-likelihood by gradient ascent (or rather a lot faster by Newton's method, if you know that).

A regression of this form is known as a **logistic regression**. It has the attractive property that it produces estimates of posterior probabilities. Another interesting property is that a logistic regression is a lot like an SVM. To see this, we replace the labels with new ones. Write $\hat{y}_i = 2y_i - 1$; this means that $\hat{y}_i$ takes the values -1 and 1, rather than 0 and 1. Now $\mathbb{I}_{[y=1]}(y_i) = \frac{\hat{y}_i+1}{2}$, so we can write

$$
\begin{aligned}
-\mathcal{L}(\beta) &= -\sum_i\left[\frac{\hat{y}_i+1}{2}\mathbf{x}_i^T\beta - \log\left(1+e^{\mathbf{x}_i^T\beta}\right)\right]\\
&= \sum_i\left[-\left(\frac{\hat{y}_i+1}{2}\mathbf{x}_i^T\beta\right) + \log\left(1+e^{\mathbf{x}_i^T\beta}\right)\right]\\
&= \sum_i\left[\log\left(\frac{1+e^{\mathbf{x}_i^T\beta}}{e^{\frac{\hat{y}_i+1}{2}\mathbf{x}_i^T\beta}}\right)\right]\\
&= \sum_i\left[\log\left(e^{\frac{-(\hat{y}_i+1)}{2}\mathbf{x}_i^T\beta} + e^{\frac{1-\hat{y}_i}{2}\mathbf{x}_i^T\beta}\right)\right]
\end{aligned}
$$

and we can interpret the term in square brackets as a loss function. If you plot it, you will notice that it behaves rather like the hinge loss. When $\hat{y}_i = 1$, if $\mathbf{x}^T\beta$ is positive, the loss is very small, but if $\mathbf{x}^T\beta$ is strongly negative, the loss grows linearly in $\mathbf{x}^T\beta$. There is similar behavior when $\hat{y}_i = -1$. The transition is smooth, unlike the hinge loss. Logistic regression should (and does) behave well for the same reasons the SVM behaves well.

Be aware that logistic regression has one annoying quirk. When the data are linearly separable (i.e., there exists some β such that $y_i \mathbf{x}_i^T \beta > 0$ for all data items), logistic regression will behave badly. To see the problem, choose the β that separates the data. Now it is easy to show that increasing the magnitude of β will increase the log-likelihood of the data; there isn't any limit. These situations arise fairly seldom in practical data.

> **Remember This:** *Logistic regression predicts the probability that a Bernoulli random variable is one using a logit link function. The result is a binary classifier whose loss is very similar to a hinge loss.*

11.3.2 Multiclass Logistic Regression

Imagine $y \in [0, 1, \ldots, C-1]$. Then it is natural to model $p(y|\mathbf{x})$ with a discrete probability distribution on these values. This can be specified by choosing $(\theta_0, \theta_1, \ldots, \theta_{C-1})$ where each term is between 0 and 1 and $\sum_i \theta_i = 1$. Our link function will need to map this constrained vector of θ values to a $\Re^{C-1}$. We can do this with a fairly straightforward variant of the logit function, too. Notice that there are $C-1$ probabilities we need to model (the C'th comes from the constraint $\sum_i \theta_i = 1$). We choose one vector β for each probability, and write β_i for the vector used to model θ_i. Then we can write

$$\mathbf{x}^T \beta_i = \log \left(\frac{\theta_i}{1 - \sum_u \theta_u} \right)$$

and this yields the model

$$P(y = 0|\mathbf{x}, \beta) = \frac{e^{\mathbf{x}^T \beta_0}}{1 + \sum_i e^{\mathbf{x}^T \beta_i}}$$

$$P(y = 1|\mathbf{x}, \beta) = \frac{e^{\mathbf{x}^T \beta_1}}{1 + \sum_i e^{\mathbf{x}^T \beta_i}}$$

$$\ldots$$

$$P(y = C - 1|\mathbf{x}, \beta) = \frac{1}{1 + \sum_i e^{\mathbf{x}^T \beta_i}}$$

and we would fit this model using maximum likelihood. The likelihood is easy to write out, and gradient descent is a good strategy for actually fitting models.

> **Remember This:** *Multiclass logistic regression predicts a multinomial distribution using a logit link function. The result is an important multiclass classifier.*

11.3.3 Regressing Count Data

Now imagine that the y_i values are counts. For example, y_i might have the count of the number of animals caught in a small square centered on $\mathbf{x}_i$ in a study region. As another example, $\mathbf{x}_i$ might be a set of features that represent a customer, and y_i might be the number of times that customer bought a particular product. The natural model for count data is a Poisson model, with parameter θ representing the intensity (reminder below).

Useful Fact: 11.2 *Definition: Poisson Distribution*

A non-negative, integer valued random variable X has a Poisson distribution when its probability distribution takes the form

$$P(\{X = k\}) = \frac{\theta^k e^{-\theta}}{k!},$$

where $\theta > 0$ is a parameter often known as the **intensity** of the distribution.

Now we need $\theta > 0$. A natural link function is to use

$$\mathbf{x}^T \beta = \log \theta$$

yielding a model

$$P(\{X = k\}) = \frac{e^{k\mathbf{x}^T \beta} e^{-e^{\mathbf{x}^T \beta}}}{k!}.$$

Now assume we have a dataset. The negative log-likelihood can be written as

$$
\begin{aligned}
-\mathcal{L}(\beta) &= -\sum_i \log \left(\frac{e^{y_i \mathbf{x}_i^T \beta} e^{-e^{\mathbf{x}_i^T \beta}}}{y_i!} \right) \\
&= -\sum_i \left(y_i \mathbf{x}_i^T \beta - e^{\mathbf{x}_i^T \beta} - \log(y_i!) \right).
\end{aligned}
$$

There isn't a closed form minimum available, but the log-likelihood is convex, and gradient descent (or Newton's method) is enough to find a minimum. Notice that the $\log(y_i!)$ term isn't relevant to the minimization, and is usually dropped.

Remember This: *You can predict count data with a GLM by predicting the parameter of a Poisson distribution with an exponential link function.*

11.3.4 Deviance

Cross-validating a model is done by repeatedly splitting a dataset into two pieces, training on one, evaluating some score on the other, and averaging the score. But we need to keep track of *what* to score. For earlier linear regression models (e.g., Sect. 11.1), we have used the squared error of predictions. This doesn't really make sense for a generalized linear model, because predictions are of quite different form. It is usual to use the **deviance** of the model. Write y_t for the true prediction at a point, $\mathbf{x}_p$ for the independent variables we want to obtain a prediction for, $\hat{\beta}$ for our estimated parameters; a generalized linear model yields $P(y|\mathbf{x}_p, \hat{\beta})$. For our purposes, you should think of the deviance as

$$-2\log P(y_t|\mathbf{x}_p, \hat{\beta})$$

(this expression is sometimes adjusted in software to deal with extreme cases, etc.). Notice that this is quite like the least squares error for the linear regression case, because there

$$-2\log P(y|\mathbf{x}_p, \hat{\beta}) = (\mathbf{x}_p^T\hat{\beta} - y_t)^2/\sigma^2 + K$$

for K some constant.

Remember This: *Evaluate a GLM with the model's deviance.*

11.4 L1 Regularization and Sparse Models

Forward and backward stagewise regression were strategies for adding independent variables to, or removing independent variables from, a model. An alternative, and very powerful, strategy is to construct a model with a method that forces some coefficients to be zero. The resulting model ignores the corresponding independent variables. Models built this way are often called **sparse models**, because (one hopes) that many independent variables will have zero coefficients, and so the model is using a sparse subset of the possible predictors.

In some situations, we are forced to use a sparse model. For example, imagine there are more independent variables than there are examples. In this case, the matrix $\mathcal{X}^T\mathcal{X}$ will be rank deficient. We could use a ridge regression (Sect. 10.4.2) and the rank deficiency problem will go away, but it would be hard to trust the resulting model, because it will likely use all the predictors (more detail below). We really want a model that uses a small subset of the predictors. Then, because the model ignores the other predictors, there will be more examples than there are predictors *that we use*.

There is now quite a strong belief among practitioners that using sparse models is the best way to deal with high dimensional problems (although there are lively debates about *which* sparse model to use, etc.). This is sometimes called the "bet on sparsity" principle: use a sparse model for high dimensional data, because dense models don't work well for such problems.

11.4.1 Dropping Variables with L1 Regularization

We have a large set of explanatory variables, and we would like to choose a small set that explains most of the variance in the independent variable. We could do this by encouraging β to have many zero entries. In Sect. 10.4.2, we saw we could regularize a regression by adding a term to the cost function that discouraged large values of β. Instead of solving for the value of β that minimized $\sum_i (y_i - \mathbf{x}_i^T \beta)^2 = (\mathbf{y} - \mathcal{X}\beta)^T (\mathbf{y} - \mathcal{X}\beta)$ (which I shall call the **error cost**), we minimized

$$\sum_i (y_i - \mathbf{x}_i^T \beta)^2 + \frac{\lambda}{2} \beta^T \beta = (\mathbf{y} - \mathcal{X}\beta)^T (\mathbf{y} - \mathcal{X}\beta) + \frac{\lambda}{2} \beta^T \beta$$

(which I shall call the **L2 regularized error**). Here $\lambda > 0$ was a constant chosen by cross-validation. Larger values of λ encourage entries of β to be small, but do not force them to be zero. The reason is worth understanding.

Write β_k for the kth component of β, and write β_{-k} for all the other components. Now we can write the L2 regularized error as a function of β_k:

$$(a + \lambda)\beta_k^2 - 2b(\beta_{-k})\beta_k + c(\beta_{-k})$$

where a is a function of the data and b and c are functions of the data and of β_{-k}. Now notice that the best value of β_k will be

$$\beta_k = \frac{b(\beta_{-k})}{(a + \lambda)}.$$

Notice that λ doesn't appear in the numerator. This means that, to force β_k to zero by increasing λ, we may have to make λ arbitrarily large. This is because the improvement in the penalty obtained by going from a small β_k to $\beta_k = 0$ is tiny—the penalty is proportional to β_k^2.

To force some components of β to zero, we need a penalty that grows linearly around zero rather than quadratically. This means we should use the **L$_1$ norm** of β, given by

$$\|\beta\|_1 = \sum_k |\beta_k|.$$

To choose β, we must now solve

$$(\mathbf{y} - \mathcal{X}\beta)^T (\mathbf{y} - \mathcal{X}\beta) + \lambda \|\beta\|_1$$

for an appropriate choice of λ. An equivalent problem is to solve a constrained minimization problem, where one minimizes

$$(\mathbf{y} - \mathcal{X}\beta)^T (\mathbf{y} - \mathcal{X}\beta) \text{ subject to } \|\beta\|_1 \le t$$

where t is some value chosen to get a good result, typically by cross-validation. There is a relationship between the choice of t and the choice of λ (with some thought, a smaller t will correspond to a bigger λ) but it isn't worth investigating in any detail.

Actually solving this system is quite involved, because the cost function is not differentiable. You should *not* attempt to use stochastic gradient descent, because this will not compel zeros to appear in $\hat{\beta}$. There are several methods, which are beyond our scope. As the value of λ increases, the number of zeros in $\hat{\beta}$ will increase too. We can choose λ in the same way we used for classification; split the training set into a training piece and a validation piece, train for different values of λ, and test the resulting regressions on the validation piece. The family of solutions $\hat{\beta}(\lambda)$ for all values of $\lambda \geq 0$ is known as the **regularization path**. One consequence of modern methods is that we can generate a very good approximation to the **regularization path** about as easily as we can get a solution for a single value of λ. As a result, cross-validation procedures for choosing λ are efficient.

Remember This: *An L_1 regularization penalty encourages models to have zero coefficients. The optimization problem that results is quite specialized. A strong approximation to the regularization path can be produced relatively easily, so cross-validation to choose λ is efficient.*

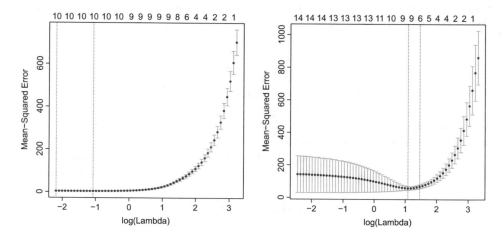

Figure 11.4: Plots of mean-squared error as a function of log regularization parameter (i.e., $\log \lambda$) for a regression of weight against all variables for the bodyfat dataset using an L1 regularizer (i.e., a lasso). These plots show mean-squared error averaged over cross-validation folds with a vertical one standard deviation bar. On the **left**, the plot for the dataset with the six outliers identified in Fig. 10.15 removed. On the **right**, the plot for the whole dataset. Notice how the outliers increase the variability of the error, and the best error. The top row of numbers gives the number of non-zero components in $\hat{\beta}$. Notice how as λ increases, this number falls (there are 15 explanatory variables, so the largest model would have 15 variables). The penalty ensures that explanatory variables with small coefficients are dropped as λ gets bigger

One way to understand the models that result is to look at the behavior of cross-validated error as λ changes. The error is a random variable, random because of the random split. It is a fair model of the error that would occur on a randomly chosen test example (assuming that the training set is "like" the test set, in a way that I do not wish to make precise yet). We could use multiple splits, and average over the splits. Doing so yields both an average error for each value of λ and an estimate of the standard deviation of error. Figure 11.4 shows the result of doing so for two datasets. Again, there is no λ that yields the smallest validation error, because the value of error depends on the random split cross-validation. A reasonable choice of λ lies between the one that yields the smallest error encountered (one vertical line in the plot) and the largest value whose mean error is within one standard deviation of the minimum (the other vertical line in the plot). It is informative to keep track of the number of zeros in $\hat{\beta}$ as a function of λ, and this is shown in Fig. 11.4.

Worked Example 11.3　*Building an L1 Regularized Regression*

Fit a linear regression to the bodyfat dataset, predicting weight as a function of all variables, and using the lasso to regularize. How good are the predictions? Do outliers affect the predictions?

Solution: I used the `glmnet` package, and I benefited a lot from example code by Trevor Hastie and Junyang Qian and published at https://web.stanford.edu/~hastie/glmnet/glmnet_alpha.html. I particularly like the R version; on my computer, the Matlab version occasionally dumps core, which is annoying. You can see from Fig. 11.4 that (a) for the case of outliers removed, the predictions are very good and (b) the outliers create problems. Note the magnitude of the error, and the low variance, for good cross-validated choices.

Another way to understand the models is to look at how $\hat{\beta}$ changes as λ changes. We expect that, as λ gets smaller, more and more coefficients become non-zero. Figure 11.5 shows plots of coefficient values as a function of $\log \lambda$ for a regression of weight against all variables for the bodyfat dataset, penalized using the L_1 norm. For different values of λ, one gets different solutions for $\hat{\beta}$. When λ is very large, the penalty dominates, and so the norm of $\hat{\beta}$ must be small. In turn, most components of $\hat{\beta}$ are zero. As λ gets smaller, the norm of $\hat{\beta}$ falls and some components of become non-zero. At first glance, the variable whose coefficient grows very large seems important. Look more carefully, this is the last component introduced into the model. But Fig. 11.4 implies that the right model has 7 components. This means that the right model has $\log \lambda \approx 1.3$, the vertical line shown in the detailed figure. In the best model, that coefficient is in fact zero.

The L_1 norm can sometimes produce an impressively small model from a large number of variables. In the UC Irvine Machine Learning repository, there is a dataset to do with the geographical origin of music (https://archive.ics.uci.edu/

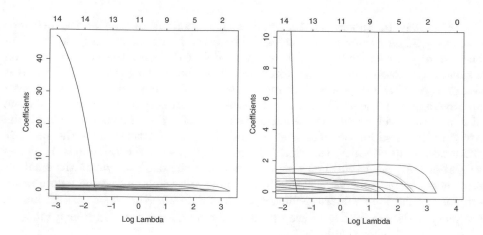

Figure 11.5: Plots of coefficient values as a function of $\log \lambda$ for a regression of weight against all variables for the bodyfat dataset, penalized using the L_1 norm. In each case, the six outliers identified in Fig. 10.15 were removed. On the **left**, the plot of the whole path for each coefficient (each curve is one coefficient). On the **right**, a detailed version of the plot. The vertical line shows the value of $\log \lambda$ that produces the model with smallest cross-validated error (look at Fig. 11.4). Notice that the variable that appears to be important, because it would have a large weight with $\lambda = 0$, does not appear in this model.

ml/datasets/Geographical+Original+of+Music). The dataset was prepared by Fang Zhou, and donors were Fang Zhou, Claire Q, and Ross D. King. Further details appear on that webpage, and in the paper: "Predicting the Geographical Origin of Music" by Fang Zhou, Claire Q, and Ross D. King, which appeared at ICDM in 2014. There are two versions of the dataset. One has 116 explanatory variables (which are various features representing music), and 2 independent variables (the latitude and longitude of the location where the music was collected). Figure 11.6 shows the results of a regression of latitude against the independent variables using L_1 regularization. Notice that the model that achieves the lowest cross-validated prediction error uses only 38 of the 116 variables.

Regularizing a regression with the L_1 norm is sometimes known as a **lasso**. A nuisance feature of the lasso is that, if several explanatory variables are correlated, it will tend to choose one for the model and omit the others (example in exercises). This can lead to models that have worse predictive error than models chosen using the L_2 penalty. One nice feature of good minimization algorithms for the lasso is that it is easy to use both an L_1 penalty and an L_2 penalty together. One can form

$$\left(\frac{1}{N}\right)\left(\sum_i (y_i - \mathbf{x}_i^T \beta)^2\right) \quad + \quad \lambda\left(\frac{(1-\alpha)}{2}\|\beta\|_2{}^2 + \alpha\|\beta\|_1\right)$$

$$\text{Error} \quad + \quad \text{Regularizer}$$

where one usually chooses $0 \le \alpha \le 1$ by hand. Doing so can both discourage large values in β and encourage zeros. Penalizing a regression with a mixed norm like this

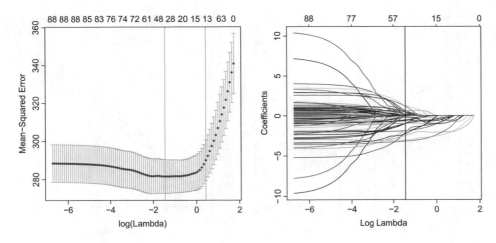

Figure 11.6: Mean-squared error as a function of log regularization parameter (i.e., $\log \lambda$) for a regression of latitude against features describing music (details in text), using the dataset at https://archive.ics.uci.edu/ml/datasets/ Geographical+Original+of+Music and penalized with the L_1 norm. The plot on the **left** shows mean-squared error averaged over cross-validation folds with a vertical one standard deviation bar. The top row of numbers gives the number of non-zero components in $\hat{\beta}$. Notice how as λ increases, this number falls. The penalty ensures that explanatory variables with small coefficients are dropped as λ gets bigger. On the **right**, a plot of the coefficient values as a function of $\log \lambda$ for the same regression. The vertical line shows the value of $\log \lambda$ that produces the model with smallest cross-validated error. Only 38 of 116 explanatory variables are used by this model

is sometimes known as **elastic net**. It can be shown that regressions penalized with elastic net tend to produce models with many zero coefficients, while not omitting correlated explanatory variables. All the computation can be done by the `glmnet` package in R (see exercises for details).

11.4.2 Wide Datasets

Now imagine we have more independent variables than examples (this is sometimes referred to as a "wide" dataset). This occurs quite often for a wide range of datasets; it's particularly common for biological datasets and natural language datasets. Unregularized linear regression must fail, because $\mathcal{X}^T \mathcal{X}$ must be rank deficient. Using an L2 (ridge) regularizer will produce an answer that should seem untrustworthy. The estimate of β is constrained by the data in some directions, but in other directions it is constrained only by the regularizer.

An estimate produced by L1 (lasso) regularization should look more reliable to you. Zeros in the estimate of β mean that the corresponding independent variables

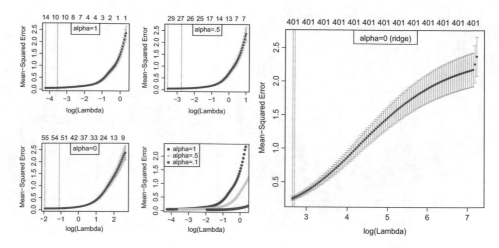

Figure 11.7: On the **left**, a comparison between three values of α in a `glmnet` regression predicting octane from NIR spectra (see Example 11.4). The plots show cross-validated error against log regularization coefficient for $\alpha = 1$ (lasso) and two elastic net cases, $\alpha = 0.5$ and $\alpha = 0.1$. I have plotted these curves separately, with error bars, and on top of each other but without error bars. The values on top of each separate plot show the number of independent variables with non-zero coefficients in the best model with that regularization parameter. On the **right**, a ridge regression for comparison. Notice that the error is considerably larger, even at the best value of the regularization parameter

are ignored. Now if there are many zeros in the estimate of β, the model is being fit with a small subset of the independent variables. If this subset is small enough, then the number of independent variables that are actually being used is smaller than the number of examples. If the model gives low enough error, it should seem trustworthy in this case. There are some hard questions to face here (e.g., does the model choose the "right" set of variables?) that we can't deal with.

> **Remember This:** *The lasso can produce impressively small models, and handles wide datasets very well.*

Worked Example 11.4   *L1 Regularized Regression for a "Wide" Dataset*

The gasoline dataset has 60 examples of near infrared spectra for gasoline of different octane ratings. The dataset is due to John H. Kalivas, and was originally described in the article "Two Data Sets of Near Infrared Spectra," in the journal *Chemometrics and Intelligent Laboratory Systems*, vol. 37, pp. 255–259, 1997. Each example has measurements at 401 wavelengths. I found this dataset in the R library pls. Fit a regression of octane against infrared spectrum using L1 regularized logistic regression.

Solution: I used the glmnet package, and I benefited a lot from example code by Trevor Hastie and Junyang Qian and published at https://web.stanford.edu/~hastie/glmnet/glmnet_alpha.html. The package will do ridge, lasso, and elastic net regressions. One adjusts a parameter in the function call, α, that balances the terms; $\alpha = 0$ is ridge and $\alpha = 1$ is lasso. Not surprisingly, the ridge isn't great. I tried $\alpha = 0.1$, $\alpha = 0.5$, and $\alpha = 1$. Results in Fig. 11.7 suggest fairly strongly that very good predictions should be available with the lasso using quite a small regularization constant; there's no reason to believe that the best ridge models are better than the best elastic net models, or vice versa. The models are very sparse (look at the number of variables with non-zero weights, plotted on the top).

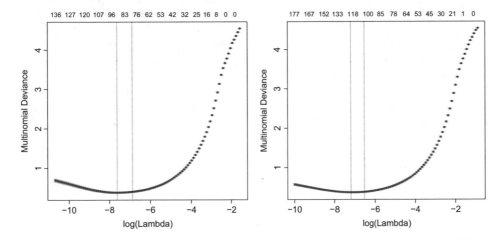

Figure 11.8: Multiclass logistic regression on the MNIST dataset, using a lasso and elastic net regularizers. On the **left**, deviance of held-out data on the digit dataset (Worked Example 11.5), for different values of the log regularization parameter in the lasso case. On the **right**, deviance of held-out data on the digit dataset (Worked Example 11.5), for different values of the log regularization parameter in the elastic net case, $\alpha = 0.5$

11.4.3 Using Sparsity Penalties with Other Models

A really nice feature of using an L1 penalty to enforce sparsity in a model is that it applies to a very wide range of models. For example, we can obtain a sparse SVM by replacing the L2 regularizer with an L1 regularizer. Most SVM packages will do this for you, although I'm not aware of any compelling evidence that this produces an improvement in most cases. All of the generalized linear models I described can be regularized with an L1 regularizer. For these cases, `glmnet` will do the computation required. The worked example shows using a multinomial (i.e., multiclass) logistic regression with an L1 regularizer.

Worked Example 11.5   *Multiclass Logistic Regression with an L1 Regularizer*

The MNIST dataset consists of a collection of handwritten digits, which must be classified into 10 classes $(0, \ldots, 9)$. There is a standard train/test split. This dataset is often called the zip code dataset because the digits come from zip codes, and has been quite widely studied. Yann LeCun keeps a record of the performance of different methods on this dataset at http://yann.lecun.com/exdb/mnist/. Obtain the Zip code dataset from http://statweb.stanford.edu/~tibs/ElemStatLearn/, and use a multiclass logistic regression with an L1 regularizer to classify it.

Solution: The dataset is rather large, and on my computer the fitting process takes a little time. Figure 11.8 shows what happens with the lasso, and with elastic net with $\alpha = 0.5$ on the training set, using `glmnet` to predict and cross-validation to select λ values. For the lasso, I found an error rate on the held-out data of 8.5%, which is OK, but not great compared to other methods. For elastic net, I found a slightly better error rate (8.2%); I believe even lower error rates are possible with these codes.

11.5 You Should

11.5.1 Remember These Terms

11.5.2 Remember These Facts

11.5.3   Remember These Procedures

Problems

Programming Exercises

11.1. *This is an extension of the previous exercise.* At https://archive.ics.uci.edu/ ml/machine-learning-databases/housing/housing.data, you will find the famous Boston Housing dataset. This consists of 506 data items. Each is 13 measurements, and a house price. The data was collected by Harrison, D. and Rubinfeld, D.L in the 1970s (a date which explains the very low house prices). The dataset has been widely used in regression exercises, but seems to be waning in popularity. At least one of the independent variables measures the fraction of population nearby that is "Black" (their word, not mine). This variable appears to have had a significant effect on house prices then (and, sadly, may still now). **Hint:** *you really shouldn't write your own code; I used* `rlm` *and* `boxcox` *in R for this.*

   (a) Use Huber's robust loss and iteratively reweighted least squares to regress the house price against all other variables. How well does this regression compare to the regression obtained by removing outliers and Box-Coxing, above?

   (b) As you should have noticed, the Box-Cox transformation can be quite strongly affected by outliers. Remove up to six outliers from this dataset using a diagnostic plot, then estimate the Box-Cox transformation. Now transform the dependent variable, and use Huber's robust loss and iteratively reweighted least squares to regress the transformed variable against all others *using all data* (i.e., put the outliers you removed to compute a Box-Cox transformation back into the regression). How does this regression compare to the regression in the previous subexercise, and to the regression obtained by removing outliers and Box-Coxing, above?

11.2. UC Irvine hosts a dataset of blog posts at https://archive.ics.uci.edu/ml/datasets/ BlogFeedback. There are 280 independent features which measure various properties of the blog post. The dependent variable is the number of comments that the blog post received in the 24 h after a base time. The zip file that you download will have training data in `blogData_train.csv`, and test data in a variety of files named `blogData_test-*.csv`.

   (a) Predict the dependent variable using all features, a generalized linear model (I'd use a Poisson model, because these are count variables), and the lasso. For this exercise, you really should use `glmnet` in R. Produce a plot of the cross-validated deviance of the model against the regularization variable (`cv.glmnet` and `plot` will do this for you). Use only the data in `blogData_train.csv`.

   (b) Your cross-validated plot of deviance likely doesn't mean all that much to you, because the deviance of a Poisson model takes a bit of getting used to. Choose a value of the regularization constant that yields a strong model, at least by the deviance criterion. Now produce a scatterplot of true values vs predicted values for data in `blogData_train.csv`. How well does this regression work? keep in mind that you are looking at predictions on the training set.

   (c) Choose a value of the regularization constant that yields a strong model, at least by the deviance criterion. Now produce a scatterplot of true values vs predicted values for data in `blogData_test-*.csv`. How well does this regression work?

   (d) Why is this regression difficult?

11.3. At http://genomics-pubs.princeton.edu/oncology/affydata/index.html, you will find a dataset giving the expression of 2000 genes in tumor and normal colon tissues. Build a logistic regression of the label (normal vs tumor) against the expression levels for those genes. There are a total of 62 tissue samples, so this is a wide regression. For this exercise, you really should use `glmnet` in R. Produce a plot of the classification error of the model against the regularization variable (`cv.glmnet`—look at the `type.measure` argument—and `plot` will do this for you). Compare the prediction of this model with the baseline of predicting the most common class.

11.4. The Jackson lab publishes numerous datasets to do with genetics and phenotypes of mice. At https://phenome.jax.org/projects/Crusio1, you can find a dataset giving the strain of a mouse, its gender, and various observations (click on the "Downloads" button). These observations are of body properties like mass, behavior, and various properties of the mouse's brain.

(a) We will predict the gender of a mouse from the body properties and the behavior. The variables you want are columns 4 through 41 of the dataset (or `bw` to `visit_time_d3_d5`; you shouldn't use the `id` of the mouse). Read the description; I've omitted the later behavioral measurements because there are many N/A's. Drop rows with N/A's (there are relatively few). How accurately can you predict gender using these measurements, using a logistic regression and the lasso? For this exercise, you really should use `glmnet` in R. Produce a plot of the classification error of the model against the regularization variable (`cv.glmnet`—look at the `type.measure` argument—and `plot` will do this for you). Compare the prediction of this model with the baseline of predicting the most common gender for all mice.

(b) We will predict the strain of a mouse from the body properties and the behavior. The variables you want are columns 4 through 41 of the dataset (or `bw` to `visit_time_d3_d5`; you shouldn't use the `id` of the mouse). Read the description; I've omitted the later behavioral measurements because there are many N/A's. Drop rows with N/A's (there are relatively few). This exercise is considerably more elaborate than the previous, because multinomial logistic regression does not like classes with few examples. You should drop strains with fewer than 10 rows. How accurately can you predict strain using these measurements, using multinomial logistic regression and the lasso? For this exercise, you really should use `glmnet` in R. Produce a plot of the classification error of the model against the regularization variable (`cv.glmnet`—look at the `type.measure` argument—and `plot` will do this for you). Compare the prediction of this model with the baseline of predicting a strain at random.

This data was described in a set of papers produced by this laboratory, and they like users to cite the papers. Papers are

- Delprato A, Bonheur B, Algéo MP, Rosay P, Lu L, Williams RW, Crusio WE. Systems genetic analysis of hippocampal neuroanatomy and spatial learning in mice. Genes Brain Behav. 2015 Nov;14(8):591–606.

- Delprato A, Algéo MP, Bonheur B, Bubier JA, Lu L, Williams RW, Chesler EJ, Crusio WE. QTL and systems genetics analysis of mouse grooming and behavioral responses to novelty in an open field. Genes Brain Behav. 2017 Nov;16(8):790–799.

- Delprato A, Bonheur B, Algéo MP, Murillo A, Dhawan E, Lu L, Williams RW, Crusio WE. A QTL on chromosome 1 modulates inter-male aggression in mice. Genes Brain Behav. 2018 Feb 19.

CHAPTER 12

Boosting

The following idea may have occurred to you after reading the chapter on regression. Imagine you have a regression that makes errors. You could try to produce a second regression that fixes those errors. You may have dismissed this idea, though, because if one uses only linear regressions trained using least squares, it's hard to see how to build a second regression that fixes the first regression's errors.

Many people have a similar intuition about classification. Imagine you have trained a classifier. You could try to train a second classifier to fix errors made by the first. There doesn't seem to be any reason to stop there, and you might try and train a third classifier to fix errors made by the first and the second, and so on. The details take some work, as you would expect. It isn't enough to just fix errors. You need some procedure to decide what the overall prediction of the system of classifiers is, and you need some way to be confident that the overall prediction will be better than the prediction produced by the initial classifier.

It is fruitful to think about correcting earlier predictions with a new set. I will start with a simple version that can be used to avoid linear algebra problems for least squares linear regression. Each regression will see different features from the previous regressions, so there is a prospect of improving the model. This approach extends easily to cover regressions that use something other than a linear function to predict values (I use a tree as an example).

Getting the best out of the idea requires some generalization. Regression builds a function that accepts features and produces predictions. So does classification. A regressor accepts features and produces numbers (or, sometimes, more complicated objects like vectors or trees, though we haven't talked about that much). A classifier accepts features and produces labels. We generalize, and call any function that accepts a feature and produces predictions a **predictor**. Predictors are trained using losses, and the main difference between a classifier and a regressor is the loss used to train the predictor.

We will build an optimal predictor as a sum of less ambitious predictors, often known as **weak learners**. We will build the optimal predictor incrementally using a greedy method, where we construct a new weak learner to improve over the sum of all previous weak learners, without adjusting the old ones. This process is called **boosting**. Setting this up will take some work, but it is worth doing, as we then have a framework makes it possible to boost a very wide range of classifiers and regressors. Boosting is particularly attractive when one has a weak learner that is simple and easy to train; one can often produce a predictor that is very accurate and can be evaluated very fast.

© Springer Nature Switzerland AG 2019
D. Forsyth, *Applied Machine Learning*,
https://doi.org/10.1007/978-3-030-18114-7_12

12.1 Greedy and Stagewise Methods for Regression

The place to start is linear regression. We will assume that we have so many features that we cannot solve the linear algebra problem resulting from least squares regression. Recall to build a linear regression of y against some high dimensional vector $\mathbf{x}$ (using the notation of Chap. 10), we will need to solve

$$\mathcal{X}^T \mathcal{X} \beta = \mathcal{X}^T \mathbf{y}$$

but this might be hard to do if $\mathcal{X}$ was really big. You're unlikely to see many problems where this really occurs, because modern software and hardware are very efficient in dealing with even enormous linear algebra problems. However, thinking about this case is very helpful. What we could do is choose some subset of the features to work with, to obtain a smaller problem, solve that, and go again.

12.1.1 Example: Greedy Stagewise Linear Regression

Write $\mathbf{x}^{(i)}$ for the i'th subset of features. For the moment, we will assume this is a small set of features and worry about how to choose the set later. Write $\mathcal{X}^{(i)}$ for the matrix constructed out of these features, etc. Now we regress $\mathbf{y}$ against $\mathcal{X}^{(1)}$. This chooses the $\hat{\beta}^{(1)}$ that minimizes the squared length of the residual vector

$$\mathbf{e}^{(1)} = \mathbf{y} - \mathcal{X}^{(1)} \hat{\beta}^{(1)}.$$

We obtain this $\hat{\beta}^{(1)}$ by solving

$$\left(\mathcal{X}^{(1)} \right)^T \mathcal{X}^{(1)} \hat{\beta}^{(1)} = \left(\mathcal{X}^{(1)} \right)^T \mathbf{y}.$$

We would now like to obtain an improved predictor by using more features in some way. We will build an improved predictor by adding some linear function of these new features to the original predictor. There are some important constraints. The improved predictor should correct errors made by the original predictor, but we do not want to change the original predictor. One reason not to is that we are building a second linear function to avoid solving a large linear algebra problem. Adjusting the original predictor at the same time will land us back where we started (with a large linear algebra problem).

To build the improved predictor, form $\mathcal{X}^{(2)}$ out of these features. The improved predictor will be

$$\mathcal{X}^{(1)} \hat{\beta}^{(1)} + \mathcal{X}^{(2)} \beta^{(2)}.$$

We do *not* want to change $\hat{\beta}^{(1)}$ and so we want to minimize

$$\left(\mathbf{y} - \left[\mathcal{X}^{(1)} \hat{\beta}^{(1)} + \mathcal{X}^{(2)} \beta^{(2)} \right] \right)^T \left(\mathbf{y} - \left[\mathcal{X}^{(1)} \hat{\beta}^{(1)} + \mathcal{X}^{(2)} \beta^{(2)} \right] \right)$$

as a function of $\beta^{(2)}$ alone. To simplify, write

$$
\begin{aligned}
\mathbf{e}^{(2)} &= \left(\mathbf{y} - \left[\mathcal{X}^{(1)} \beta^{(1)} + \mathcal{X}^{(2)} \beta^{(2)} \right] \right) \\
&= \mathbf{e}^{(1)} - \mathcal{X}^{(2)} \beta^{(2)}
\end{aligned}
$$

and we must choose $\beta^{(2)}$ to minimize

$$\left(\mathbf{e}^{(1)} - \mathcal{X}^{(2)}\beta^{(2)}\right)^T \left(\mathbf{e}^{(1)} - \mathcal{X}^{(2)}\beta^{(2)}\right).$$

This follows a familiar recipe. We obtain this $\hat{\beta}^{(1)}$ by solving

$$\left(\mathcal{X}^{(2)}\right)^T \mathcal{X}^{(2)}\hat{\beta}^{(2)} = \left(\mathcal{X}^{(2)}\right)^T \mathbf{e}^{(1)}.$$

Notice this is a linear regression of $\mathbf{e}^{(1)}$ against the features in $\mathcal{X}^{(2)}$. This is extremely convenient. The linear function that improves the original predictor is obtained using the same procedure (linear regression) as the original predictor. We just regress the residual (rather than $\mathbf{y}$) against the new features.

The new linear function is not guaranteed to make the regression better, but it will not make the regression worse. Because our choice of $\hat{\beta}^{(2)}$ minimizes the squared length of $\mathbf{e}^{(2)}$, we have that

$$\mathbf{e}^{(2)^T}\mathbf{e}^{(2)} \leq \mathbf{e}^{(1)^T}\mathbf{e}^{(1)}$$

with equality only if $\mathcal{X}^{(2)}\hat{\beta}^{(2)} = \mathbf{0}$. In turn, the second round did not make the residual worse. If the features in $\mathcal{X}^{(2)}$ aren't all the same as those in $\mathcal{X}^{(1)}$, it is very likely to have made the residual better.

Extending all this to an R'th round is just a matter of notation; you can write an iteration with $\mathbf{e}^{(0)} = \mathbf{y}$. Then you regress $\mathbf{e}^{(j-1)}$ against the features in $\mathcal{X}^{(j)}$ to get $\hat{\beta}^{(j)}$, and

$$\mathbf{e}^{(j)} = \mathbf{e}^{(j-1)} - \mathcal{X}^{(j)}\hat{\beta}^{(j)} = \mathbf{e}^{(0)} - \sum_{u=1}^{j} \mathcal{X}^{(u)}\hat{\beta}^{(u)}.$$

The residual never gets bigger (at least if your arithmetic is exact). This procedure is referred to as **greedy stagewise linear regression**. It's stagewise, because we build up the model in steps. It's greedy, because we do not adjust our estimate of $\hat{\beta}^{(1)}, \ldots, \hat{\beta}^{(j-1)}$ when we compute $\hat{\beta}^{(j)}$, etc.

This process won't work for a linear regression when we use all the features in $\mathcal{X}^{(1)}$. It's worth understanding why. Consider the first step. We will choose β to minimize $(\mathbf{y} - \mathcal{X}\beta)^T(\mathbf{y} - \mathcal{X}\beta)$. But there's a closed form solution for this β (which is $\hat{\beta} = (\mathcal{X}^T\mathcal{X})^{-1}\mathcal{X}^T\mathbf{y}$; remind yourself if you've forgotten by referring to Chap. 10), and this is a global minimizer. So to minimize

$$\left(\left[\mathbf{y} - \mathcal{X}\hat{\beta}\right] - \mathcal{X}\gamma\right)^T \left(\left[\mathbf{y} - \mathcal{X}\hat{\beta}\right] - \mathcal{X}\gamma\right)$$

by choice of γ, we'd have to have $\mathcal{X}\gamma = 0$, meaning that the residual wouldn't improve.

At this point, greedy stagewise linear regression may look like nothing more than a method of getting otherwise unruly linear algebra under control. But it's actually a model recipe, exposed in the box below. This recipe admits very substantial generalization.

Procedure: 12.1 *Greedy Stagewise Linear Regression*

We choose to minimize the squared length of the residual vector. Write

$$\mathcal{L}^{(j)}(\beta) = \left\| (\mathbf{e}^{(j-1)} - \mathcal{X}^{(j)}\beta) \right\|^2.$$

Start with $\mathbf{e}^{(0)} = \mathbf{y}$ and $j = 1$. Now iterate:

- choose a set of features to form $\mathcal{X}^{(j)}$;
- construct $\hat{\beta}^{(j)}$ by minimizing $\mathcal{L}^{(j)}(\beta)$; do so by solving the linear system

$$\left(\mathcal{X}^{(j)}\right)^T \mathcal{X}^{(j)} \hat{\beta}^{(j)} = \left(\mathcal{X}^{(j)}\right)^T \mathbf{e}^{(j-1)}.$$

- form $\mathbf{e}^{(j)} = \mathbf{e}^{(j-1)} - \mathcal{X}^{(j)}\hat{\beta}^{(j)}$;
- increment j to be $j + 1$.

The prediction for the training data is

$$\sum_j \mathcal{X}^{(j)}\hat{\beta}^{(j)}.$$

Write $\mathbf{x}$ for a test point for which you want a prediction, and $\mathbf{x}^{(j)}$ for the j'th set of features from that test point. The prediction for $\mathbf{x}$ is

$$\sum_j \mathbf{x}^{(j)}\hat{\beta}^{(j)}.$$

It is natural to choose the features at random, as more complicated strategies might be hard to execute. There isn't an obvious criterion for stopping, but looking at a plot of the test error with iterations will be informative.

Remember This: *A linear regression that has a very large number of features could result in a linear algebra problem too large to be conveniently solved. In this case, there is an important strategy. Choose a small subset of features and fit a model. Now choose a small random subset of features and use them to fit a regression that predicts the residual of the current model. Add the regression to the current model, and go again. This recipe can be aggressively generalized, and is extremely powerful.*

12.1.2 Regression Trees

I haven't seen the recipe in Box 12.1 used much for linear regressions, but as a model recipe it's extremely informative. It becomes much more interesting when applied to regressions that don't use linear functions. It is straightforward to coopt machinery we saw in the context of classification to solve regression problems, too. A **regression tree** is defined by analogy with a decision tree (Sect. 2.2). One builds a tree by splitting on coordinates, so each leaf represents a cell in space where the coordinates satisfy some inequalities. For the simplest regression tree, each leaf contains a single value representing the value the predictor takes in that cell (one can place other prediction methods in the leaves; we won't bother). The splitting process parallels the one we used for classification, but now we can use the error in the regression to choose the split instead of the information gain.

Worked Example 12.1   *Regressing Prawn Scores Against Location*

At http://www.statsci.org/data/oz/reef.html you will find a dataset describing prawns caught between the coast of northern Queensland and the Great Barrier Reef. There is a description of the dataset at that URL; the data was collected and analyzed by Poiner et al, cited at that URL. Build a regression tree predicting prawn score 1 (whatever that is!) against latitude and longitude using this dataset.

Solution: This dataset is nice, because it is easy to visualize interesting predictors. Figure 12.1 shows a 3D scatterplot of score 1 against latitude and longitude. There are good packages for building such trees (I used R's `rpart`). Figure 12.1 shows a regression tree fitted with that package, as an image. This makes it easy to visualize the function. The darkest points are the smallest values, and the lightest points are the largest. You can see what the tree does: carve space into boxes, then predict a constant inside each.

Remember This: *A regression tree is like a classification tree, but a regression tree stores values rather than labels in the leaves. The tree is fitted using a procedure quite like fitting a classification tree.*

12.1.3 Greedy Stagewise Regression with Trees

We wish to regress y against $\mathbf{x}$ using many regression trees. Write $f(\mathbf{x}; \theta^{(j)})$ for a regression tree that accepts $\mathbf{x}$ and produces a prediction (here $\theta^{(j)}$ are parameters

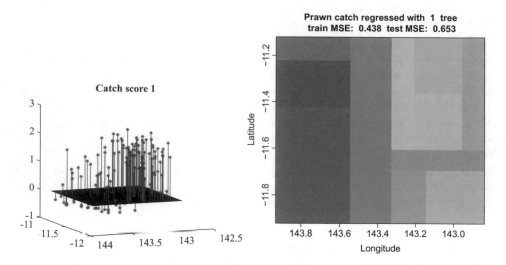

Figure 12.1: On the **left**, a 3D scatterplot of score 1 of the prawn trawls data from http://www.statsci.org/data/oz/reef.html, plotted as a function of latitude and longitude. On the **right**, a regression using a single regression tree, to help visualize the kind of predictor these trees produce. You can see what the tree does: carve space into boxes, then predict a constant inside each. The intensity scale is chosen so that the range is symmetric; because there are no small negative numbers, there are no very dark boxes. The odd axes (horizontal runs from bigger to smaller) are chosen so that you can register the left to the right by eye

internal to the tree: where to split; what is in the leaves; and so on). Write the regression as

$$F(\mathbf{x}) = \sum_j f(\mathbf{x}; \theta^{(j)})$$

where there might be quite a lot of trees indexed by j. Now we must fit this regression model to the data by choosing values for each $\theta^{(j)}$. We could fit the model by minimizing

$$\sum_i (y_i - F(\mathbf{x}_i))^2$$

as a function of all the $\theta^{(j)}$'s. This is unattractive, because it isn't clear how to solve this optimization problem.

The recipe for greedy stagewise linear regression applies here, with very little change. The big difference is that there is no need to choose a new subset of independent variables each time (the regression tree fitting procedure will do this). The recipe looks (roughly) like this: regress y against $\mathbf{x}$ using a regression tree; construct the residuals; and regress the residuals against $\mathbf{x}$; and repeat until some termination criterion.

In notation, start with an $F^{(0)} = 0$ (the initial model) and $j = 0$. Write $e_i^{(j)}$ for the residual at the j'th round and the i'th example. Set $e_i^{(0)} = y_i$. Now iterate the following steps:

- Choose $\hat{\theta}^{(j)}$ to minimize

$$\sum_i \left(e_i^{(j-1)} - f(\mathbf{x}_i; \theta) \right)^2$$

  as a function of θ using some procedure to fit regression trees.
- Set
$$e_i^{(j)} = e_i^{(j-1)} - f(\mathbf{x}_i; \hat{\theta}^{(j)}).$$

- Increment j to $j + 1$.

This is sometimes referred to as **greedy stagewise regression**. Notice that there is no particular reason to stop, unless (a) the residual is zero at all data points or (b) for some reason, it is clear that no future progress would be possible. For reference, I have put this procedure in a box below.

Worked Example 12.2   *Greedy Stagewise Regression for Prawns*

Construct a stagewise regression of score 1 against latitude and longitude, using the prawn trawls dataset from http://www.statsci.org/data/oz/reef.html. Use a regression tree.

Solution: There are good packages for building regression trees (I used R's `rpart`). Stagewise regression is straightforward. I started with a current prediction of zero. Then I iterated: form the current residual (score 1—current prediction); regress that against latitude and longitude; then update the current residual. Figures 12.2 and 12.3 show the result. For this example, I used a function of two dimensions so I could plot the regression function in a straightforward way. It's easy to visualize a regression tree in 2D. The root node of the tree splits the plane in half, usually with an axis aligned line. Then each node splits its parent into two pieces, so each leaf is a rectangular cell on the plane (which might stretch to infinity). The value is constant in each leaf. You can't make a smooth predictor out of such trees, but the regressions are quite good (Fig. 12.3).

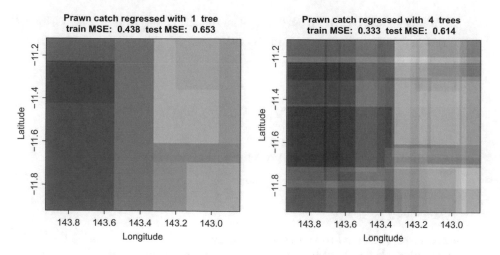

Figure 12.2: Score 1 of the prawn trawls data from http://www.statsci.org/data/oz/ reef.html, regressed against latitude and longitude (I did not use depth, also in that dataset; this means I could plot the regression easily). The axes (horizontal runs from bigger to smaller) are chosen so that you can register with the plot in Fig. 12.1 by eye. The intensity scale is chosen so that the range is symmetric; because there are no small negative numbers, there are no very dark boxes. The figure shows results using 1 and 4 trees. Notice the model gets more complex as we add trees. Further stages appear in Fig. 12.3, which uses the same intensity scale

---

Procedure: 12.2 *Greedy Stagewise Regression with Trees*

Write $f(\mathbf{x}; \theta)$ for a regression tree, where θ encodes internal parameters (where to split, thresholds, and so on). We will build a regression that is a sum of trees, so the regression is

$$F(\mathbf{x}; \theta) = \sum_j f(\mathbf{x}; \theta^{(j)}).$$

We choose to minimize the squared length of the residual vector, so write

$$\mathcal{L}^{(j)}(\theta) = \sum_i \left(e_i^{(j-1)} - f(\mathbf{x}_i; \theta) \right)^2.$$

Start with $e_i^{(0)} = y_i$ and $j = 0$. Now iterate:

- construct $\hat{\theta}^{(j)}$ by minimizing $\mathcal{L}^{(j)}(\theta)$ using regression tree software (which should give you an approximate minimizer);
- form $e_i^{(j)} = e_i^{(j-1)} - f(\mathbf{x}_i; \hat{\theta}^{(j)})$;
- increment j to $j + 1$.

The prediction for a data item **x** is

$$\sum_j f(\mathbf{x}; \hat{\theta}^{(j)})$$

There isn't an obvious criterion for stopping, but looking at a plot of the test error with iterations will be informative.

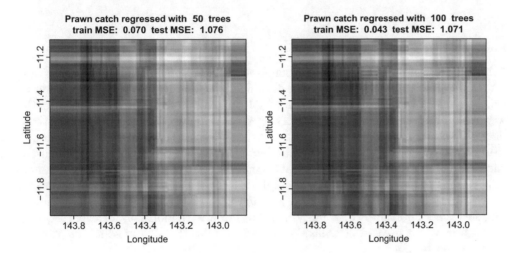

Figure 12.3: Score 1 of the prawn trawls data from http://www.statsci.org/data/oz/ reef.html, regressed against latitude and longitude (I did not use depth, also in that dataset; this means I could plot the regression easily). The axes (horizontal runs from bigger to smaller) are chosen so that you can register with the plot in Fig. 12.1 by eye. The intensity scale is chosen so that the range is symmetric; because there are no small negative numbers, there are no very dark boxes. The figure shows results of a greedy stagewise regression using regression trees using 50 and 100 trees. Notice that both train and test error go down, and the model gets more complex as we add trees

None of this would be helpful if the regression using trees $1 \ldots j$ is worse than the regression using trees $1 \ldots j{-}1$. Here is an argument that establishes that greedy stagewise regression should make progress in the training error. Assume that, if there is any tree that reduces the residual, the software will find one such tree; if not, it will return a tree that is a single leaf containing 0. Then $\left\| \mathbf{e}^{(j)} \right\|^2 \leq \left\| \mathbf{e}^{(j-1)} \right\|^2$, because the tree was chosen to minimize $\left\| \mathbf{e}^{(j)} \right\|^2 = \mathcal{L}^{(j)}(\theta) = \left\| (\mathbf{e}^{(j-1)} - f(\mathbf{x}; \theta)) \right\|$. If the tree is successful at minimizing this expression, the error will not go up.

In practice, greedy stagewise regression is well-behaved. One could reasonably fear overfitting. Perhaps only the training error goes down as you add trees, but the test error might go up. This can happen, but it tends not to happen (see the examples). We can't go into the reasons here (and they have some component of mystery, anyhow).

> **Remember This:** *It is difficult to fit a weighted sum of regression trees to data directly. Greedy stagewise methods offer a straightforward procedure. Fit a tree to the data to obtain an initial model. Now repeat: Fit a tree to predict the residual of the current model; add that tree to the current model. Stop by looking at validation error.*

12.2 Boosting a Classifier

The recipes I have given above are manifestations of a general approach. This approach applies to both regression and classification. The recipes seem more natural in the context of regression (which is why I did those versions first). But in both regression and classification we are trying to build a **predictor**—a function that accepts features and reports either a number (regression) or a label (classification). Notice we can encode the label as a number, meaning we could classify with regression machinery. In particular, we have some function $F(\mathbf{x})$. For both regression and classification, we apply F, for example, $\mathbf{x}$ to obtain a prediction. The regressor or classifier is learned by choosing a function that gets good behavior on a training set. This notation is at a fairly high level of abstraction (so, for example, the procedure we've used in classification where we take the sign of some function is represented by F).

12.2.1 The Loss

In early chapters, it seemed as though we used different kinds of predictor for classification and regression. But you might have noticed that the predictor used for linear support vector machines bore a strong similarity to the predictor used for linear regression, though we trained these two in quite different ways. There are many kinds of predictor—linear functions; trees; and so on. We now take the view that the kind of predictor you use is just a matter of convenience (what package you have available; what math you feel like doing; etc.). Once you know what kind of predictor you will use, you must choose the parameters of that predictor. In this new view, the really important difference between classification and regression is the **loss** that you use to choose these parameters. The loss is the cost function used to evaluate errors, and so to train the predictor. Training a classifier involves using a loss that penalizes errors in class prediction in some way, and training a regressor means using a loss that penalizes prediction errors.

The **empirical loss** is the average loss on the training set. Different predictors F produce different losses at different examples, so the loss depends on the predictor F. Notice the kind of predictor isn't what's important; instead, the loss scores the difference between what a predictor produced and what it should have produced. Now write $\mathcal{L}(F)$ for this empirical loss. There are many plausible losses that apply to different prediction problems. Here are some examples:

- For least squares regression, we minimized the least squares error:

$$\mathcal{L}_{ls}(F) = \frac{1}{N} \sum_i (y_i - F(\mathbf{x}_i))^2$$

(though the $1/N$ term sometimes was dropped as irrelevant; Sect. 10.2.2).
- For a linear SVM, we minimized the hinge loss:

$$\mathcal{L}_h(F) = \frac{1}{N} \sum_i \max(0, 1 - y_i F(\mathbf{x}_i))$$

(assuming that labels are 1 or −1; Sect. 2.1.1).
- For logistic regression, we minimized the logistic loss:

$$\mathcal{L}_{lr}(F) = \frac{1}{N} \sum_i \left[\log \left(e^{\frac{-(y_i+1)}{2} F(\mathbf{x}_i)} + e^{\frac{1-y_i}{2} F(\mathbf{x}_i)} \right) \right]$$

(again, assuming that labels are 1 or −1; Sect. 11.3.1).

We construct a loss by taking the average over the training data of a **pointwise loss**—a function ℓ that accepts three arguments: a y-value, a vector $\mathbf{x}$, and a prediction $F(\mathbf{x})$. This average is an estimate of the expected value of that pointwise loss over all data.

- For least squares regression,

$$\ell_{ls}(y, \mathbf{x}, F) = (y - F(\mathbf{x}))^2.$$

- For a linear SVM,

$$\ell_h(y, \mathbf{x}, F) = \max(0, 1 - yF(\mathbf{x})).$$

- For logistic regression,

$$\ell_{lr}(y, \mathbf{x}, F) = \left[\log \left(e^{\frac{-(y+1)}{2} F(\mathbf{x})} + e^{\frac{1-y}{2} F(\mathbf{x})} \right) \right].$$

We often used a regularizer with these losses. It is quite common in boosting to ignore this regularization term, for reasons I will explain below.

Remember This: *Models are predictors that accept a vector and predict some value. All our models are scored using an average of a pointwise loss function that compares the prediction at each data point with the training value at that point. The important difference between classification and regression is the pointwise loss function that is used.*

12.2.2 Recipe: Stagewise Reduction of Loss

We have used a predictor that was a sum of weak learners (equivalently, individual linear regressions; regression trees). Now generalize by noticing that scaling each weak learner could possibly produce a better result. So our predictor is

$$F(\mathbf{x}; \theta, \mathbf{a}) = \sum_j a_j f(\mathbf{x}; \theta^{(j)})$$

where a_j is the scaling for each weak learner.

Assume we have some F, and want to compute a weak learner that improves it. Whatever the particular choice of loss $\mathcal{L}$, we need to minimize

$$\frac{1}{N} \sum_i \ell(y_i, \mathbf{x}_i, F(\mathbf{x}_i) + a_j f(\mathbf{x}_i; \theta^{(j)})).$$

For most reasonable choices of loss, we can differentiate ℓ and we write

$$\left. \frac{\partial \ell}{\partial F} \right|_i$$

to mean the partial derivative of that function with respect to the F argument, evaluated at the point $(y_i, \mathbf{x}_i, F(\mathbf{x}_i))$. Then a Taylor series gives us

$$\frac{1}{N} \sum_i \ell(y_i, \mathbf{x}_i, F(\mathbf{x}_i) + a_j f(\mathbf{x}_i; \theta^{(j)})) \quad \approx \quad \frac{1}{N} \sum_i \ell(y_i, \mathbf{x}_i, F(\mathbf{x}_i))$$
$$+ a_j \frac{1}{N} \sum_i \left[\left(\left. \frac{\partial \ell}{\partial F} \right|_i \right) f(\mathbf{x}_i; \theta^{(j)}) \right].$$

In turn, this means that we can minimize by finding parameters $\hat{\theta}^{(j)}$ such that

$$\frac{1}{N} \sum_i \left(\left. \frac{\partial \ell}{\partial F} \right|_i \right) f(\mathbf{x}_i; \hat{\theta}^{(j)})$$

is negative. This predictor should cause the loss to go down, at least for small values of a_j. Now assume we have chosen an appropriate predictor, represented by $\hat{\theta}^{(j)}$ (the estimate of the predictor's parameters). Then we can obtain a_j by minimizing

$$\Phi(a_j) = \frac{1}{N} \sum_i \ell(y_i, \mathbf{x}_i, F(\mathbf{x}_i) + a_j f(\mathbf{x}_i; \hat{\theta}^{(j)}))$$

which is a one-dimensional problem (remember, F and $\hat{\theta}^{(j)}$ are known; only a_j is unknown).

This is quite a special optimization problem. It is one-dimensional (which simplifies many important aspects of optimization). Furthermore, from the Taylor series argument, we expect that the best choice of a_j is greater than zero. A problem with these properties is known as a **line search** problem, and there are strong and effective procedures in any reasonable optimization package for line search problems. You could use a line search method from an optimization package, or just minimize this function with an optimization package, which should recognize it as line search. The overall recipe, which is extremely general, is known as **gradient boost**; I have put it in a box, below.

Procedure: 12.3 *Gradient Boost*

We wish to choose a predictor F that minimizes a loss

$$\mathcal{L}(F) = \frac{1}{N} \sum_j \ell(y_j, \mathbf{x}_j, F).$$

We will do so iteratively by searching for a predictor of the form

$$F(\mathbf{x}; \theta) = \sum_j \alpha_j f(\mathbf{x}; \theta^{(u)}).$$

Start with $F = 0$ and $j = 0$. Now iterate:

- form a set of weights, one per example, where

$$w_i^{(j)} = \frac{\partial \ell}{\partial F}\bigg|_i$$

  (this means the partial derivative of $\ell(y, \mathbf{x}, F)$ with respect to the F argument, evaluated at the point $(y_i, \mathbf{x}_i, F(\mathbf{x}_i))$);
- choose $\hat{\theta}^{(j)}$ (and so the predictor f) so that

$$\sum_i w_i^{(j)} f(\mathbf{x}_i; \hat{\theta}^{(j)})$$

  is negative;
- now form $\Phi(a_j) = \mathcal{L}(F + a_j f(\cdot; \hat{\theta}^{(j)}))$ and search for the best value of a_j using a line search method.

The prediction for any data item $\mathbf{x}$ is

$$\sum_j \alpha_j f(\mathbf{x}; \hat{\theta}^{(j)}).$$

There isn't an obvious criterion for stopping, but looking at a plot of the test error with iterations will be informative.

The important problem here is finding parameters $\hat{\theta}^{(j)}$ such that

$$\sum_i w_i^{(j)} f(\mathbf{x}_i; \hat{\theta}^{(j)})$$

is negative. For some predictors, this can be done in a straightforward way. For others, this problem can be rearranged into a regression problem. We will do an example of each case.

Remember This: *Gradient boosting builds a sum of predictors using a greedy stagewise method. Fit a predictor to the data to obtain an initial model. Now repeat: Compute the appropriate weight at each data point; fit a predictor using these weights; search for the best weight with which to add this predictor to the current model; and add the weighted predictor to the current model. Stop by looking at validation error. The weight is a partial derivative of the loss with respect to the predictor, evaluated at the current value of the predictor.*

12.2.3 Example: Boosting Decision Stumps

The name "weak learner" comes from the considerable body of theory covering when and how boosting should work. An important fact from that theory is that the predictor $f(\cdot; \hat{\theta}^{(j)})$ needs only to be a descent direction for the loss—i.e., we need to ensure that adding some positive amount of $f(\cdot; \hat{\theta}^{(j)})$ to the prediction will result in an improvement in the loss. This is a very weak constraint in the two-class classification case (it boils down to requiring that the learner can do slightly better than a 50% error rate on a weighted version of the dataset), so that it is reasonable to use quite a simple classifier for the predictor.

One very natural classifier is a **decision stump**, which tests one linear projection of the features against a threshold. The name follows, rather grossly, because this is a highly reduced decision tree. There are two common strategies. In one, the stump tests a single feature against a threshold. In the other, the stump projects the features onto some vector chosen during learning, and tests that against a threshold.

Decision stumps are useful because they're easy to learn, though not in themselves a particularly strong classifier. We have examples $(\mathbf{x}_i, y_i)$. We will assume that y_i are 1 or -1. Write $f(\mathbf{x}; \theta)$ for the stump, which will predict -1 or 1. For gradient boost, we will receive a set of weights h_i (one per example), and try to learn a decision stump that *minimizes* the sum $\sum_i h_i f(\mathbf{x}_i; \theta)$ by choice of θ. We use a straightforward search, looking at each feature and for each, checking a set of thresholds to find the one that maximizes the sum. If we seek a stump that projects features, we project the features onto a set of random directions first. The box below gives the details.

Procedure: 12.4 *Learning a Decision Stump*

We have examples $(\mathbf{x}_i, y_i)$. We will assume that y_i are 1 or -1, and $\mathbf{x}_i$ have dimension d. Write $f(\mathbf{x}; \theta)$ for the stump, which will predict -1 or 1. We receive a set of weights h_i (one per example), and wish to learn a decision stump that *minimizes* the sum $\sum_i h_i f(\mathbf{x}_i; \theta)$. If the dataset is too large for your computational resources, obtain a subset by sampling uniformly at random without replacement. The parameters will be a projection, a threshold and a sign. Now for $j = 1 : d$

- Set $\mathbf{v}_j$ to be either a random d-dimensional vector *or* the j'th basis vector (i.e., all zeros, except a one in the j'th component).
- Compute $r_i = \mathbf{v}_j^T \mathbf{x}_i$.
- Sort these r's; now construct a collection of thresholds t from the sorted r's where each threshold is halfway between the sorted values.
- For each t, construct two predictors. One reports 1 if $r > t$, and -1 otherwise; the other reports -1 if $r > t$ and 1 otherwise. For each of these predictors, compute the value $\sum_i h_i f(\mathbf{x}_i; \theta)$. If this value is smaller than any seen before, keep $\mathbf{v}_j$, t, and the sign of the predictor.

Now report the $\mathbf{v}_j$, t, and sign that obtained the best value.

Remember This: *Decision stumps are very small decision trees. They are easy to fit, and have a particularly good record with gradient boost.*

12.2.4 Gradient Boost with Decision Stumps

We will work with two-class classification, as boosting multiclass classifiers can be tricky. One can apply gradient boost to any loss that appears convenient. However, there is a strong tradition of using the **exponential loss**. Write y_i for the true label for the i'th example. We will label examples with 1 or -1 (it is easy to derive updates for the case when the labels are 1 or 0 from what follows). Then the exponential loss is

$$\ell_e(y, \mathbf{x}, F(\mathbf{x})) = e^{[-yF(\mathbf{x})]}.$$

Notice if $F(\mathbf{x})$ has the right sign, the loss is small; if it has the wrong sign, the loss is large.

We will use a decision stump. Decision stumps report a label (i.e., 1 or -1). Notice this doesn't mean that F reports only 1 or -1, because F is a weighted sum of predictors. Assume we know F_{r-1}, and seek a_r and f_r. We then form

$$w_i^{(j)} = \left.\frac{\partial \ell}{\partial F}\right|_i = -y_i e^{[-y_i F(\mathbf{x}_i)]}.$$

Notice there is one weight per example. The weight is negative if the label is positive, and positive if the label is negative. If F gets the example right, the weight will have small magnitude, and if F gets the example wrong, the weight will have large magnitude. We want to choose parameters $\hat{\theta}^{(j)}$ so that

$$C(\theta^{(j)}) = \sum_i w_i^{(j)} f(\mathbf{x}_i; \hat{\theta}^{(j)}).$$

is negative. Assume that the search described in Box 12.4 is successful in producing such an $f(\mathbf{x}_i; \hat{\theta}^{(j)})$. To get a negative value, $f(\cdot; \hat{\theta}^{(j)})$ should try to report the same sign as the example's label (recall the weight is negative if the label is positive, and positive if the label is negative). This means that (mostly) if F gets a positive example right, $f(\cdot; \hat{\theta}^{(j)})$ will try to increase the value that F takes, etc.

The search produces an $f(\cdot; \hat{\theta}^{(j)})$ that has a large absolute value of $C(\theta^{(j)})$. Such an $f(\cdot; \hat{\theta}^{(j)})$ should have large absolute values, for example, where $w_i^{(j)}$ has large magnitude. But these are examples that F got very badly wrong (i.e., produced a prediction of large magnitude, but the wrong sign).

It is easy to choose a decision stump that minimizes this expression $C(\theta)$. The weights are fixed, and the stump reports either 1 or -1, so all we need to do is search for a split that achieves a minimum. You should notice that the minimum is always negative (unless all weights are zero, which can't happen). This is because you can multiply the stump's prediction by -1 and so flip the sign of the score.

12.2.5 Gradient Boost with Other Predictors

A decision stump makes it easy to construct a predictor such that

$$\sum_i w_{r-1,i} f_r(\mathbf{x}_i; \theta_r)$$

is negative. For other predictors, it may not be so easy. It turns out that this criterion can be modified, making it straightforward to use other predictors. There are two ways to think about these modifications, which end up in the same place: choosing $\hat{\theta}^{(j)}$ to minimize

$$\sum_i \left(\left[-w_i^{(j)} \right] - f(\mathbf{x}_i; \theta^{(j)}) \right)^2$$

is as good (or good enough) for gradient boost to succeed. This is an extremely convenient result, because many different regression procedures can minimize this loss. I will give both derivations, as different people find different lines of reasoning easier to accept.

Reasoning About Minimization: Notice that

$$\sum_i \left(\left[-w_i^{(j)} \right] - f(\mathbf{x}_i; \theta^{(j)}) \right)^2 = \sum_i \left[\begin{array}{c} \left(w_i^{(j)} \right)^2 \\ + (f(\mathbf{x}_i; \theta^{(j)}))^2 \\ + 2 (w_i^{(j)} f(\mathbf{x}_i; \theta^{(j)})) \end{array} \right].$$

Now assume that $\sum_i (f(\mathbf{x}_i; \theta^{(j)}))^2$ is not affected by $\theta^{(j)}$. For example, f could be a decision tree that reports either 1 or -1. In fact, it is usually sufficient that $\sum_i (f(\mathbf{x}_i; \theta^{(j)}))^2$ is not much affected by $\theta^{(j)}$. Then if you have a small value of

$$\sum_i \left(\left[-w_i^{(j)} \right] - f(\mathbf{x}_i; \theta^{(j)}) \right)^2,$$

that must be because $\sum_i w_{r-1,i} f(\mathbf{x}_i; \theta_r)$ is negative. So we seek $\theta^{(j)}$ that minimizes

$$\sum_i \left(\left[-w_i^{(j)} \right] - f(\mathbf{x}_i; \theta^{(j)}) \right)^2.$$

Reasoning About Descent Directions: You can think of $\mathcal{L}$ as a function that accepts a vector of prediction values, one at each data point. Write $\mathbf{v}$ for this vector. The values are produced by the current predictor. In this model, we have that

$$\nabla_{\mathbf{v}} \mathcal{L} \propto w_i^{(j)}.$$

In turn, this suggests we should minimize $\mathcal{L}$ by obtaining a new predictor f which takes values as close as possible to $-\nabla_{\mathbf{v}} \mathcal{L}$—that is, choose f_r that minimizes

$$\sum_i \left(\left[-w_i^{(j)} \right] - f(\mathbf{x}_i; \theta^{(j)}) \right)^2.$$

Remember This: *The original fitting criterion for predictors in gradient boost is awkward. An easy argument turns this into a familiar regression problem.*

12.2.6 Example: Is a Prescriber an Opiate Prescriber?

You can find a dataset of prescriber behavior focusing on opiate prescriptions at https://www.kaggle.com/apryor6/us-opiate-prescriptions. One column of this data is a 0-1 answer, giving whether the individual prescribed opiate drugs more than 10 times in the year. The question here is: does a doctor's pattern of prescribing predict whether that doctor will predict opiates?

You can argue this question either way. It is possible that doctors who see many patients who need opiates also see many patients who need other kinds of drug for similar underlying conditions. This would mean the pattern of drugs prescribed would suggest whether the doctor prescribed opiates. It is possible that there are prescribers engaging in deliberate fraud (e.g., prescribing drugs that aren't necessary, for extra money). Such prescribers would tend to prescribe drugs that have informal uses that people are willing to pay for, and opiates are one such drug, so the pattern of prescriptions would be predictive. The alternative possibility is that patients who need opiates attend doctors randomly, so that the pattern of drugs prescribed isn't predictive.

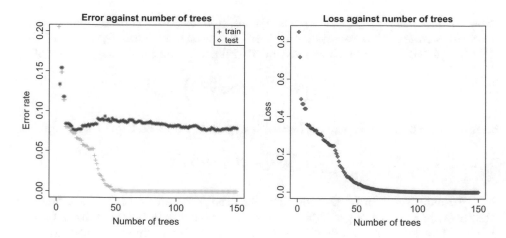

Figure 12.4: Models for a boosted decision tree classifier predicting whether a given prescriber will write more than 10 opioid prescriptions in a year, using the data of https://www.kaggle.com/apryor6/us-opiate-prescriptions. **Left:** train and test error against number of trees; **right:** exponential loss against number of trees. Notice that both test and train error go down, but there is a test–train gap. Notice also a characteristic property of boosting; continuing to boost after the training error is zero (about 50 trees in this case) still results in improvements in the test error. Note also that lower exponential loss doesn't guarantee lower training error

We will predict the `Opioid.Prescriber` column from the other entries, using a boosted decision tree and the exponential loss function. Confusingly, the column is named `Opioid.Prescriber` but all the pages, etc. use the term "opiate"; the internet suggests that "opiates" come from opium, and "opioids" are semi-synthetic or synthetic materials that bind to the same receptors. Quite a lot of money rides on soothing the anxieties of internet readers about these substances, so I'm inclined to assume that easily available information is unreliable; for us, they will mean the same thing.

This is a fairly complicated classification problem. It is natural to try gradient boost using a regression tree. To fit the regression tree, I used R's `rpart`; for line search, I used Newton's method. Doing so produces quite good classification (Fig. 12.4). This figure illustrates two very useful and quite typical feature of a boosted classifier.

- **The test error usually declines even after the training error is zero.** Look at Fig. 12.4, and notice the training error hits zero shortly after 50 trees. The loss is not zero there—exponential loss can never be zero—and continues to decline as more trees are added, even when the training error hits zero. Better, the test error continues to decline, though slowly. The exponential loss has the property that, even if the training error is zero, the predictor tries to have larger magnitude at each training point (i.e., boosting tries to make $F(\mathbf{x}_i)$ larger if y_i is positive, and smaller if y_i is negative). In turn, this means that, even after the training error is zero, changes to F might cause some test examples to change sign.

- **The test error doesn't increase sharply, however, far boosting proceeds.** You could reasonably be concerned that adding new weak learners to a boosted predictor would eventually cause overfitting problems. This can happen, but doesn't happen very often. It's also quite usual that the overfitting is mild. For this reason, it was believed until relatively recently that overfitting could never happen. Mostly, adding weak learners results in slow improvements to test error. This effect is most reliable when the weak learners are relatively simple, like decision stumps. The predictor we learn is regularized by the fact that a collection of decision stumps is less inclined to overfit than one might reasonably expect. In turn, this justifies the usual practice of not incorporating explicit regularizers in a boosting loss.

12.2.7 Pruning the Boosted Predictor with the Lasso

You should notice there is a relatively large number of predictors here, and it's reasonable to wonder if one could get good results with fewer. This is a good question. When you construct a set of boosted predictors, there is no guarantee they are all necessary to achieve a particular error rate. Each new predictor is constructed to cause the loss to go down. But the loss could go down without causing the error rate to go down. There is a reasonable prospect that some of the predictors are redundant.

Whether this matters depends somewhat on the application. It may be important to evaluate the minimum number of predictors. Furthermore, having many predictors might (but doesn't usually) create generalization problems. One strategy to remove redundant predictors is to use the lasso. For a two-class classifier, one uses a generalized linear model (logistic regression) applied to the values of the predictors at each example. Figure 12.5 shows the result of using a lasso (from `glmnet`) to the predictors used to make Fig. 12.4. Notice that reducing the size of the model seems not to result in significant loss of classification accuracy here.

There is one point to be careful about. You should not compute a cross-validated estimate of error on all data. That estimate of error will be biased low, because you are using some data on which the predictors were trained (you must have a training set to fit the boosted model and obtain the predictors in the first place). There are two options: you could fit a lasso on the training data, then evaluate on test; or you could use cross-validation to evaluate a fitted lasso on the test set alone. Neither strategy is perfect. If you fit a lasso to the training data, you may not make the best estimate of coefficients, because you are not taking into account variations caused by test–train splits (Fig. 12.5). But if you use cross-validation on the test set alone, you will be omitting quite a lot of data. This is a large dataset (25,000 prescribers) so I tried both approaches (compare Fig. 12.5 with Fig. 12.6). A better option might be to apply the Lasso *during* the boosting process, but this is beyond our scope.

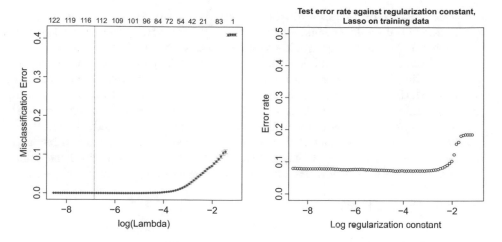

Figure 12.5: **Left:** A cross-validation plot from `cv.glmnet` for the lasso applied to predictions on training data obtained from all 150 trees from the boosted model of Fig. 12.4. You should not believe this cross-validated error, because the predictors were trained on this data. This means that data on both sides of the cross-validation split have been seen by the model (though not by the lasso). This plot suggests zero error is attainable by quite small models. But the cross-validated error here is biased low as the plot on the right confirms. **Right:** The error of the best model for each value of the regularization constant in the plot on the left, now evaluated on a held-out set. If you have enough data, you could break it into train (for training predictors and the lasso), validation (to select a model, using a plot like this), and test (to evaluate the resulting model)

> **Remember This:** *You can prune boosted models with the lasso. It's often very effective, but you need to be careful about how you choose a model—it's easy to accidentally evaluate on training data.*

12.2.8 Gradient Boosting Software

Up to this point, the examples shown have used simple loops I wrote with R. This is fine for small datasets, but gradient boost can be applied very successfully to extremely large datasets. For example, many recent Kaggle competitions have been won with gradient boosting methods. Quite a lot of the work in boosting methods admits parallelization across multiple threads or across multiple machines. Various clever speedups are also available. When the dataset is large, you need to have software that can exploit these tricks. As of writing, the standard is **XGBoost**, which can be obtained from https://xgboost.ai. This is the work of a large open source developer community, based around code by Tianqi Chen and a paper by Tianqi Chen and Carlos Guestrin. The paper is *XGBoost: A Scalable Tree Boosting*

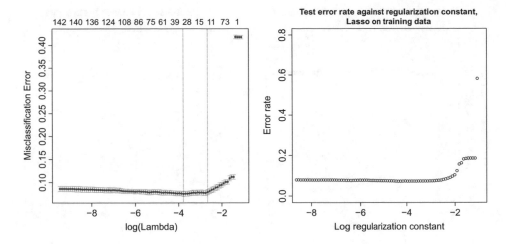

Figure 12.6: Here I have split the data into two. I used a training set to produce predictors using gradient boosting applied to decision stumps. I then applied `cv.glmnet` to a separate test set. **Left:** A cross-validation plot from `cv.glmnet` for the lasso applied to predictions on *test* data passed through each of the 150 trees made by the boosted model of Fig. 12.4. This gives an accurate estimate of the error, as you can see by comparing to the test error of the best model for each value of the regularization constant (**right**). This approach gives a better estimate of what the model will do, but may present problems if you have little data

System, which you can find in Proc. SIGKDD 2016, or at https://arxiv.org/abs/1603.02754.

XGBoost has a variety of features to notice (see the tutorials at https://xgboost.readthedocs.io/en/latest/tutorials/index.html). XGBoost doesn't do line search. Instead, one sets a parameter `eta`—the value of α in Procedure 12.3—which is fixed. Generally, larger `eta` values result in larger changes of the model with each new tree, but a greater chance of overfitting. There is an interaction between this parameter and the maximum depth of the tree you use. Generally, the larger the maximum depth of a tree (which can be selected), the more likely you will see overfitting, unless you set `eta` small.

XGBoost offers early stopping. If properly invoked, it can monitor error on an appropriate set of data (training or validation, your choice) and, if there is insufficient progress, it will stop training. Using this requires care. If you stop early using test data, the estimate of model performance that XGBoost returns must be biased. This is because it chooses a model (when to stop) using test data. You should follow the recipe of splitting data into three (train, validation, test), then train on training data, use validation for early stopping, and evaluate on test data.

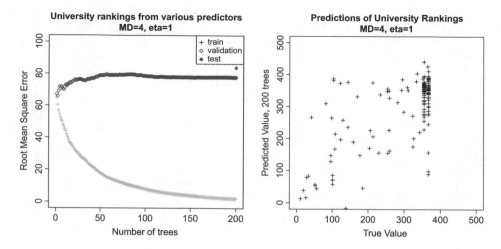

Figure 12.7: On the **left**, training and test error reported by XGBoost for the university ranking data as in example 12.3. The error is RMS error, because I modelled the rank as a continuous variable. The test error is slightly larger than validation error, likely a result of having a small test set. On the **right**, plots of predicted value against true value for the final regression. A regression is not a particularly good way of predicting rankings, because it does not know there needs to be a distinct value for each university and it does not know that there can be no ranking greater than the number of universities. Despite these disadvantages, the method can generally predict when a university will be highly ranked but tends to mix up the rankings of lower ranked (= larger number) universities

---

Remember This: *Very good, very fast, very scalable gradient boosting software is available.*

Worked Example 12.3   *Predicting the Quality of Education of a University*

You can find a dataset of measures of universities at https://www.kaggle.com/ mylesoneill/world-university-rankings/data. These measures are used to predict rankings. From these measures, but not using the rank or the name of the university, predict the quality of education using a stagewise regression. Use XGBoost.

Solution:

Ranking universities is a fertile source of light entertainment for assorted politicians, bureaucrats, and journalists. I have no idea what any of the numbers in this dataset mean (and I suspect I may not be the only one). Anyhow, one could get some sense of how reasonable they are by trying to predict the quality of education score from the others. This is a nice model problem for getting used to XGBoost. I modelled the rank as a continuous variable (which isn't really the best way to produce learned rankers—but we're just trying to see what a new tool does with a regression problem). This means that root-mean-square error is a natural loss. Figure 12.7 shows plots of a simple model. This is trained with trees whose maximum depth is 4. For Fig. 12.7, I used $\eta = 1$, which is quite aggressive. The scatterplot of predictions against true values for held-out data (in Fig. 12.7) suggests the model has a fairly good idea whether a university is strong or weak, but isn't that good at predicting the rank of universities where the rank is quite large (i.e., there are many stronger universities). For Fig. 12.8, I used a maximum depth of 8 and $\eta = 0.1$, which is much more conservative. I allowed the training procedure to stop early if it saw 100 trees without an improvement on a validation set. This model is distinctly better than the model of Fig. 12.7. The scatterplot of predictions against true values for held-out data (in Fig. 12.7) suggests the model has a fairly good idea whether a university is strong or weak, but isn't that good at predicting the rank of universities where the rank is quite large (i.e., there are many stronger universities).

Worked Example 12.4   *Opioid Prescribers with XGBoost*

Use XGBoost to obtain the best test accuracy you can on the dataset of Sect. 12.2.6. Investigate how accurate a model you can build by varying the parameters.

Solution: XGBoost is very fast (somewhere between 10 and 100 times faster than my homebrew gradient booster using rpart), so one can fiddle with hyperparameters to see what happens. Figure 12.9 shows models trained with depth 1 trees (so decision stumps, and comparable with the models of Fig. 12.4). The `eta` values were 1 and 0.5. You should notice distinct signs of overfitting—the validation error is drifting upwards quite early in training and continues to do so. There is a very large number of stumps (800) so that all effects are visible. Figure 12.10 shows a model trained with max depth 1 and `eta` of 0.2; again there are notable signs of overfitting. Training conservatively with a deeper tree (max depth 4, `eta` of 0.1, and early stopping) leads to a somewhat better behaved model. All these models are more accurate than those of Fig. 12.4.

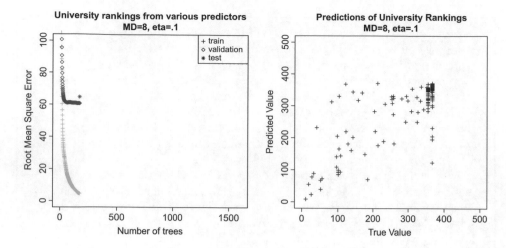

Figure 12.8: On the **left**, training and test error for reported by `xgboost` for the university ranking data as in example 12.3. This model uses a deeper tree (maximum depth of 8) and a smaller `eta` (0.1). It stops once adding 100 trees hasn't changed the validation error much. On the **right**, plots of predicted value against true value for the final regression. This model is notably more accurate than that of Fig. 12.7

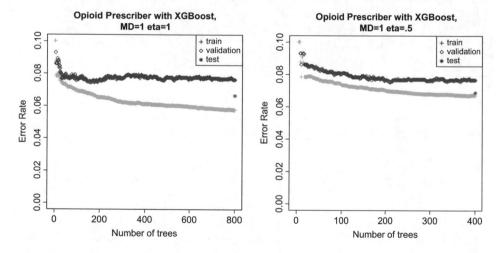

Figure 12.9: On the **left**, training, validation, and test error reported by `xgboost` for the opioid data as in example 12.2.6. Validation error is not strictly needed, as I did not apply early stopping. But the plot is informative. Notice how the validation error drifts up as the number of trees increases. Although this effect is slow and small, it's a sign of overfitting. Decreasing the `eta` (**right**) does not cure this trend (see also Fig. 12.10)

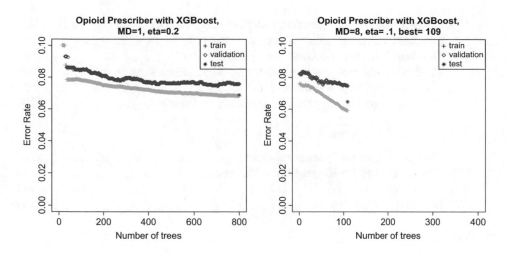

Figure 12.10: On the **left**, training, validation, and test error reported by `xgboost` for the opioid data as in example 12.2.6. This figure should be compared to Fig. 12.9. As in that figure, I used a fixed number of trees but now a rather small `eta`—this still doesn't cure the problem. On the **right**, I used deeper trees, and an even smaller `eta`, with early stopping. This produces the strongest model so far

12.3 You Should

12.3.1 Remember These Definitions

12.3.2 Remember These Terms

12.3.3 Remember These Facts

12.3.4 Remember These Procedures

12.3.5 Be Able to

- Set up and solve a regression problem using

Problems

12.1. Show that you cannot improve the training error of a linear regression using all features by a stagewise step.

(a) First, write $\hat{\beta}$ for the value that minimizes

$$(\mathbf{y} - \mathcal{X}\beta)^T(\mathbf{y} - \mathcal{X}\beta).$$

Now show that for $\overline{\beta} \neq \hat{\beta}$, we have

$$(\mathbf{y} - \mathcal{X}\overline{\beta})^T(\mathbf{y} - \mathcal{X}\overline{\beta}) \geq (\mathbf{y} - \mathcal{X}\hat{\beta})^T(\mathbf{y} - \mathcal{X}\hat{\beta}).$$

(b) Now explain why this means the residual can't be improved by regressing it against the features.

12.2. This exercise compares regression trees to linear regression. Mostly, one can improve the training error of a regression tree model by a stagewise step. Write $f(\mathbf{x}; \theta)$ for a regression tree, where θ encodes internal parameters (where to split, thresholds, and so on).

(a) Write $\hat{\theta}$ for the parameters of the regression tree that minimizes

$$\mathcal{L}(\theta) = \sum_i (y_i - f(\mathbf{x}_i; \theta))^2.$$

over all possible depths, splitting variables, and splitting thresholds. Why is $\mathcal{L}(\hat{\theta}) = 0$?

(b) How many trees achieve this value? Why would you not use that tree (those trees) in practice?

(c) A regression tree is usually regularized by limiting the maximum depth. Why (roughly) should this work?

12.3. We will fit a regression model to N one-dimensional points x_i. The value at the i'th point is y_i. We will use regression stumps. These are regression trees that have two leaves. A regression stump can be written as

$$f(x; t, v_1, v_2) = \begin{cases} v_1 & \text{for } x > t \\ v_2 & \text{otherwise} \end{cases}$$

(a) Assume that each data point is distinct (so you don't have $x_i = x_j$ for $i \neq j$). Show that you can build a regression with zero error with N stumps.

(b) Is it possible to build a regression with zero error with fewer than N stumps?

(c) Is there a procedure for fitting stumps that guarantees that gradient boosting results in a model that has zero error when you use exactly N stumps? **Warning:** *This might be quite hard.*

12.4. We will fit a classifier to N data points $\mathbf{x}_i$ with labels y_i (which are 1 or -1). The data points are distinct. We use the exponential loss, and use decision stumps identified using Procedure 12.4. Write

$$F_r(\mathbf{x}; \theta, \mathbf{a}) = \sum_{j=1}^{r} a_j f(\mathbf{x}; \theta^{(j)})$$

for the predictor that uses r decision stumps, $F_0 = 0$, and $L(F_r)$ for the exponential loss evaluated for that predictor on the dataset.

(a) Show that there is some α_1 so that $L(F_1) < L(F_0)$ when you use this procedure for fitting stumps.

(b) Show that there is some α_i so that $L(F_i) < L(F_{i-1})$ when you use this procedure for fitting stumps.

(c) All this means that the loss must continue to decline through an arbitrary number of rounds of boosting. Why does it not stop declining?

(d) If the loss declines at every round of boosting, does the training error do so as well? Why?

Programming Exercises

General Remark: *These exercises are suggested activities, and are rather open ended. Installing multi-threaded XGBoost—which you'll need—on a Mac can get quite exciting, but nothing that can't be solved with a bit of searching.*

12.5. Reproduce the example of Sect. 12.2.6, using a decision stump. You should write your own code for this stump and gradient boost. Prune the boosted predictor with the lasso. What test accuracy do you get?

12.6. Reproduce the example of Sect. 12.2.6, using XGBoost and adjusting hyperparameters (**eta**; the maximum depth of the tree; and so on) to get the best result. What test accuracy do you get?

12.7. Use XGBoost to classify MNIST digits, working directly with the pixels. This means you will have a 784-dimensional feature set. What test accuracy do you get? (mine was surprisingly high compared to the example of Sect. 17.2.1).

12.8. Investigate feature constructions for using XGBoost to classify MNIST digits. The subexercises suggest feature constructions. What test accuracy do you get?

(a) One natural construction is to project the images onto a set of principal components (50 is a good place to start, yielding a 50- dimensional feature vector).

(b) Another natural construction is to project the images each of the perclass principal components (50 is a good place to start, yielding a 500-dimensional feature vector).

(c) Yet another natural construction is to use vector quantization for windows on a grid in each image.

12.9. Use XGBoost to classify CIFAR-10 images, working directly with the pixels. This means you will have a 3072-dimensional feature set. What test accuracy do you get?

12.10. Investigate feature constructions for using XGBoost to classify CIFAR-10 images. The subexercises suggest feature constructions. What test accuracy do you get?

    (a) One natural construction is to project the images onto a set of principal components (50 is a good place to start, yielding a 50- dimensional feature vector).

    (b) Another natural construction is to project the images each of the per-class principal components (50 is a good place to start, yielding a 500-dimensional feature vector).

    (c) Yet another natural construction is to use vector quantization for windows on a grid in each image.

PART FIVE

Graphical Models

CHAPTER 13

Hidden Markov Models

There are many situations where one must work with sequences. Here is a simple, and classical, example. We see a sequence of words, but the last word is missing. I will use the sequence "I had a glass of red wine with my grilled xxxx." What is the best guess for the missing word? You could obtain one possible answer by counting word frequencies, then replacing the missing word with the most common word. This is "the," which is not a particularly good guess because it doesn't fit with the previous word. Instead, you could find the most common pair of words matching "grilled xxxx," and then choose the second word. If you do this experiment (I used Google Ngram viewer, and searched for "grilled *"), you will find mostly quite sensible suggestions (I got "meats," "meat," "fish," "chicken," in that order). If you want to produce random sequences of words, the next word should depend on some of the words you have already produced. A model with this property that is very easy to handle is a Markov chain (defined below).

It is really common to see a noisy sequence, and want to recover the noise free version. You should think of this recipe in a very broad way. So, for example, the recipe applies when one hears sound and would like to turn it into text. Here, the sound is the noisy sequence, and the text is the noise free version. You might see handwriting (noisy sequence) and want to recover text (noise free version). You might see video of a person moving (noisy sequence) and want to recover joint angles (noise free version). The standard model for this recipe is a hidden Markov model. We assume the noise free sequence is generated by a known Markov model, and the procedure that produced observed items from the sequence is known. In this case, a straightforward inference algorithm yields the noise free sequence from the observations. Furthermore, a hidden Markov model can be learned from examples using EM.

13.1 Markov Chains

A sequence of random variables X_n is a **Markov chain** if it has the property that,

$$P(X_n = j|\text{values of all previous states}) = P(X_n = j|X_{n-1}),$$

or, equivalently, only the last state matters in determining the probability of the current state. The probabilities $P(X_n = j|X_{n-1} = i)$ are the **transition probabilities**. We will always deal with discrete random variables here, and we will assume that there is a finite number of states. For all our Markov chains, we will assume that

$$P(X_n = j|X_{n-1} = i) = P(X_{n-1} = j|X_{n-2} = i).$$

© Springer Nature Switzerland AG 2019
D. Forsyth, *Applied Machine Learning*,
https://doi.org/10.1007/978-3-030-18114-7_13

Formally, we focus on *discrete time, time homogenous Markov chains in a finite state space*. With enough technical machinery one can construct many other kinds of Markov chain.

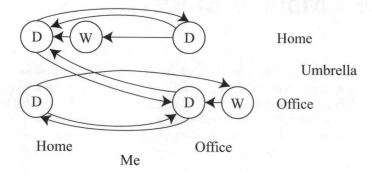

Figure 13.1:   A directed graph representing the umbrella example. Notice you can't arrive at the office wet with the umbrella at home (you'd have taken it), and so on. Labelling the edges with probabilities is left to the reader

One natural way to build Markov chains is to take a finite directed graph and label each directed edge from node i to node j with a probability. We interpret these probabilities as $P(X_n = j | X_{n-1} = i)$ (so the sum of probabilities over *outgoing* edges at any node must be 1). The Markov chain is then a **biased random walk** on this graph. A bug (or any other small object you prefer) sits on one of the graph's nodes. At each time step, the bug chooses one of the outgoing edges at random. The probability of choosing an edge is given by the probabilities on the drawing of the graph (equivalently, the transition probabilities). The bug then follows that edge. The bug keeps doing this until it hits an end state.

Worked Example 13.1   *Umbrellas*

I own one umbrella, and I walk from home to the office each morning, and back each evening. If it is raining (which occurs with probability p, and my umbrella is with me), I take it; if it is not raining, I leave the umbrella where it is. We exclude the possibility that it starts raining while I walk. Where I am, and whether I am wet or dry, forms a Markov chain. Draw a state machine for this Markov chain.

Solution: Figure 13.1 gives this chain. A more interesting question is with what probability I arrive at my destination wet? Again, we will solve this with simulation.

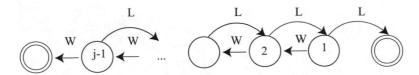

Figure 13.2:   A directed graph representing the gambler's ruin example. I have labelled each state with the amount of money the gambler has at that state. There are two end states, where the gambler has zero (is ruined), or has j and decides to leave the table. The problem we discuss is to compute the probability of being ruined, given the start state is s. This means that any state except the end states could be a start state. I have labelled the state transitions with "W" (for win) and "L" for lose, but have omitted the probabilities

---

Worked Example 13.2   *The Gambler's Ruin*

Assume you bet 1 a tossed coin will come up heads. If you win, you get 1 and your original stake back. If you lose, you lose your stake. But this coin has the property that $P(H) = p < 1/2$. You have s when you start. You will keep betting until either (a) you have 0 (you are ruined; you can't borrow money) or (b) the amount of money you have accumulated is j, where $j > s$. The coin tosses are independent. The amount of money you have is a Markov chain. Draw the underlying state machine. Write $P(\text{ruined, starting with } s|p) = p_s$. It is straightforward that $p_0 = 1$, $p_j = 0$. Show that

$$p_s = pp_{s+1} + (1-p)p_{s-1}.$$

Solution: Figure 13.2 illustrates this example. The recurrence relation follows because the coin tosses are independent. If you win the first bet, you have $s+1$ and if you lose, you have $s - 1$.

Notice an important difference between Examples 13.1 and 13.2. For the gambler's ruin, the sequence of random variables can end (and your intuition likely tells you it should do so reliably). We say the Markov chain has an **absorbing state**—a state that it can never leave. In the example of the umbrella, there is an infinite sequence of random variables, each depending on the last. Each state of this chain is **recurrent**—it will be seen repeatedly in this infinite sequence. One way to have a state that is not recurrent is to have a state with outgoing but no incoming edges.

The gambler's ruin example illustrates some points that are quite characteristic of Markov chains. You can often write recurrence relations for the probability of various events. Sometimes you can solve them in the closed form, though we will not pursue this thought further. It is often very helpful to think creatively about what the random variable is (Example 13.3).

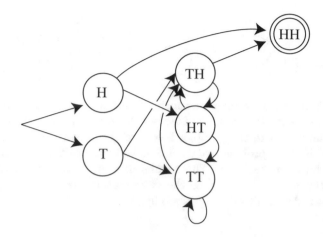

Figure 13.3: A directed graph representing the coin flip example, using the pairs of random variables described in Worked Example 13.3. A sequence "HTHTHH" (where the last two H's are the last two flips) would be generated by transitioning to H, then to HT, then to TH, then to HT, then to TH, then to HH. By convention, the end state is a double circle. Each edge has probability 1/2

Worked Example 13.3 *Multiple Coin Flips*

You choose to flip a fair coin until you see two heads in a row, and then stop. Represent the resulting sequence of coin flips with a Markov chain. What is the probability that you flip the coin four times?

Solution: You could think of the chain as being a sequence of independent coin flips. This is a Markov chain, but it isn't very interesting, and it doesn't get us anywhere. A better way to think about this problem is to have the X's be *pairs* of coin flips. The rule for changing state is that you flip a coin, then append the result to the state and drop the first item. Then you need a special state for stopping, and some machinery to get started. Figure 13.3 shows a drawing of the directed graph that represents the chain. The last three flips must have been THH (otherwise you'd go on too long, or end too early). But, because the second flip must be a T, the first could be either H or T. This means there are two sequences that work: $HTHH$ and $TTHH$. So $P(4 \text{ flips}) = 2/16 = 1/8$. We might want to answer significantly more interesting questions. For example, what is the probability that we must flip the coin more than 10 times? It is often possible to answer these questions by analysis, but we will use simulations.

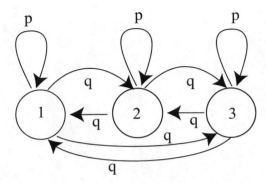

Figure 13.4: A virus can exist in one of 3 strains. At the end of each year, the virus mutates. With probability α, it chooses uniformly and at random from one of the 2 other strains, and turns into that; with probability $1 - \alpha$, it stays in the strain it is in. For this figure, we have transition probabilities $p = (1 - \alpha)$ and $q = (\alpha/2)$

Useful Facts: 13.1 *Markov Chains*

A Markov chain is a sequence of random variables X_n with the property that:

$$P(X_n = j | \text{values of all previous states}) = P(X_n = j | X_{n-1}).$$

13.1.1 Transition Probability Matrices

Define the matrix $\mathcal{P}$ with $p_{ij} = P(X_n = j | X_{n-1} = i)$. Notice that this matrix has the properties that $p_{ij} \geq 0$ and

$$\sum_j p_{ij} = 1$$

because at the end of each time step the model must be in some state. Equivalently, the sum of transition probabilities for outgoing arrows is one. Non-negative matrices with this property are **stochastic matrices**. By the way, you should look very carefully at the i's and j's here—Markov chains are usually written in terms of *row* vectors, and this choice makes sense in that context.

> **Worked Example 13.4**   *Viruses*
>
> Write out the transition probability matrix for the virus of Fig. 13.4, assuming that $\alpha = 0.2$.
>
> **Solution:** We have $P(X_n = 1 | X_{n-1} = 1) = (1 - \alpha) = 0.8$, and $P(X_n = 2 | X_{n-1} = 1) = \alpha/2 = P(X_n = 3 | X_{n-1} = 1)$; so we get
>
> $$\begin{pmatrix} 0.8 & 0.1 & 0.1 \\ 0.1 & 0.8 & 0.1 \\ 0.1 & 0.1 & 0.8 \end{pmatrix}$$

Now imagine we do not know the initial state of the chain, but instead have a probability distribution. This gives $P(X_0 = i)$ for each state i. It is usual to take these k probabilities and place them in a k-dimensional *row vector*, which is usually written π. From this information, we can compute the probability distribution over the states at time 1 by

$$
\begin{aligned}
P(X_1 = j) &= \sum_i P(X_1 = j, X_0 = i) \\
&= \sum_i P(X_1 = j | X_0 = i) P(X_0 = i) \\
&= \sum_i p_{ij} \pi_i.
\end{aligned}
$$

If we write $\mathbf{p}^{(n)}$ for the row vector representing the probability distribution of the state at step n, we can write this expression as

$$\mathbf{p}^{(1)} = \pi \mathcal{P}.$$

Now notice that

$$
\begin{aligned}
P(X_2 = j) &= \sum_i P(X_2 = j, X_1 = i) \\
&= \sum_i P(X_2 = j | X_1 = i) P(X_1 = i) \\
&= \sum_i p_{ij} \left(\sum_{ki} p_{ki} \pi_k \right).
\end{aligned}
$$

so that

$$\mathbf{p}^{(n)} = \pi \mathcal{P}^n.$$

This expression is useful for simulation, and also allows us to deduce a variety of interesting properties of Markov chains.

Useful Facts: 13.2   *Transition Probability Matrices*

A finite state Markov chain can be represented with a matrix $\mathcal{P}$ of transition probabilities, where the i, j'th element $p_{ij} = P(X_n = j | X_{n-1} = i)$. This matrix is a stochastic matrix. If the probability distribution of state X_{n-1} is represented by π_{n-1}, then the probability distribution of state X_n is given by $\pi_{n-1}^T \mathcal{P}$.

13.1.2 Stationary Distributions

Worked Example 13.5   *Viruses*

We know that the virus of Fig. 13.4 started in strain 1. After two-state transitions, what is the distribution of states when $\alpha = 0.2$? when $\alpha = 0.9$? What happens after 20 state transitions? If the virus starts in strain 2, what happens after 20 state transitions?

Solution: If the virus started in strain 1, then $\pi = [1, 0, 0]$. We must compute $\pi(\mathcal{P}(\alpha))^2$. This yields $[0.66, 0.17, 0.17]$ for the case $\alpha = 0.2$ and $[0.4150, 0.2925, 0.2925]$ for the case $\alpha = 0.9$. Notice that, because the virus with small α tends to stay in whatever state it is in, the distribution of states after 2 years is still quite peaked; when α is large, the distribution of states is quite uniform. After 20 transitions, we have $[0.3339, 0.3331, 0.3331]$ for the case $\alpha = 0.2$ and $[0.3333, 0.3333, 0.3333]$ for the case $\alpha = 0.9$; you will get similar numbers even if the virus starts in strain 2. After 20 transitions, the virus has largely "forgotten" what the initial state was.

In Example 13.5, the distribution of virus strains after a long interval appears not to depend much on the initial strain. This property is true of many Markov chains. Assume that our chain has a finite number of states. Assume that any state can be reached from any other state, by some sequence of transitions. Such chains are called **irreducible**. Notice this means there is no absorbing state, and the chain cannot get "stuck" in a state or a collection of states. Then there is a unique vector **s**, usually referred to as the **stationary distribution**, such that for *any* initial state distribution π,

$$\lim_{n \to \infty} \pi \mathcal{P}^{(n)} = \mathbf{s}.$$

Equivalently, if the chain has run through many steps, it no longer matters what the initial distribution is. The probability distribution over states will be **s**.

The stationary distribution can often be found using the following property. Assume the distribution over states is **s**, and the chain goes through one step. Then the new distribution over states must be **s** too. This means that

$$\mathbf{s}\mathcal{P} = \mathbf{s}$$

so that **s** is an eigenvector of $\mathcal{P}^T$, with eigenvalue 1. It turns out that, for an irreducible chain, there is exactly one such eigenvector.

The stationary distribution is a useful idea in applications. It allows us to answer quite natural questions, without conditioning on the initial state of the chain. For example, in the umbrella case, we might wish to know the probability I arrive home wet. This could depend on where the chain starts (Example 13.6). If you look at the figure, the Markov chain is irreducible, so there is a stationary distribution and (as long as I've been going back and forth long enough for the chain to "forget" where it started), the probability it is in a particular state doesn't depend on where it started. So the most sensible interpretation of this probability is the probability of a particular state in the stationary distribution.

Worked Example 13.6   *Umbrellas, but Without a Stationary Distribution*

This is a different version of the umbrella problem, but with a crucial difference. When I move to town, I decide randomly to buy an umbrella with probability 0.5. I then go from office to home and back. If I have bought an umbrella, I behave as in Example 13.1. If I have not, I just get wet. Illustrate this Markov chain with a state diagram.

Solution: Figure 13.5 does this. Notice this chain *isn't* irreducible. The state of the chain in the far future depends on where it started (i.e., did I buy an umbrella or not).

Useful Facts: 13.3   *Many Markov Chains Have Stationary Distributions*

If a Markov chain has a finite set of states, and if it is possible to get from any state to any other state, then the chain will have a stationary distribution. A sample state of the chain taken after it has been running for a long time will be a sample from that stationary distribution. Once the chain has run for long enough, it will visit states with a frequency corresponding to that stationary distribution, though it may take many state transitions to move from state to state.

13.1.3 Example: Markov Chain Models of Text

Imagine we wish to model English text. The very simplest model would be to estimate individual letter frequencies (most likely, by counting letters in a large body of example text). We might count spaces and punctuation marks as letters. We regard the frequencies as probabilities, then model a sequence by repeatedly drawing a letter from that probability model. You could even punctuate with this model by regarding punctuation signs as letters, too. We expect this model will produce sequences that are poor models of English text—there will be very long strings of "a"s, for example. This is clearly a (rather dull) Markov chain. It is sometimes referred to as a 0-th order chain or a 0-th order model, because each letter depends on the 0 letters behind it.

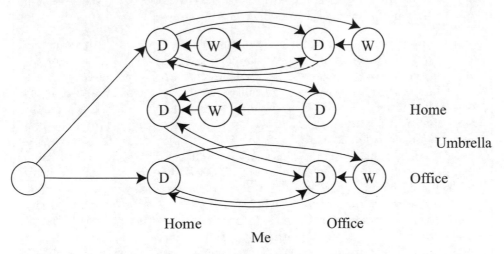

Figure 13.5: In this umbrella example, there can't be a stationary distribution; what happens depends on the initial, random choice of buying/not buying an umbrella

A slightly more sophisticated model would be to work with pairs of letters. Again, we would estimate the frequency of pairs by counting letter pairs in a body of text. We could then draw a first letter from the letter frequency table. Assume this is an "a." We would then draw the second letter by drawing a sample from the conditional probability of encountering each letter after "a," which we could compute from the table of pair frequencies. Assume this is an "n." We get the third letter by drawing a sample from the conditional probability of encountering each letter after "n," which we could compute from the table of pair frequencies, and so on. This is a first order chain (because each letter depends on the one letter behind it).

Second and higher order chains (or models) follow the general recipe, but the probability of a letter depends on more of the letters behind it. You may be concerned that conditioning a letter on the two (or k) previous letters means we don't have a Markov chain, because I said that the n'th state depends on only the $n-1$'th state. The cure for this concern is to use states that represent two (or k) letters, and adjust transition probabilities so that the states are consistent. So for a second order chain, the string "abcde" is a sequence of four states, "ab," "bc," "cd," and "de."

Worked Example 13.7   *Modelling Short Words*

Obtain a text resource, and use a trigram letter model to produce four letter words. What fraction of bigrams (resp. trigrams) do not occur in this resource? What fraction of the words you produce are actual words?

Solution: I used the text of a draft of this chapter. I ignored punctuation marks, and forced capital letters to lower case letters. I found 0.44 of the bigrams and 0.90 of the trigrams were not present. I built two models. In one, I just used counts to form the probability distributions (so there were many zero probabilities). In the other, I split a probability of 0.1 between all the cases that had not been observed. A list of 20 word samples from the first model is "ngen," "ingu," "erms," "isso," "also," "plef," "trit," "issi," "stio," "esti," "coll," "tsma," "arko," "llso," "bles," "uati," "namp," "call," "riat," "eplu"; two of these are real English words (three if you count "coll," which I don't; too obscure), so perhaps 10% of the samples are real words. A list of 20 word samples from the second model is "hate," "ther," "sout," "vect," "nces," "ffer," "msua," "ergu," "blef," "hest," "assu," "fhsp," "ults," "lend," "lsoc," "fysj," "uscr," "ithi," "prow," "lith"; four of these are real English words (you might need to look up "lith," but I refuse to count "hest" as being too archaic), so perhaps 20% of the samples are real words. In each case, the samples are too small to take the fraction estimates all that seriously.

Letter models can be good enough for (say) evaluating communication devices, but they're not great at producing words (Example 13.7). More effective language models are obtained by working with words. The recipe is as above, but now we use words in place of letters. It turns out that this recipe applies to such domains as protein sequencing, DNA sequencing, and music synthesis as well, but now we use amino acids (resp. base pairs; notes) in place of letters. Generally, one decides what the basic item is (letter, word, amino acid, base pair, note, etc.). Then individual items are called **unigrams** and 0'th order models are **unigram models**; pairs are **bigrams** and first order models are **bigram models**; triples are **trigrams**, second order models **trigram models**; and for any other n, groups of n in sequence are **n-grams** and $n-1$'th order models are **n-gram models**.

Worked Example 13.8   *Modelling Text with n-Grams of Words*

Build a text model that uses bigrams (resp. trigrams, resp. n-grams) of words, and look at the paragraphs that your model produces.

Solution: This is actually a fairly arduous assignment, because it is hard to get good bigram frequencies without working with enormous text resources. Google publishes n-gram models for English words with the year in which the n-gram occurred and information about how many different books it occurred in. So, for example, the word "circumvallate" appeared 335 times in 1978, in 91 distinct books—some books clearly felt the need to use this term more than once. This information can be found starting at http://storage.googleapis.com/books/ngrams/books/datasetsv2.html. The raw dataset is huge, as you would expect. There are numerous n-gram language models on the web. Jeff Attwood has a brief discussion of some models at https://blog.codinghorror.com/markov-and-you/; Sophie Chou has some examples, and pointers to code snippets and text resources, at http://blog.sophiechou.com/2013/how-to-model-markov-chains/. Fletcher Heisler, Michael Herman, and Jeremy Johnson are authors of RealPython, a training course in Python, and give a nice worked example of a Markov chain language generator at https://realpython.com/blog/python/lyricize-a-flask-app-to-create-lyrics-using-markov-chains/. Markov chain language models are effective tools for satire. Garkov is Josh Millard's tool for generating comics featuring a well-known cat (at http://joshmillard.com/garkov/). There's a nice Markov chain for reviewing wines by Tony Fischetti at http://www.onthelambda.com/2014/02/20/how-to-fake-a-sophisticated-knowledge-of-wine-with-markov-chains/.

It is usually straightforward to build a unigram model, because it is usually easy to get enough data to estimate the frequencies of the unigrams. There are many more bigrams than unigrams, many more trigrams than bigrams, and so on. This means that estimating frequencies can get tricky. In particular, you might need to collect an immense amount of data to see every possible n-gram several times. Without seeing every possible n-gram several times, you will need to deal with estimating the probability of encountering rare n-grams *that you haven't seen*. Assigning these n-grams a probability of zero is unwise, because that implies that they *never* occur, as opposed to occur seldom.

There are a variety of schemes for **smoothing** data (essentially, estimating the probability of rare items that have not been seen). The simplest one is to assign some very small fixed probability to every n-gram that has a zero count. It turns out that this is not a particularly good approach, because, for even quite small n, the fraction of n-grams that have zero count can be very large. In turn, you can find that most of the probability in your model is assigned to n-grams you have never seen. An improved version of this model assigns a fixed probability to unseen n-grams, then divides that probability up between all of the n-grams that have never been seen before. This approach has its own characteristic problems. It ignores evidence that some of the unseen n-grams are more common than others. Some of the unseen n-grams have $(n-1)$ leading terms that are $(n-1)$-grams that we *have* observed. These $(n-1)$-grams likely differ in frequency, suggesting that n-grams involving them should differ in frequency, too. More sophisticated schemes are beyond our scope, however.

13.2 Hidden Markov Models and Dynamic Programming

Imagine we wish to build a program that can transcribe speech sounds into text. Each small chunk of text can lead to one, or some, sounds, and some randomness is involved. For example, some people pronounce the word "fishing" rather like "fission." As another example, the word "scone" is sometimes pronounced rhyming with "stone," sometimes rhyming with "gone," and very occasionally rhyming with "toon" (really!). A Markov chain supplies a model of all possible text sequences, and allows us to compute the probability of any particular sequence. We will use a Markov chain to model text sequences, but what we observe is sound. We must have a model of how sound is produced by text. With that model and the Markov chain, we want to produce text that (a) is a likely sequence of words and (b) is likely to have produced the sounds we hear.

Many applications contain the main elements of this example. We might wish to transcribe music from sound. We might wish to understand American sign language from video. We might wish to produce a written description of how someone moves from video observations. We might wish to break a substitution cipher. In each case, what we want to recover is a sequence that can be modelled with a Markov chain, but we don't see the states of the chain. Instead, we see noisy measurements that *depend* on the state of the chain, and we want to recover a state sequence that is (a) likely under the Markov chain model and (b) likely to have produced the measurements we observe.

13.2.1 Hidden Markov Models

Assume we have a finite state, time homogenous Markov chain, with S states. This chain will start at time 1, and the probability distribution $P(X_1 = i)$ is given by the vector π. At time u, it will take the state X_u, and its transition probability matrix is $p_{ij} = P(X_{u+1} = j | X_u = i)$. We do not observe the state of the chain. Instead, we observe some Y_u. We will assume that Y_u is also discrete, and there are a total of O possible states for Y_u for any u. We can write a probability distribution for these observations $P(Y_u | X_u = i) = q_i(Y_u)$. This distribution is the **emission distribution** of the model. For simplicity, we will assume that the emission distribution does not change with time.

We can arrange the emission distribution into a matrix $\mathcal{Q}$. A **hidden Markov model** consists of the transition probability distribution for the states, the relationship between the state and the probability distribution on Y_u, and the initial distribution on states, that is, $(\mathcal{P}, \mathcal{Q}, \pi)$. These models are often dictated by an application. An alternative is to build a model that best fits a collection of observed data, but doing so requires technical machinery we cannot expound here.

I will sketch how one might build a model for transcribing speech, but you should keep in mind this is just a sketch of a very rich area. We can obtain the probability of a word following some set of words using n-gram resources, as in Sect. 13.1.3. We then build a model of each word in terms of small chunks of word that are likely to correspond to common small chunks of sound. We will call these chunks of sound **phonemes**. We can look up the different sets of phonemes that correspond to a word using a pronunciation dictionary. We can combine these two resources into a model of how likely it is one will pass from one phoneme inside a word to another, which might either be inside this word or inside another word. We now have $\mathcal{P}$. We will not spend much time on π, and might even model it as a uniform distribution. We can use a variety of strategies to build $\mathcal{Q}$. One is to build discrete features of a sound signal, then count how many times a particular set of features is produced when a particular phoneme is played.

13.2.2 Picturing Inference with a Trellis

Assume that we have a sequence of N measurements Y_i that we believe to be the output of a known hidden Markov model. We wish to recover the "best" corresponding sequence of X_i. Doing so is **inference**. We will choose the sequence that maximizes the posterior probability of $X_1, \ldots, X_N$, conditioned on the observations and the model, which is

$$P(X_1, X_2, \ldots, X_N | Y_1, Y_2, \ldots, Y_N, \mathcal{P}, \mathcal{Q}, \pi).$$

This is **maximum a posteriori** inference (or **MAP** inference).

It is equivalent to recover a sequence X_i that minimizes

$$- \log P(X_1, X_2, \ldots, X_N | Y_1, Y_2, \ldots, Y_N, \mathcal{P}, \mathcal{Q}, \pi).$$

This is more convenient, because (a) the log turns products into sums, which will be convenient and (b) minimizing the negative log probability gives us a formulation that is consistent with algorithms in Chap. 14. The negative log probability factors as

$$- \log \left(\frac{P(X_1, X_2, \ldots, X_N, Y_1, Y_2, \ldots, Y_N | \mathcal{P}, \mathcal{Q}, \pi)}{P(Y_1, Y_2, \ldots, Y_N)} \right)$$

and this is

$$- \log P(X_1, X_2, \ldots, X_N, Y_1, Y_2, \ldots, Y_N | \mathcal{P}, \mathcal{Q}, \pi) + \log P(Y_1, Y_2, \ldots, Y_N).$$

Notice that $P(Y_1, Y_2, \ldots, Y_N)$ doesn't depend on the sequence of X_u we choose, and so the second term can be ignored. What is important here is that we can decompose $- \log P(X_1, X_2, \ldots, X_N, Y_1, Y_2, \ldots, Y_N | \mathcal{P}, \mathcal{Q}, \pi)$ in a very useful way, because the X_u form a Markov chain. We want to minimize

$$- \log P(X_1, X_2, \ldots, X_N, Y_1, Y_2, \ldots, Y_N | \mathcal{P}, \mathcal{Q}, \pi)$$

but this is

$$- \left[\begin{array}{c} \log P(X_1) + \log P(Y_1 | X_1) + \\ \log P(X_2 | X_1) + \log P(Y_2 | X_2) + \\ \cdots \\ \log P(X_N | X_{n-1}) + \log P(Y_N | X_N). \end{array} \right]$$

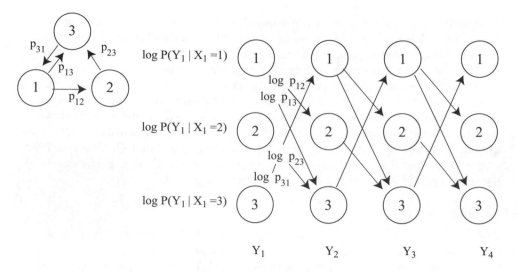

Figure 13.6: At the **top left**, a simple state transition model. Each outgoing edge has some probability, though the topology of the model forces two of these probabilities to be 1. Below, the trellis corresponding to that model. Each path through the trellis corresponds to a legal sequence of states, for a sequence of three measurements. We weight the arcs with the log of the transition probabilities, and the nodes with the log of the emission probabilities. I have shown some weights

Notice that this cost function has an important structure. It is a sum of terms. There are terms that depend on a single X_i (unary terms) and terms that depend on two (binary terms). Any state X_i appears in at most two binary terms.

We can illustrate this cost function in a structure called a **trellis**. This is a weighted, directed graph consisting of N copies of the state space, which we arrange in columns. There is a column corresponding to each measurement. We add a directed arrow from any state in the u'th column to any state in the $u+1$'th column if the transition probability between the states isn't 0. This represents the fact that there is a possible transition between these states. We then label the trellis with weights. We weight the node representing the case that state $X_u = j$ in the column corresponding to Y_u with $-\log P(Y_u | X_u = j)$. We weight the arc from the node representing $X_u = i$ to that representing $X_{u+1} = j$ with $-\log P(X_{u+1} = j | X_u = i)$.

The trellis has two crucial properties. Each directed path through the trellis from the start column to the end column represents a legal sequence of states. Now for some directed path from the start column to the end column, sum all the weights for the nodes and edges along this path. This sum is the negative log of the joint probability of that sequence of states with the measurements. You can verify each of these statements easily by reference to a simple example (try Fig. 13.6)

There is an efficient algorithm for finding the path through a trellis which maximizes the sum of terms. The algorithm is usually called **dynamic programming** or the **Viterbi algorithm**. I will describe this algorithm both in narrative and as a recursion. We could proceed by finding, for each node in the first column,

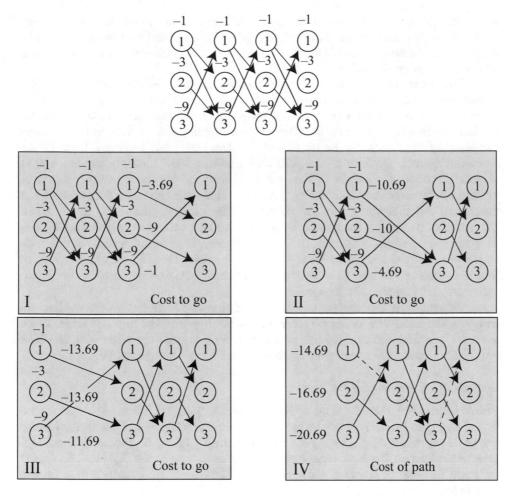

Figure 13.7: An example of finding the best path through a trellis. The probabilities of leaving a node are uniform (and remember, $\ln 2 \approx -0.69$). Details in the text

the best path from that node to any node in the last. There are S such paths, one for each node in the first column. Once we have these paths, we can choose the one with highest log joint probability. Now consider one of these paths. It passes through the i'th node in the u'th column. The path segment from this node to the end column must, itself, be the best path from this node to the end. If it wasn't, we could improve the original path by substituting the best. This is the key insight that gives us an algorithm.

Start at the *final* column of the tellis. We can evaluate the best path from each node in the final column to the final column, because that path is just the node, and the value of that path is the node weight. Now consider a two-state path, which will start at the second-last column of the trellis (look at panel I in Fig. 13.7). We can easily obtain the value of the best path leaving each node in this column. Consider a node: we know the weight of each arc leaving the node and the weight of the node at the far end of the arc, so we can choose the path segment with the largest value of the sum; this arc is the best we can do leaving that node. This sum is the best value obtainable on leaving that node—which is often known as the **cost to go function**.

Now, because we know the best value obtainable on leaving each node in the second-last column, we can figure out the best value obtainable on leaving each node in the third-last column (panel II in Fig. 13.7). At each node in the third-last column, we have a choice of arcs. Each of these reaches a node *from which we know the value of the best path*. So we can choose the best path leaving a node in the third-last column by finding the path that has the best value of: the arc weight leaving the node; the weight of the node in the second-last column the arc arrives at; and the value of the path leaving that node. This is much more easily done than described. All this works just as well for the third-last column, etc. (panel III in Fig. 13.7) so we have a recursion. To find the value of the best path with $X_1 = i$, we go to the corresponding node in the first column, then add the value of the node to the value of the best path leaving that node (panel IV in Fig. 13.7). Finally, to find the value of the best path leaving the first column, we compute the minimum value over all nodes in the first column.

We can also get the path with the minimum likelihood value. When we compute the value of a node, we erase all but the best arc leaving that node. Once we reach the first column, we simply follow the path from the node with the best value. This path is illustrated by dashed edges in Fig. 13.7 (panel IV).

13.2.3 Dynamic Programming for HMMs: Formalities

We will formalize the recursion of the previous section with two ideas. First, we define $C_w(j)$ to be the cost of the best path segment to the end of the trellis *leaving* the node representing $X_w = j$. Second, we define $B_w(j)$ to be the node in column $w + 1$ that lies on the best path *leaving* the node representing $X_w = j$. So $C_w(j)$ tells you the cost of the best path, and $B_w(j)$ tells you what node is next on the best path.

Now it is straightforward to find the cost of the best path leaving each node in the second-last column, and also the path. In symbols, we have

$$C_{N-1}(j) = \min_u \left[-\log P(X_N = u | X_{N-1} = j) - \log P(Y_N | X_N = u) \right]$$

and

$$B_{N-1}(j) = \operatorname*{argmin}_u \left[-\log P(X_N = u | X_{N-1} = j) - \log P(Y_N | X_N = u) \right].$$

You should check this against step I of Fig. 13.7

Once we have the best path leaving each node in the $w + 1$'th column and its cost, it's straightforward to find the best path leaving the w'th column and its cost. In symbols, we have

$$C_w(j) = \min_u \left[-\log P(X_{w+1} = u | X_w = j) - \log P(Y_{w+1} | X_{w+1} = u) - C_{w+1}(u) \right]$$

and

$$B_w(j) = \underset{u}{\mathrm{argmin}} \; [-\log P(X_{w+1} = u | X_w = j) - \log P(Y_{w+1} | X_{w+1} = u) - C_{w+1}(u)].$$

Check this against steps II and III in Fig. 13.7.

Now finding the best path is easy. We run the recursion until we have $C_1(j)$ for each j. This gives the cost of the best path leaving the j'th node in column 1. We choose the node with the best cost, say $\hat{j}$. The next node on the best path is $B_1(\hat{j})$; and the path is $B_1(\hat{j}), B_2(B_1(\hat{j})), \ldots$.

13.2.4   Example: Simple Communication Errors

Hidden Markov models can be used to correct text errors. We will simplify somewhat, and assume we have text that has no punctuation marks, and no capital letters. This means there are a total of 27 symbols (26 lower case letters, and a space). We send this text down some communication channel. This could be a telephone line, a fax line, a file saving procedure, or anything else. This channel makes errors independently at each character. For each location, with probability $1-p$ the output character at that location is the same as the input character. With probability p, the channel chooses randomly between the character one ahead or one behind in the character set, and produces that instead. You can think of this as a simple model for a mechanical error in one of those now ancient printers where a character strikes a ribbon to make a mark on the paper. We must reconstruct the transmission from the observations.

*	e	t	i	a	o	s	n	r	h
1.9e−1	9.7e−2	7.9e−2	6.6e−2	6.5e−2	5.8e−2	5.5e−2	5.2e−2	4.8e−2	3.7e−2

TABLE 13.1: The most common single letters (unigrams) that I counted from a draft of this chapter, with their probabilities. The "*" stands for a space. Spaces are common in this text, because I have tended to use short words (from the probability of the "*", average word length is between five and six letters)

I built a unigram model, a bigram model, and a trigram model. I stripped the text of this chapter of punctuation marks and mapped the capital letters to lower case letters. I used an HMM package (in my case, for Matlab; but there's a good one for R as well) to perform inference. The main programming here is housekeeping to make sure the transition and emission models are correct. About 40% of the bigrams and 86% of the trigrams did not appear in the text. I smoothed the bigram and trigram probabilities by dividing the probability 0.01 evenly between all unobserved bigrams (resp. trigrams). The most common unigrams, bigrams, and trigrams appear in Tables 13.1, 13.2, and 13.3. As an example sequence, I used

> the trellis has two crucial properties each directed path through the trellis from the start column to the end column represents a legal sequence of states now for some directed path from the start column to the end column sum all the weights for the nodes and edges along this path this sum is the log of the joint probability of that sequence of states

Lead char					
*	*t (2.7e−2)	*a (1.7e−2)	*i (1.5e−2)	*s (1.4e−2)	*o (1.1e−2)
e	e* (3.8e−2)	er (9.2e−3)	es (8.6e−3)	en (7.7e−3)	el (4.9e−3)
t	th (2.2e−2)	t* (1.6e−2)	ti (9.6e−3)	te (9.3e−3)	to (5.3e−3)
i	in (1.4e−2)	is (9.1e−3)	it (8.7e−3)	io (5.6e−3)	im (3.4e−3)
a	at (1.2e−2)	an (9.0e−3)	ar (7.5e−3)	a* (6.4e−3)	al (5.8e−3)
o	on (9.4e−3)	or (6.7e−3)	of (6.3e−3)	o* (6.1e−3)	ou (4.9e−3)
s	s* (2.6e−2)	st (9.4e−3)	se (5.9e−3)	si (3.8e−3)	su (2.2e−3)
n	n* (1.9e−2)	nd (6.7e−3)	ng (5.0e−3)	ns (3.6e−3)	nt (3.6e−3)
r	re (1.1e−2)	r* (7.4e−3)	ra (5.6e−3)	ro (5.3e−3)	ri (4.3e−3)
h	he (1.4e−2)	ha (7.8e−3)	h* (5.3e−3)	hi (5.1e−3)	ho (2.1e−3)

TABLE 13.2: The most common bigrams that I counted from a draft of this chapter, with their probabilities. The "*" stands for a space. For each of the 10 most common letters, I have shown the five most common bigrams with that letter in the lead. This gives a broad view of the bigrams, and emphasizes the relationship between unigram and bigram frequencies. Notice that the first letter of a word has a slightly different frequency than letters (top row: bigrams starting with a space are first letters). About 40% of the possible bigrams do not appear in the text

*th	the	he*	is*	*of	of*	on*	es*	*a*	ion
1.7e−2	1.2e−2	9.8e−3	6.2e−3	5.6e−3	5.4e−3	4.9e−3	4.9e−3	4.9e−3	4.9e−3

tio	e*t	in*	*st	*in	at*	ng*	ing	*to	*an
4.6e−3	4.5e−3	4.2e−3	4.1e−3	4.1e−3	4.0e−3	3.9e−3	3.9e−3	3.8e−3	3.7e−3

TABLE 13.3: The most frequent 10 trigrams in a draft of this chapter, with their probabilities. Again, "*" stands for space. You can see how common "the" and "*a*" are; "he*" is common because "*the*" is common. About 80% of possible trigrams do not appear in the text

with the measurements you can verify each of these statements easily by reference to a simple example

(which is text you could find in a draft of this chapter). There are 456 characters in this sequence.

When I ran this through the noise process with $p = 0.0333$, I got

theztrellis has two crucial properties each directed path through the tqdllit from the start column to the end coluln represents a legal sequencezof states now for some directed path from the start column to thf end column sum aml the veights for the nodes and edges along this path this sum is the log of the joint probability oe that sequence of states wish the measurements youzcan verify each of these statements easily by reference to a simple examqle

which is mangled but not too badly (13 of the characters are changed, so 443 locations are the same).

The unigram model produces

the trellis has two crucial properties each directed path through the tqdllit from the start column to the end column represents a legal sequence of states now for some directed path from the start column to thf end column sum aml the veights for the nodes and edges along this path this sum is the log of the joint probability oe that sequence of states wish the measurements you can verify each of these statements easily by reference to a simple examqle

which fixes three errors. The unigram model only changes an observed character when the probability of encountering that character on its own is less than the probability it was produced by noise. This occurs only for "z," which is unlikely on its own and is more likely to have been a space. The bigram model produces

she trellis has two crucial properties each directed path through the trellit from the start column to the end coluln represents a legal sequence of states now for some directed path from the start column to the end column sum aml the veights for the nodes and edges along this path this sum is the log of the joint probability oe that sequence of states wish the measurements you can verify each of these statements easily by reference to a simple example

This is the same as the correct text in 449 locations, so somewhat better than the noisy text. The trigram model produces

the trellis has two crucial properties each directed path through the trellit from the start column to the end column represents a legal sequence of states now for some directed path from the start column to the end column sum all the weights for the nodes and edges along this path this sum is the log of the joint probability of that sequence of states with the measurements you can verify each of these statements easily by reference to a simple example

which corrects all but one of the errors (look for "trellit").

13.3 Learning an HMM

There are two very distinct cases for learning an HMM. In the first case, the hidden states have known and important semantics. For example, the hidden states could be words or letters. In this case, we want any model we learn to respect the semantics of the hidden states. For example, the model should recover the right words to go with ink or sound or whatever. This case is straightforward (Sect. 13.3.1).

In the second case, we want to model sequences, but the hidden states are just a modelling device. One example is motion capture data. Various devices can be used to measure the position of human joints in space while a person is moving around. These devices report position as a function of time. This kind of data is extremely useful in making computer generated imagery (CGI) for films and computer games. We might observe some motion capture data, and try to make more (HMMs actually do this quite poorly). As another example, we might observe stock price data, and try to make more. As yet another example, we might observe encrypted text and want to make more encrypted text (for example, to confuse the people encrypting the text). This case isn't like the examples I've used to describe or justify HMMs, but it occurs fairly often. In this case, EM is an effective learning algorithm (Sect. 13.3.2).

13.3.1 When the States Have Meaning

There are two interesting versions of this case. In one, we see example sequences of X_i with corresponding Y_i. Since everything is discrete, building models of $P(X_{i+1}|X_i)$ and of $P(Y|X)$ is straightforward—one simply counts. This assumes that there is enough example data. When there is not—usually signalled by zero counts for some cases—one must use smoothing methods that are beyond our scope. If one has this kind of data, it is possible to build other kinds of sequence model; I describe these models in the following chapter.

In the second case, we see example sequences of X_i but do not see the Y_i corresponding to those sequences. A standard example is a substitution cipher for English text. Here it is easy to get a lot of data for $P(X_{i+1}|X_i)$ (one looks up text resources in English), but we have no data for $P(Y|X)$ because we do not know what X corresponds to observed Y's. Learning in this case is a straightforward variant of the EM algorithm for learning when there is no X data (below; for the variant, see the exercises).

13.3.2 Learning an HMM with EM

We have a dataset $\mathbf{Y}$ for which we believe a hidden Markov model is an appropriate model. This dataset consists of R sequences of visible states. The u'th sequence has $N(u)$ elements. We will assume that the observed values lie in a discrete space (i.e., there are O possible values that the Y's can take, and no others). We wish to choose a model that best represents a set of data. Assume, for the moment, that we knew each hidden state corresponding to each visible state. Write $Y_t^{(u)}$ is the observed value for the t'th observed state in the u'th sequence; write $X_t^{(u)}$ for the random variable representing the hidden value for the t'th observed state in the u'th sequence; write s_k for the hidden state values (where k is in the range $1\ldots S$); and write y_k for the possible values for Y (where k is in the range $1\ldots O$).

The hidden Markov model is given by three sets of parameters, π, $\mathcal{P}$, and $\mathcal{Q}$. We will assume that these parameters are not affected by where we are in the sequence (i.e., the model is homogeneous). First, π is an S-dimensional vector. The i'th element, π_i, of this vector gives the probability that the model starts in state s_i, i.e., $\pi_i = P(X_1 = s_i|\theta)$. Second, $\mathcal{P}$ is an $S \times S$-dimensional table. The i, j'th element of this table gives $P(X_{t+1} = s_j|X_t = s_i)$. Finally, $\mathcal{Q}$ is an $O \times S$-dimensional table. We will write $q_j(y_i) = P(Y_t = y_i|X_t = s_j)$ for the i, j'th element of this table. Note I will write θ to represent all of these parameters together.

Now assume that we *know* the values of $X_t^{(u)}$ for all t, u, (i.e., for each $Y_t^{(u)}$ we know that $X_t^{(u)} = s_i$). Then estimating the parameters is straightforward. We can estimate each by counting. For example, we estimate π_i by counting the number of sequences where $X_1 = s_i$, then dividing by the total number of sequences. We will encapsulate this knowledge in a function $\delta_t^{(u)}(i)$, where

$$\delta_t^{(u)}(i) = \begin{cases} 1 & \text{if } X_t^{(u)} = s_i \\ 0 & \text{otherwise} \end{cases}.$$

If we know $\delta_t^{(u)}(i)$, we have

$$
\begin{aligned}
\pi_i &= \frac{\text{number of times in } s_i \text{ at time } 1}{\text{number of sequences}} \\[2mm]
&= \frac{\sum_{u=1}^{R} \delta_1^{(u)}(i)}{R} \\[4mm]
\mathcal{P}_{ij} &= \frac{\text{number of transitions from } s_j \text{ to } s_i}{\text{total number of transitions}} \\[2mm]
&= \frac{\sum_{u=1}^{R} \sum_{t=1}^{N(u)-1} \delta_t^{(u)}(j) \delta_{t+1}^{(u)}(i)}{\sum_{u=1}^{R} [N(u) - 1]} \\[4mm]
q_j(y_i) &= \frac{\text{number of times in } s_j \text{ and observe } Y = y_i}{\text{number of times in } s_j} \\[2mm]
&= \frac{\sum_{u=1}^{R} \sum_{t=1}^{N(u)} \delta_t^{(u)}(j) \delta(Y_t^{(u)}, y_i)}{\sum_{u=1}^{R} \sum_{t=1}^{N(u)} \delta_t^{(u)}(j)}
\end{aligned}
$$

where $\delta(u, v)$ is one if its arguments are equal and zero otherwise.

The problem (of course) is that we *don't know* $\delta_t^{(u)}(i)$. But we have been here before (Sects. 9.2.1 and 9.2.3). The situation follows the recipe for EM: we have missing variables (the $X_t^{(u)}$; or, equivalently, the $\delta_t^{(u)}(i)$) where the log-likelihood can be written out cleanly in terms of the missing variables. We assume we know an estimate of the parameters $\hat{\theta}^{(n)}$. We construct

$$Q(\theta; \hat{\theta}^{(n)}) = \mathbb{E}_{P(\delta|Y, \hat{\theta}^{(n)})}[\log P(\delta, Y | \theta)]$$

(the E-step). Then we compute

$$\hat{\theta}^{(n+1)} = \underset{\theta}{\text{argmin}} \; Q(\theta; \hat{\theta}^{(n)})$$

(the M-step). As usual, the problem is the E-step. I will not derive this in detail (enthusiasts can easily reconstruct the derivation from what follows together with Chap. 9). The essential point is that we need to recover

$$\xi_t^{(u)}(i) = \mathbb{E}_{P(\delta|Y, \hat{\theta}^{(n)})}\left[\delta_t^{(u)}(i)\right] = P(X_t^{(u)} = s_i | Y, \hat{\theta}^{(n)}).$$

For the moment, assume we know these. Then we have

$$
\begin{aligned}
\hat{\pi}_i^{(n+1)} &= \text{expected frequency of being in } s_i \text{ at time 1} \\
&= \frac{\sum_{u=1}^{R} \xi_1^{(u)}(i)}{R}
\end{aligned}
$$

$$
\begin{aligned}
\hat{P}_{ij}^{(n+1)} &= \frac{\text{expected number of transitions from } s_j \text{ to } s_i}{\text{expected number of transitions from state } s_j} \\
&= \frac{\sum_{u=1}^{R} \sum_{t=1}^{N(u)} \xi_t^{(u)}(j)\xi_{t+1}^{(u)}(i)}{\sum_{u=1}^{R} \sum_{t=1}^{N(u)} \xi_t^{(u)}(j)}
\end{aligned}
$$

$$
\begin{aligned}
\hat{q}_j^{(n+1)}(k) &= \frac{\text{expected number of times in } s_j \text{ and observing } Y = y_k}{\text{expected number of times in state } s_j} \\
&= \frac{\sum_{u=1}^{R} \sum_{t=1}^{N(u)} \xi_t^{(u)}(j)\delta(Y_t^{(u)}, y_k)}{\sum_{u=1}^{R} \sum_{t=1}^{N(u)} \xi_t^{(u)}(j)}
\end{aligned}
$$

where $\delta(u, v)$ is one if its arguments are equal and zero otherwise.

To evaluate $\xi_t^{(u)}(i)$, we need two intermediate variables: a **forward variable** and a **backward variable**. The forward variable is

$$
\alpha_t^{(u)}(j) = P(Y_1^{(u)}, \ldots, Y_t^{(u)}, X_t^{(u)} = s_j | \hat{\theta}^{(n)}).
$$

The backward variable is

$$
\beta_t^{(u)}(j) = P(\{Y_{t+1}^{(u)}, Y_{t+2}^{(u)}, \ldots, Y_{N(u)}^{(u)}\} | X_t^{(u)} = s_j | \hat{\theta}^{(n)}).
$$

Now assume that we know the values of these variables, we have that

$$
\begin{aligned}
\xi_t^{(u)}(i) &= P(X_t^{(u)} = s_i | \hat{\theta}^{(n)}, \mathbf{Y}^{(u)}) \\
&= \frac{P(\mathbf{Y}^{(u)}, X_t^{(u)} = s_i | \hat{\theta}^{(n)})}{P(\mathbf{Y}^{(u)} | \hat{\theta}^{(n)})} \\
&= \frac{\alpha_t^{(u)}(i)\beta_t^{(u)}(i)}{\sum_{i=1}^{S} \alpha_t^{(u)}(i)\beta_t^{(u)}(i)}
\end{aligned}
$$

Both the forward and backward variables can be evaluated by induction. We get $\alpha_t^{(u)}(j)$ by observing that:

$$
\begin{aligned}
\alpha_1^{(u)}(j) &= P(Y_1^{(u)}, X_1^{(u)} = s_j | \hat{\theta}^{(n)}) \\
&= \pi_j^{(n)} q_j^{(n)}(Y_1).
\end{aligned}
$$

Now for all other t's, we have

$$
\begin{aligned}
\alpha_{t+1}^{(u)}(j) &= P(Y_1^{(u)}, \ldots, Y_{t+1}^{(u)}, X_{t+1}^{(u)} = s_j | \hat{\theta}^{(n)}) \\
&= \sum_{l=1}^{S} P(Y_1^{(u)}, \ldots, Y_t^{(u)}, Y_{t+1}^{(u)}, X_t^{(u)} = s_l, X_{t+1}^{(u)} = s_j | \hat{\theta}^{(n)}) \\
&= \left(\sum_{l=1}^{S} \left[\begin{array}{c} P(Y_1^{(u)}, \ldots, Y_t^{(u)}, X_t^{(u)} = s_l | \hat{\theta}^{(n)}) \times \\ P(X_{t+1}^{(u)} = s_j | X_t^{(u)} = s_l, \hat{\theta}^{(n)}) \end{array} \right] \right) \\
&\quad \times P\left(Y_{t+1}^{(u)} | X_{t+1}^{(u)} = s_j, \hat{\theta}^{(n)}\right) \\
&= \left[\sum_{l=1}^{S} \alpha_t^{(u)}(l) p_{lj}^{(n)} \right] q_j^{(n)}(Y_{t+1})
\end{aligned}
$$

We get $\beta_t^{(u)}(j)$ by observing that:

$$
\begin{aligned}
\beta_{N(u)}^{(u)}(j) &= P(\text{no further output} | X_{N(u)}^{(u)} = s_j, \hat{\theta}^{(n)}) \\
&= 1.
\end{aligned}
$$

Now for all other t we have

$$
\begin{aligned}
\beta_t^{(u)}(j) &= P\left(Y_{t+1}^{(u)}, Y_{t+2}^{(u)}, \ldots, Y_{N(u)}^{(u)} | X_t^{(u)} = s_j, \hat{\theta}^{(n)}\right) \\
&= \sum_{l=1}^{S} \left[P\left(Y_{t+1}^{(u)}, Y_{t+2}^{(u)}, \ldots, Y_{N(u)}^{(u)}, X_{t+1}^{(u)} = s_l | X_t^{(u)} = s_j, \hat{\theta}^{(n)}\right) \right] \\
&= \sum_{l=1}^{S} \left[\begin{array}{c} P\left(Y_{t+2}^{(u)}, \ldots, Y_{N(u)}^{(u)} | X_{t+1}^{(u)} = s_j, \hat{\theta}^{(n)}\right) \\ \times P\left(Y_{t+1}^{(u)}, X_{t+1}^{(u)} = s_l | X_t^{(u)} = s_j, \hat{\theta}^{(n)}\right) \end{array} \right] \\
&= P\left(Y_{t+2}^{(u)}, \ldots, Y_{N(u)}^{(u)} | X_{t+1}^{(u)} = s_j, \hat{\theta}^{(n)}\right) \\
&\quad \left(\sum_{l=1}^{S} \left[\begin{array}{c} P\left(X_{t+1}^{(u)} = s_l | X_t^{(u)} = s_j, \hat{\theta}^{(n)}\right) \\ \times P\left(Y_{t+1}^{(u)} | X_{t+1}^{(u)} = s_l \hat{\theta}^{(n)}\right) \end{array} \right] \right) \\
&= \beta_{t+1}(j) \left(\sum_{l=1}^{S} \left[q_l^{(n)}\left(Y_{t+1}^{(u)}\right) p_{lj}^{(n)} \right] \right)
\end{aligned}
$$

As a result, we have a simple fitting algorithm, collected in Algorithm 13.1.

Procedure: 13.1 *Fitting Hidden Markov Models with EM*

We fit a model to a data sequence $\mathbf{Y}$ is achieved by a version of EM.
We seek the values of parameters $\theta = (\mathcal{P}, \mathcal{Q}, \pi)_i$. We assume we have
an estimated $\hat{\theta}^{(n)}$, and then compute the coefficients of a new model;
this iteration is guaranteed to converge to a local maximum of $P(\mathbf{Y}|\hat{\theta})$.

Until $\hat{\theta}^{(n+1)}$ is the same as $\hat{\theta}^{(n)}$
   compute the forward variables α and β
   using the procedures of Algorithms 13.2 and 13.3

   compute $\xi_t^{(u)}(i) = \dfrac{\alpha_t^{(u)}(i)\beta_t^{(u)}(i)}{\sum_{i=1}^{S}\alpha_t^{(u)}(i)\beta_t^{(u)}(i)}$

   compute the updated parameters using the procedures of Procedure 13.4
end

Procedure: 13.2 *Computing the Forward Variable for Fitting an HMM*

$$\alpha_1^{(u)}(j) \;=\; \pi_j^{(n)} q_j^{(n)}(Y_1)$$

$$\alpha_{t+1}^{(u)}(j) \;=\; \left[\sum_{l=1}^{S}\alpha_t^{(u)}(l)p_{lj}^{(n)}\right] q_j^{(n)}(Y_{t+1})$$

Procedure: 13.3 *Computing the Backward Variable for Fitting an HMM*

$$\beta_{N(u)}^{(u)}(j) \;=\; 1$$

$$\beta_t^{(u)}(j) \;=\; \beta_{t+1}(j)\left(\sum_{l=1}^{S}\left[q_l^{(n)}(Y_{t+1}^{(u)})p_{lj}^{(n)}\right]\right)$$

Procedure: 13.4 *Updating Parameters for Fitting an HMM*

$$\hat{\pi}_i^{(n+1)} = \frac{\sum_{u=1}^{R} \xi_1^{(u)}(i)}{R}$$

$$\hat{\mathcal{P}}_{ij}^{(n+1)} = \frac{\sum_{u=1}^{R} \sum_{t=1}^{N(u)} \xi_t^{(u)}(j)\xi_{t+1}^{(u)}(i)}{\sum_{u=1}^{R} \sum_{t=1}^{N(u)} \xi_t^{(u)}(j)}$$

$$\hat{q}_j(k)^{(n+1)} = \frac{\sum_{u=1}^{R} \sum_{t=1}^{N(u)} \xi_t^{(u)}(j)\delta(Y_t^{(u)}, y_k)}{\sum_{u=1}^{R} \sum_{t=1}^{N(u)} \xi_t^{(u)}(j)}$$

where $\delta(u, v)$ is one if its arguments are equal and zero otherwise.

13.4 You Should

13.4.1 Remember These Terms

13.4.2   Remember These Facts

13.4.3   Be Able to

- Set up a simple HMM and use it to solve problems.
- Learn a simple HMM from data using EM

Problems

13.1. Multiple die rolls: You roll a fair die until you see a 5, then a 6; after that, you stop. Write $P(N)$ for the probability that you roll the die N times.

  (a) What is $P(1)$?

  (b) Show that $P(2) = (1/36)$.

  (c) Draw a directed graph encoding all the sequences of die rolls that you could encounter. Don't write the events on the edges; instead, write their probabilities. There are 5 ways not to get a 5, but only one probability, so this simplifies the drawing.

  (d) Show that $P(3) = (1/36)$.

  (e) Now use your directed graph to argue that $P(N) = (5/6)P(N-1) + (25/36)P(N-2)$.

13.2. More complicated multiple coin flips: You flip a fair coin until you see either HTH or THT, and then you stop. We will compute a recurrence relation for $P(N)$.

  (a) Draw a directed graph for this chain.

  (b) Think of the directed graph as a finite state machine. Write Σ_N for some string of length N accepted by this finite state machine. Use this finite state machine to argue that $Sigma_N$ has one of four forms:

     1. $TT\Sigma_{N-2}$
     2. $HH\Sigma_{N-3}$
     3. $THH\Sigma_{N-2}$
     4. $HTT\Sigma_{N-3}$

  (c) Now use this argument to show that $P(N) = (1/2)P(N-2)+(1/4)P(N-3)$.

13.3. For the umbrella example of worked Example 13.1, assume that with probability 0.7 it rains in the evening, and 0.2 it rains in the morning. I am conventional, and go to work in the morning, and leave in the evening.

  (a) Write out a transition probability matrix.

  (b) What is the stationary distribution? (you should use a simple computer program for this).

  (c) What fraction of evenings do I arrive at home wet?

  (d) What fraction of days do I arrive at my destination dry?

Programming Exercises

13.4. A dishonest gambler has two dice and a coin. The coin and one die are both fair. The other die is unfair. It has $P(n) = [0.5, 0.1, 0.1, 0.1, 0.1, 0.1]$ (where n is the number displayed on the top of the die). At the start, the gambler chooses a die uniformly and at random. At each subsequent step, the gambler chooses a die by flipping a weighted coin. If the coin comes up heads (probability p), the gambler changes the die, otherwise, the gambler keeps the same die. The gambler rolls the chosen die.

   (a) Model this process with a hidden Markov model. The emitted symbols should be $1, \ldots, 6$. Doing so requires only two hidden states (which die is in hand). Simulate a long sequence of rolls using this model for the case $p = 0.01$ and $p = 0.5$. What difference do you see?

   (b) Use your simulation to produce 10 sequences of 100 symbols for the case $p = 0.1$. Record the hidden state sequence for each of these. Now recover the hidden state using dynamic programming (you should likely use a software package for this; there are many good ones for R and Matlab). What fraction of the hidden states is correctly identified by your inference procedure?

13.5. A dishonest gambler has two dice and a coin. The coin and one die are both fair. The other die is unfair. It has $P(n) = [0.5, 0.1, 0.1, 0.1, 0.1, 0.1]$ (where n is the number displayed on the top of the die). At the start, the gambler chooses a die uniformly and at random. At each subsequent step, the gambler chooses a die by flipping a weighted coin. If the coin comes up heads (probability p), the gambler changes the die, otherwise, the gambler keeps the same die. The gambler rolls the chosen die.

   (a) Model this process with a hidden Markov model. The emitted symbols should be $1, \ldots, 6$. Doing so requires only two hidden states (which die is in hand). Produce one sequence of 1000 symbols for the case $p = 0.2$ with your simulator.

   (b) Use the sequence of symbols and EM to learn a hidden Markov model with two states.

   (c) It turns out to be difficult with the tools at our disposal to compare your learned model with the true model. Can you do this by inferring a sequence of hidden states using each model, then comparing the inferred sequences? Explain. *Hint:* No—but the reason requires a moment's thought.

   (d) Simulate a sequence of 1000 states using the learned model and also using the true model. For each sequence compute the fraction of 1's, 2's, etc. observed in the sequence. Does this give you any guide as to how good the learned model is? *Hint:* can you use the chi-squared test to tell if any differences you see are due to chance?

13.6. **Warning: this exercise is fairly elaborate, though straightforward.** We will correct text errors using a hidden Markov model.

   (a) Obtain the text of a copyright-free book in plain characters. One natural source is Project Gutenberg, at https://www.gutenberg.org. Simplify this text by dropping all punctuation marks except spaces, mapping capital letters to lower case, and mapping groups of many spaces to a single space. The result will have 27 symbols (26 lower case letters and a space). From this text, count unigram, bigram, and trigram letter frequencies.

   (b) Use your counts to build models of unigram, bigram, and trigram letter probabilities. You should build both an unsmoothed model and at least one smoothed model. For the smoothed models, choose some small amount of probability ϵ and split this between all events with zero count. Your models should differ only by the size of ϵ.

(c) Construct a corrupted version of the text by passing it through a process that, with probability p_c, replaces a character with a randomly chosen character, and otherwise reports the original character.

(d) For a reasonably sized block of corrupted text, use an HMM inference package to recover the best estimate of your true text. Be aware that your inference will run more slowly as the block gets bigger, but you won't see anything interesting if the block is (say) too small to contain any errors.

(e) For $p_c = 0.01$ and $p_c = 0.1$, estimate the error rate for the corrected text for different values of ϵ. Keep in mind that the corrected text could be worse than the corrupted text.

13.7. Warning: this exercise is fairly elaborate, though straightforward. We will break a substitution cipher using a hidden Markov model.

(a) Obtain the text of a copyright-free book in plain characters. One natural source is Project Gutenberg, at https://www.gutenberg.org. Simplify this text by dropping all punctuation marks except spaces, mapping capital letters to lower case, and mapping groups of many spaces to a single space. The result will have 27 symbols (26 lower case letters and a space). From this text, count unigram, bigram, and trigram letter frequencies.

(b) Use your counts to build models of unigram, bigram, and trigram letter probabilities. You should build both an unsmoothed model and at least one smoothed model. For the smoothed models, choose some small amount of probability ϵ and split this between all events with zero count. Your models should differ only by the size of ϵ.

(c) Construct a ciphered version of the text. We will use a substitution cipher, which you can represent as randomly chosen permutation of 27 points. You should represent this permutation as a 27×27 permutation matrix. Remember a permutation matrix contains only zeros and ones. Each column and each row contains exactly one 1. You can now represent each character with a 27-dimensional one-hot vector. This is a 27-dimensional vector. One component is 1, and the others are 0. For the i'th character, the i'th component is 1 (so an "a" is represented by a vector with 1 in the first component, etc.). The document becomes a sequence of these vectors. Now you can get a representation of the ciphered document by multiplying the representation of the original document by the permutation matrix.

(d) Using at least 10,000 ciphered characters apply EM, rather like Sect. 13.3.2, to estimate an HMM. You should use the unigram model of the previous subexercise as the transition model, and you should *not* re-estimate the transition model. Instead, you should estimate the emission model and prior only. This is straightforward; plug the known transition model, the estimated emission model, and the prior into the E-step, and then in the M-step update only the emission model and the prior.

(e) For a reasonably sized block of ciphered text, use an HMM inference package and your learned model to recover the best estimate of your true text. Be aware that your inference will run more slowly as the block gets bigger, but you won't see anything interesting if the block is (say) too small to contain any errors.

(f) Now perform the last two steps using your bigram and trigram models. Which model deciphers with the lowest error rate?

C H A P T E R   14

Learning Sequence Models Discriminatively

In this chapter, I resolve two problems that you might not have noticed in the previous chapter. First, HMMs aren't that natural for many sequences, because a model that represents (say) ink conditioned on (say) a letter is odd. Generative models like this must often do much more work than is required to solve a problem, and modelling the letter conditioned on the ink is usually much easier (this is why classifiers work). Second, in many applications you would want to learn a model that produces the right sequence of hidden states given a set of observed states, as opposed to maximizing likelihood.

Resolving these issues requires some generalization. Hidden Markov models have two very nice properties. First, they can be represented as graphs; second, inference is easy and efficient. In this chapter, we will look at other models which can be represented on graphs and which allow easy inference. Inference is efficient if the graph representation is a forest. One apparently natural model for sequences meets our criteria, is discriminative, and has a nasty hidden bug. The better model can't be interpreted in terms of joint probabilities of pairs of hidden states, but still allows easy inference.

Now we want our model to accept (say) sequences of ink and produce (say) sequences of characters. One approach to training is to choose a model that maximizes the joint probability of training data. But this may not be what we really seek. Assume the training data consists of (say) sequences of ink and (say) sequences of characters. Then we what we really want is that, when our model receives a ground truth sequence of ink, it produces corresponding ground truth sequence of characters. This view is much more like training a classifier (which is trained to produce the ground truth label when it gets a training input, or mostly). Training a sequence model to produce the sequences you want turns out to be a straightforward, but interesting, generalization of our reasoning about classifiers.

14.1 Graphical Models

What made an HMM an attractive model is that inference is easy. We could search an exponential space of paths in polynomial time to find the best path. I showed this by transferring the cost function for an HMM to a trellis, then reasoning about paths on that trellis. But nothing I did required the cost function that we were minimizing to come from log probabilities (you should revisit Sect. 13.2 to check this point, which is important). I could put arbitrary node and edge weights on the trellis, and still use my algorithm to recover the path with the smallest sum of weights.

© Springer Nature Switzerland AG 2019
D. Forsyth, *Applied Machine Learning*,
https://doi.org/10.1007/978-3-030-18114-7_14

14.1.1 Inference and Graphs

This opens the possibility that we could assemble other kinds of model—not HMMs—for which inference is easy. This idea turns out to be fruitful, but requires thinking about what made inference easy. For an HMM, I could factor the log of the joint probability by

$$
\begin{aligned}
-\log P(Y_1, Y_2, \ldots, Y_N, X_1, X_2, \ldots, X_N) = \ & -\log P(X_1) - \log P(Y_1|X_1) - \\
& \log P(X_2|X_1) - \log P(Y_2|X_2) - \\
& \cdots \\
& \log P(X_N|X_{N-1}) - \log P(Y_N|X_N).
\end{aligned}
$$

Inference requires choosing the $X_1, \ldots, X_N$ values that maximize this objective function. Notice the objective function is a sum of two kinds of term. There are **unary terms** which are functions that take one argument, and which we write $V(X_i)$. Notice that the variable identifies *which* vertex function we are talking about; the convention follows probability notation. There are **binary terms** which are functions that take two arguments, and which we write $E(X_i, X_j)$, using the same convention. For example, in the case of the HMM, the variables would be the hidden states, the unary terms would be the negative logs of emission probabilities, and binary terms would be the negative logs of transition probabilities.

It is quite natural to draw this objective function as a graph. There is one vertex for each variable (and the unary terms are sometimes called **vertex terms**), and one edge for each binary term (binary terms are sometimes called **edge terms**). It turns out that if this graph is a forest, then optimization will be easy, which I shall sketch here. By the way, this does not mean that if the graph is not a forest, optimization is necessarily hard; there is a large and interesting collection of details to be navigated here, all of which we shall ignore.

The simplest case is a **chain graph**. A chain graph looks like a chain (hence the name), and is the graph that arises from an HMM. There is one vertex term for each X_i (for the HMM, this is $-\log P(Y_i|X_i)$). There is an edge term for X_i and X_{i+1} for each i where both exist (for the HMM, this is $-\log P(X_{i+1}|X_i)$). We could write the objective function for inference as

$$
f(X_1, \ldots, X_n) = \sum_{i=1}^{i=N} V(X_i) + \sum_{i=1}^{i=N-1} E(X_i, X_{i+1})
$$

and we wish to minimize this function. Now we define a new function, the **cost-to-go function**, with a recursive definition. Write

$$
f_{\text{cost-to-go}}^{(N-1)}(X_{N-1}) = \min_{X_N} \left[E(X_{N-1}, X_N) + V(X_N) \right].
$$

This function represents the effect of a choice of value for X_{N-1} on the terms that involve X_N, where one chooses the best possible choice of X_N. This means that

$$
\min_{X_1, \ldots, X_N} f(X_1, \ldots, X_N)
$$

is equal to

$$\min_{X_1, \ldots, X_{N-1}} \left(f(X_1, \ldots, X_{N-1}) + f_{\text{cost-to-go}}^{(N-1)}(X_{N-1}) \right),$$

which means that we can eliminate the Nth variable from the optimization by replacing the term

$$E(X_{N-1}, X_N) + V(X_N)$$

with

$$f_{\text{cost-to-go}}^{(N-1)}(X_{N-1}),$$

which is a function of X_{N-1} alone.

Equivalently, assume we must choose a value for X_{N-1}. The cost-to-go function tells us the value of $E(X_{N-1}, X_N) + V(X_N)$ obtained by making the best choice of X_N conditioned on our choice of X_{N-1}. Because any other choice would not lead to a minimum, if we know the cost-to-go function at X_{N-1}, we can now compute the best choice of X_{N-1} conditioned on our choice of X_{N-2}. This yields that

$$\min_{X_{N-1}, X_N} \left[E(X_{N-2}, X_{N-1}) + V(X_{N-1}) + E(X_{N-1}, X_N) + V(X_N) \right]$$

is equal to

$$\min_{X_{N-1}} \left[E(X_{N-2}, X_{N-1}) + V(X_{N-1}) + \left(\min_{X_N} E(X_{N-1}, X_N) + V(X_N) \right) \right].$$

But all this can go on recursively, yielding

$$f_{\text{cost-to-go}}^{(k)}(X_k) = \min_{X_{k+1}} E(X_k, X_{k+1}) + V(X_k) + f_{\text{cost-to-go}}^{(k+1)}(X_{k+1}).$$

This is basically what we did with a trellis in Sect. 13.2.2. Notice that

$$\min_{X_1, \ldots, X_N} f(X_1, \ldots, X_N)$$

is equal to

$$\min_{X_1, \ldots, X_{N-1}} \left(f(X_1, \ldots, X_{N-1}) + f_{\text{cost-to-go}}^{(N-1)}(X_{N-1}) \right)$$

which is equal to

$$\min_{X_1, \ldots, X_{N-2}} \left(f(X_1, \ldots, X_{N-2}) + f_{\text{cost-to-go}}^{(N-2)}(X_{N-2}) \right),$$

and we can apply the recursive definition of the cost-to-go function to get

$$\min_{X_1, \ldots, X_N} f(X_1, \ldots, X_N) = \min_{X_1} \left(f(X_1) + f_{\text{cost-to-go}}^{1}(X_1) \right).$$

All this gives another description of the trellis maximization process. We start at X_N, and construct $f_{\text{cost-to-go}}^{(N-1)}(X_{N-1})$. We can represent this function as a table, giving the value of the cost-to-go function for each possible value of X_{N-1}. We build a second table giving the optimum X_N for each possible value of X_{N-1}. From this, we can build $f_{\text{cost-to-go}}^{(N-2)}(X_{N-2})$, again as a table, and also the best X_{N-1} as a function of X_{N-2}, again as a table, and so on. Now we arrive at X_1. We obtain the solution for X_1 by choosing the X_1 that yields the best value of $\left(f_{\text{chain}}(X_1) + f_{\text{cost-to-go}}^1(X_1)\right)$. But from this solution, we can obtain the solution for X_2 by looking in the table that gives the best X_2 as a function of X_1, and so on. It should be clear that this process yields a solution in polynomial time; if each X_i can take one of k values, then the time is $O(NK^2)$.

This strategy will work for a model with the structure of a forest. The proof is an easy induction. If the forest has no edges (i.e., consists entirely of nodes), then it is obvious that a simple strategy applies (choose the best value for each X_i independently). This is clearly polynomial. Now assume that the algorithm yields a result in polynomial time for a forest with e edges, and show that it works for a forest with $e+1$ edges. There are two cases. The new edge could link two existing trees, in which case we could reorder the trees so the nodes that are linked are roots, construct a cost-to-go function for each root, and then choose the best pair of states for these roots from the cost-to-go functions. Otherwise, one tree had a new edge added, joining the tree to an isolated node. In this case, we reorder the tree so that this new node is the root and build a cost-to-go function from the leaves to the root. The fact that the algorithm works is a combinatorial insight. In Sect. 15.1, we will see graphical models that do not admit easy inference because their graph is not a forest. In those cases, we will need to use approximation strategies for inference.

Remember This: *Dynamic programming inference doesn't require the weights to be log probabilities to work. All that is required is a cost function on a forest that is a sum of charges for vertices and edges. In this general form, dynamic programming works by repeatedly eliminating a node and computing cost-to-go function.*

14.1.2 Graphical Models

A probability model where the inference problem can be drawn as a graph is called a **graphical model**. There are good reasons to use models with this property. It is known when inference is easy. If inference is hard (as we shall see), there are often quite good approximation procedures. The models often fit interesting problems quite naturally, and can be quite easy to construct (Fig. 14.1).

Here is one quite useful way to write the general class of graphical models. Assume we wish to construct a probability distribution over a collection of R vari-

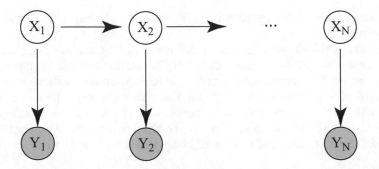

Figure 14.1: An HMM is an example of a graphical model. The joint probability distribution of the hidden variables X_i and the observations Y_i factors as in the text. Each variable appears as a vertex. There is an edge between pairs of variables that appear in the factorization. These edges have an arrowhead drawn according to conventions about conditional probability. Finally, the observed values are shaded

ables, $U_1, \ldots, U_R$. We want to draw a graph, so this distribution must be factored into a set of terms where each term depends on at most two variables. This means that there are some functions ψ_i, ϕ_{ij} so that

$$-\log P(U_1, \ldots, U_R) = \sum_{i=1}^{R} \psi_i(U_i) + \sum_{(i,j) \in \text{pairs}} \phi_{ij}(U_i, U_j) + K$$

where K is the log of the **normalizing constant** (it ensures the distribution sums to one), and is of no interest to us at present.

Now assume we have a probability distribution that can be written in this form. We partition the U variables into two groups. The X_i are unknown and we need to recover them by inference, and the Y_j are known. In turn, this means that some of the ψ (resp. ϕ) become constants because the argument is (resp. both arguments are) known; these are of no interest, and we can drop them from the drawing. Some of the ϕ become effectively functions of a single argument, because the value of the other argument is known and fixed. It is usual to keep these edges and relevant nodes in the drawing, but shade the known argument. In turn, inference involves solving for the values of a set of discrete variables to minimize the value of an objective function $f(X_1, \ldots, X_n)$.

For some graphical models (HMMs are a good example), $\phi_{ij}(U_i, U_j)$ can be interpreted in terms of conditional probability. In the case of an HMM, if you make the right choices, you can think of $\phi_{ii+1}(U_i, U_{i+1})$ as $\log P(U_{i+1}|U_i)$. In models like this, it is usual to add an arrowhead to the edge pointing to the variable that is in front of the conditioning bar (i.e., U_{j+1} in the example).

14.1.3 Learning in Graphical Models

Graphical models can be tricky to learn. We have seen that we can use EM to learn an HMM that models sequences without knowing the hidden states corresponding to the observations. This approach doesn't extend to all graphical models, by any manner of means.

The rest of this chapter focuses on a natural and important learning strategy which is very different from using EM. I will describe this for sequence models, but it can be used for other models too. Assume we have a collection of example sequences of observations (write $\mathbf{Y}^{(u)}$ for the u'th such sequence) *and* of hidden states (write $\mathbf{X}^{(u)}$ for the u'th such sequence). We construct a family of cost functions $C(X, Y; \theta)$ parametrized by θ. We then choose the θ so that inference applied to the cost function yields the right answer. So we want to choose θ so that

$$\operatorname*{argmin}_{\mathbf{X}} C(\mathbf{Y}^{(u)}, \mathbf{X}; \theta)$$

is $\mathbf{X}^{(u)}$ or "close to" it. The details require quite a lot of work, which we do below. What is important now is that this strategy applies to *any* model where easy inference is available. This means we can generalize quite significantly from HMMs, which is the next step.

14.2 Conditional Random Field Models for Sequences

HMM models have been widely used, but have one odd feature that is inconsistent with practical experience. Recall X_i are the hidden variables, and Y_i are the observations. HMMs model

$$P(Y_1, \ldots, Y_n | X_1, \ldots, X_n) \propto P(Y_1, \ldots, Y_n, X_1, \ldots, X_n),$$

which is the probability of observations given the hidden variables. This is modelled using the factorization

$$
\begin{aligned}
P(Y_1, Y_2, \ldots, Y_N, X_1, X_2, \ldots, X_N) \;=\;\; & P(X_1)P(Y_1|X_1) \\
& P(X_2|X_1)P(Y_2|X_2) \\
& \ldots \\
& P(X_N|X_{N-1})P(Y_N|X_N).
\end{aligned}
$$

In much of what we will do, this seems unnatural. For example, in the case of reading written text, we would be modelling the probability of the observed ink given the original text. But we would not attempt to find a single character by modelling the probability of the observed ink given the character (I will call this a **generative** strategy). Instead, we would search using a classifier, which is a model of the probability of the character conditioned on the observed ink (I will call this a **discriminative** strategy). The two strategies are quite different in practice. A generative strategy would need to explain all possible variants of the ink that a character could produce, but a discriminative strategy just needs to judge whether the ink observed is the character or not.

> **Remember This:** *HMMs have an odd feature that is often inconvenient. Generating observations from labels can require a much more detailed model than choosing a label from an observation.*

14.2.1 MEMMs and Label Bias

One alternative would be to look for a model that factors in a different way. For example, we could consider

$$
\begin{aligned}
P(X_1, X_2, \ldots, X_N | Y_1, Y_2, \ldots, Y_N) \;=\; & P(X_1 | Y_1) \times \\
& P(X_2 | Y_2, X_1) \times \\
& P(X_3 | X_2, Y_2) \times \\
& \ldots \times \\
& P(X_N | X_{N-1}, Y_N).
\end{aligned}
$$

This means that

$$
\begin{aligned}
-\log P(X_1, X_2, \ldots, X_N | Y_1, Y_2, \ldots, Y_N) \;=\; & -\log P(X_1 | Y_1) \\
& -\log P(X_2 | Y_2, X_1) \\
& -\log P(X_3 | X_2, Y_2) \\
& \ldots \\
& -\log P(X_N | X_{N-1}, Y_N).
\end{aligned}
$$

This is still a set of edge and vertex functions, but notice there is only one vertex function $(-\log P(X_1 | Y_1))$. All the remaining terms are edge functions. Models of this form are known as **maximum entropy Markov models** or **MEMMs**.

These models are deprecated. You should not use one without a special justification. Rather than just ignoring these models, I have described them because the reason they are deprecated is worth understanding, and because if I don't, it's quite likely you'll invent them on your own.

The problem is this: these models very often ignore measurements, as a result of their structure. To see this, assume we have fitted a model, and wish to recover the best sequence of X_i corresponding to a given sequence of observations Y_i. We must minimize

$$
\begin{aligned}
-\log P(X_1, X_2, \ldots, X_N | Y_1, Y_2, \ldots, Y_N) \;=\; & -\log P(X_1 | Y_1) \\
& -\log P(X_2 | Y_2, X_1) \\
& -\log P(X_3 | X_2, Y_2) \\
& \ldots \\
& -\log P(X_N | X_{N-1}, Y_N).
\end{aligned}
$$

by choice of $X_1, \ldots, X_N$. We can represent this cost function on a trellis, as for the HMM, but now notice that the costs on the trellis behave differently. For an HMM,

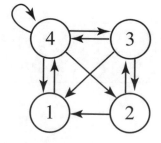

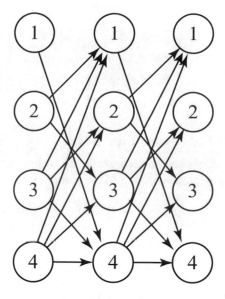

Figure 14.2: On the **left**, a Markov chain used in an MEMM. On the **right**, the resulting trellis. Notice that, in state 1, there is only one outgoing edge. In turn, this means that $-\log P(X_{i+1} = 2 | X_i = 1, Y_i) = 0$, *whatever* the value of Y_i. This guarantees mischief, detailed in the text

each state (circle) in a trellis had a cost, corresponding to $-\log P(Y_i | X_i)$, and each edge had a cost $(-\log P(X_{i+1} | X_i))$, and the cost of a particular sequence was the sum of the costs along the implied path. But for an MEMM, the representation is slightly different. There is no term associated with each state in the trellis; instead, we associate the edge going from the state $X_i = U$ to the state $X_{i+1} = V$ with the cost $-\log P(X_{i+1} = V | X_i = U, Y_i)$. Again, the cost of a sequence of states is represented by the sum of costs along the corresponding path. This may look to you like a subtle change, but it has nasty effects.

Look at the example of Fig. 14.2. Notice that when the model is in state 1, it can only transition to state 4. In turn, this means that $-\log P(X_{i+1} = 4 | X_i = 1, Y_i) = 0$ *whatever* the measurement Y_i is. Furthermore, either $P(X_{i+1} = 3 | X_i = 2, Y_i) \geq 0.5$ or $P(X_{i+1} = 1 | X_i = 2, Y_i) \geq 0.5$ (because there are only two options leaving state 2). Here the measurement can determine which of the two options has higher probability. That figure shows a trellis corresponding to three measurements. In this trellis, the path 2 1 4 will be the lowest cost path *unless* the first measurement overwhelmingly disfavors the transition $2 \to 1$. This is because most other paths must share weights between many outgoing edges; but $1 \to 4$ is very cheap, and $2 \to 1$ will be cheap unless there is an unusual measurement. Paths which pass through many states with few outgoing edges are strongly favored. This is known as the **label bias problem**. There are some fixes that can be applied, but it is better to reengineer the model.

> **Remember This:** *The obvious fix to the generative property of an HMM doesn't work, because the model can ignore or discount measurements.*

14.2.2 Conditional Random Field Models

We want a model of sequences that is discriminative, but doesn't have the label bias problem. We'd also like that model to be as tractable as an HMM for sequences. We can achieve this by ensuring that the graphical representation of the model is a chain graph. We'd like the resulting model to be discriminative in form—i.e., we should be able to interpret the vertex functions as $-\log P(X_i|Y_i)$—but we don't want an MEMM.

We start with the cost functions. Write $E_i(a, b)$ for the cost of the edge from $X_i = a$ to $X_{i+1} = b$, and write $V_i(a)$ for the cost of assigning $X_i = a$. We will use $V_i(a) = -\log P(X_i = a|Y_i)$ as a vertex cost function, because we want the model to be discriminative. We will assume that $E_i(a, b)$ and $V_i(a)$ are bounded (straightforward, because the variables are all discrete anyhow), but we will not apply any direct probabilistic interpretation to this function. Instead, we interpret the whole model by stating that

$$-\log P(X_1 = x_1, X_2 = x_2, \ldots, X_N = x_N | Y_1, Y_2, \ldots, Y_N)$$

is

$$[V_1(x_1) + E_1(x_1, x_2) + V_2(x_2) + E_2(x_2, x_3) + \ldots V_N(x_N)] + K$$

where K is the log of the normalizing constant, chosen to ensure the probability distribution sums to 1. There is a crucial difference with the MEMM; there are now node as well as edge costs *but* we can't interpret the edge costs as transition probabilities. A model with this structure is known as a **conditional random field**.

Notice the minus sign. This means that the best sequence has the smallest value of

$$\begin{bmatrix} V_1(x_1) + E_1(x_1, x_2) + \\ V_2(x_2) + E_2(x_2, x_3) + \\ \ldots \\ V_N(x_N) \end{bmatrix},$$

and we can think of this expression as a cost. Inference is straightforward by dynamic programming, as above, if we have known E_i and V_i terms.

> **Remember This:** *In a CRF, the weights on edges cannot be directly interpreted as conditional probabilities. Instead, we write out a joint probability model. If the model forms a forest, dynamic programming will work.*

14.2.3 Learning a CRF Takes Care

What is much more interesting is *learning* a CRF. We don't have a probabilistic interpretation of E_i, so we can't reconstruct an appropriate table of values by (say) counting. There must be some set of parameters, θ, that we are trying to adjust. However, we need some principle to drive the choice of θ. One strategy is to maximize

$$-\log P(X_1 = x_1, X_2 = x_2, \ldots, X_N = x_N | Y_1, Y_2, \ldots, Y_N, \theta)$$

but this is difficult to do in practice. The problem is that the normalizing constant K depends on θ. This means that we need to be able to compute K. It is possible to do this, but the computation is moderately expensive.

We have that

$$K = \log \left(\sum_{u_1, u_2, \ldots, u_n} \exp - [V_1(u_1) + E_1(u_1, u_2) + V_2(u_2) + E_2(u_2, u_3) + \ldots V_N(u_N)] \right)$$

and this sum is over an exponential space (all possible combinations of $u_1, u_2, \ldots, u_N$). Dynamic programming will yield this sum for a graph that is a forest. Notice that

$$\sum_{u_1, u_2, \ldots, u_N} \exp - [V_1(u_1) + E_1(u_1, u_2) + V_2(u_2) + E_2(u_2, u_3) + \ldots V_N(u_N)]$$

is

$$\sum_{u_1, u_2, \ldots, u_{N-1}} \left(\begin{array}{c} \exp - [V_1(u_1) + E_1(u_1, u_2) + V_2(u_2) + E_2(u_2, u_3) + \ldots V_{N-1}] \\ \times \\ \sum_{u_N} \exp - [E_{N-1}(u_{N-1}, u_N) + V_N(u_N)] \end{array} \right).$$

From this, we get the usual recursion. Write

$$f_{\text{sum-prod to } i}(u_i) = \sum_{u_{i+1}, \ldots, u_N} \exp - \left[\begin{array}{c} E_i(u_i, u_{i+1}) + \\ V_{i+1}(u_{i+1}) + \\ E_{i+1}(u_{i+1}, u_{i+2}) + \\ \ldots \\ V_N(u_N) \end{array} \right]$$

and notice that

$$f_{\text{sum-prod to } N-1}(u_{N-1}) = \sum_{u_N} \exp - [E_{N_1}(u_{N-1}, u_N) + V_N(u_N)].$$

Now

$$f_{\text{sum-prod to } i}(u_i) = \sum_{u_{i+1}} (\exp - [E_i(u_i, u_{i+1}) + V_{i+1}(u_{i+1})]$$

$$\times f_{\text{sum-prod to } i+1}(u_{i+1}))$$

and

$$K = \log \sum_{u_1} f_{\text{sum-prod to } 1}(u_1).$$

This procedure is sometimes known as the **sum-products algorithm**. I have described this procedure because it gives me a chance to re-emphasize just how powerful and useful dynamic programming and its underlying reasoning are. But computing K like this to learn θ is unattractive. You'd need the gradient of K, which can also be computed from the recursion, but you'd need to apply this procedure for each step of the descent process. Furthermore, this approach is somewhat indirect, because it constructs the probability model that maximizes the log-posterior of the observed data. That isn't necessarily what we want—instead, we'd like the model to produce the right answer at inference time. In the next section, I expound a more direct training principle.

> **Remember This:** *You can learn a CRF for sequences using maximum likelihood, but doing so takes care. You must compute the normalizing constant, which can be done with the sum-products algorithm.*

14.3 Discriminative Learning of CRFs

A really powerful strategy for learning a CRF follows by obtaining both the observations and the state for a set of example sequences. Different sequences in this set might have different lengths; this doesn't matter. Now write $Y_i^{(k)}$ for the i'th element of the k'th sequence of observations, etc. For any set of parameters, we can recover a solution from the observations using dynamic programming (write Inference $(Y_1, \ldots, Y_N^{(k)}, \theta)$ for this). Now we will choose a set of parameters $\hat{\theta}$ so that

$$\text{Inference}\left(Y_1^{(k)}, \ldots, Y_N^{(k)}, \hat{\theta}\right) \text{ is close to } X_1^{(k)} \ldots X_N^{(k)}.$$

In words, the principle is this: Choose parameters so that, if you infer a sequence of hidden states from a set of training observations, you will get the hidden states that are (about) the same as those observed.

14.3.1 Representing the Model

We need a parametric representation of V_i and E_i; we'll then search for the parameters that yield the right model. We will simplify notation somewhat, by assuming that each vertex function is the same function and that each edge function is the same function. This isn't required, but it is the most usual case, and we have enough to deal with. We now construct a set of functions $\phi_j(U, V)$ (one for each j), and a vector of parameters $\theta_j^{(v)}$. Finally, we choose V to be a weighted sum of these basis functions, so that $V(u) = \sum_j \theta_j^{(v)} \phi_j(u, Y_i)$. Similarly, for E_i we will construct a set of functions $\psi_j(U, V)$ (one for each j) and a vector of parameters $\theta_j^{(e)}$. We choose E to be a weighted sum of these basis functions, so that $E(U, V) = \sum_j \theta_j^{(e)} \psi_j(U, V)$. Modelling involves choosing $\phi_j(U, V)$ and $\psi_j(U, V)$. Learning involves choosing $\theta = (\theta^{(v)}, \theta^{(e)})$.

I give some sample constructions below, but you may find them somewhat involved at first glance. What is important is that (a) we have some parameters θ so that, for different choices of parameter, we get different cost functions; and (b) for any choice of parameters (say $\hat{\theta}$) and any sequence of Y_i we can label each vertex and each edge on the trellis with a number representing the cost of that vertex or edge. Assume we have these properties. Then for any particular $\hat{\theta}$ we can construct the best sequence of X_i using dynamic programming (as above). Furthermore, we can try to adjust the choice of $\hat{\theta}$ so that for the i'th training sequence, $\mathbf{y}^{(i)}$, inference yields $\mathbf{x}^{(i)}$ or something close by.

14.3.2 Example: Modelling a Sequence of Digits

Here is a massive simplification of the problem of reading a sequence of digits. Assume we see a sequence of digits, where the observations are inked numerals, like MNIST, which appear in a window of fixed size, like MNIST. Assume also we know the location of each digit's window, which is the important simplification.

Each $\phi_j(U, V)$ accepts two arguments. One is the ink pattern on the paper (the V), and the other is a label to be attached to the ink (the U). Here are three possible approaches for building ϕ_j.

- Multinomial logistic regression works quite well on MNIST using just the pixel values as features. This means that you can compute 10 linear functions of the pixel values (one for each numeral) such that the linear function corresponding to the right numeral is smaller than any other of the functions, at least most of the time. Write $L_u(V)$ for the linear function of the ink (V) that corresponds to the u'th digit. Then we could use $\sum_u \mathbb{I}_{[U=u]} L_u(V)$ as a ϕ.
- For each possible numeral u and each pixel location p build a feature function $\phi(U, V) = \mathbb{I}_{[U=u]} \mathbb{I}_{[V(p)=0]}$. This is 1 if $U = u$ (i.e., for a particular numeral, u) *and* the ink at pixel location p is dark, and otherwise zero. We index these feature functions in any way that seems convenient to get $\phi_j(U, V)$.
- For each class x, we will build several different classifiers each of which can tell that class from all the others. We obtain different classifiers by using different training sets; or different features; or different classifier architectures; or all of these. Write $g_{i,u}(V)$ for the i'th classifier for class u. We ensure that $g_{i,u}(V)$ is small if V is of class u, and large otherwise. Then for each classifier and each class we can build a feature function by $\phi(U, V) = g_{i,U}(V)$. We index these feature functions in any way that seems convenient.

We must now build the $\psi_j(U, V)$. Here U and V can take the value of any state. I will assume the states are labelled with counting numbers, without any loss of generality, and will write a, b for particular values of the state. One simple construction is to build one ψ for each a, b pair, yielding $\psi_j(U, V) = \mathbb{I}_{[U=a]} \mathbb{I}_{[V=b]}$. If there are S possible states, there will be S^2 of these feature functions. Each one takes the value 1 when U and V take the corresponding state values, otherwise is 0. This construction will allow us to represent any possible cost for any transitions, as long as the cost doesn't depend on the observations.

We now have a model of the cost. I will write sequences like vectors, so $\mathbf{x}$ is a sequence, and x_i is the i'th element of that sequence. Write $C(\mathbf{x}; \mathbf{y}, \theta)$ for the cost of a sequence $\mathbf{x}$ of hidden variables, conditioned on observed values $\mathbf{y}$ and parameters θ. I'm suppressing the number of items in this sequence for conciseness, but will use N if I need to represent it. We have

$$C(\mathbf{x}; \mathbf{y}, \theta) = \sum_{i=1}^{N} \left[\sum_{j} \theta_j^{(v)} \phi_j(x_i, y_i) \right] + \sum_{i=1}^{N-1} \left[\left(\sum_{l} \theta_l^{(e)} \psi_l(x_i, x_{i+1}) \right) \right].$$

Notice that this cost function is linear in θ. We will use this to build a search for the best setting of θ.

14.3.3 Setting Up the Learning Problem

I will write $\mathbf{x}^{(i)}$ for the i'th training sequence of hidden states, and $\mathbf{y}^{(i)}$ for the i'th training sequence of observations. I will write $x_j^{(i)}$ for the hidden state at step j in the i'th training sequence, etc. The general principle we will adopt is that we should train a model by choosing θ such that, if we apply inference to $\mathbf{y}^{(i)}$, we will recover $\mathbf{x}^{(i)}$ (or something very similar).

For *any* sequence $\mathbf{x}$, we would like to have $C(\mathbf{x}^{(i)}; \mathbf{y}^{(i)}, \theta) \leq C(\mathbf{x}; \mathbf{y}^{(i)}, \theta)$. This inequality is much more general than it seems, because it covers *any* available sequence. Assume we engage in inference on the model represented by θ, using $\mathbf{y}^{(i)}$ as observed variables. Write $\mathbf{x}^{+,i}$ for the sequence recovered by inference, so that

$$\mathbf{x}^{+,i} = \underset{\mathbf{x}}{\text{argmin}}\ C(\mathbf{x}; \mathbf{y}^{(i)}, \theta)$$

(i.e., $\mathbf{x}^{+,i}$ is the sequence recovered from the model by inference if the parameters take the value θ). In turn, the inequality means that

$$C(\mathbf{x}^{(i)}; \mathbf{y}^{(i)}, \theta) \leq C(\mathbf{x}^{+,i}; \mathbf{y}^{(i)}, \theta).$$

It turns out that this is not good enough; we would also like the cost of solutions that are further from the true solution to be higher. So we want to ensure that the cost of a solution grows at least as fast as its distance from the true solution. Write $d(\mathbf{u}, \mathbf{v})$ for some appropriate distance between two sequences $\mathbf{u}$ and $\mathbf{v}$. We want to have

$$C(\mathbf{x}^{(i)}; \mathbf{y}^{(i)}, \theta) + d(\mathbf{x}, \mathbf{x}^{(i)}) \leq C(\mathbf{x}; \mathbf{y}^{(i)}, \theta).$$

Again, we want this inequality to be true for *any* sequence $\mathbf{x}$. This means that

$$C(\mathbf{x}^{(i)}; \mathbf{y}^{(i)}, \theta) \leq C(\mathbf{x}; \mathbf{y}^{(i)}, \theta) - d(\mathbf{x}, \mathbf{x}^{(i)})$$

for *any* $\mathbf{x}$. Now write

$$\mathbf{x}^{(*,i)} = \underset{\mathbf{x}}{\text{argmin}}\ C(\mathbf{x}; \mathbf{y}^{(i)}, \theta) - d(\mathbf{x}, \mathbf{x}^{(i)}).$$

The inequality becomes

$$C(\mathbf{x}^{(i)}; \mathbf{y}^{(i)}, \theta) \leq C(\mathbf{x}^{(*,i)}; \mathbf{y}^{(i)}, \theta) - d(\mathbf{x}^{(*,i)}, \mathbf{x}^{(i)}).$$

This constraint is likely to be violated in practice. Assume that

$$\xi_i = \max(C(\mathbf{x}^{(i)}; \mathbf{y}^{(i)}, \theta) - C(\mathbf{x}^{(*,i)}; \mathbf{y}^{(i)}, \theta) + d(\mathbf{x}^{(*,i)}, \mathbf{x}^{(i)}), 0)$$

so that ξ_i measures the extent to which the constraint is violated. We would like to choose θ so that we have the smallest possible set of constraint violations. It is natural to want to minimize the sum of ξ_i over all training data. But we also want to ensure that θ is not "too large," for the same reasons we regularized a support vector machine. Choose a regularization constant λ. Then we want to choose θ to minimize the regularized cost

$$\sum_{i \in \text{examples}} \xi_i + \lambda \theta^T \theta$$

where ξ_i is defined as above. This problem is considerably harder than it might look, because each ξ_i is a (rather strange) function of θ.

14.3.4 Evaluating the Gradient

We will solve the learning problem by stochastic gradient descent, as usual. First, we obtain an initial value of θ. Then we repeatedly choose a minibatch of examples at random, evaluate the gradient for that minibatch, update the estimate of θ, and go again. There is the usual nuisance of choosing a steplength, etc. which is handled in the usual way. The important question is evaluating the gradient.

Imagine we have chosen the u'th example. We must evaluate $\nabla_\theta \xi_u$. Recall

$$\xi_u = \max(C(\mathbf{x}^{(u)}; \mathbf{y}^{(u)}, \theta) - C(\mathbf{x}^{(*,u)}; \mathbf{y}^{(u)}, \theta) + d(\mathbf{x}^{(*,u)}, \mathbf{x}^{(u)}), 0)$$

and assume that we know $\mathbf{x}^{(*,u)}$. We will ignore the concern that ξ_u may not be differentiable in θ as a result of the max. If $\xi_u = 0$, we will say the gradient is zero. For the other case, recall that

$$C(\mathbf{x}; \mathbf{y}, \theta) = \sum_{i=1}^{N} \left[\sum_j \theta_j^{(v)} \phi_j^{(v)}(x_i, y_i) \right] + \sum_{i=1}^{N-1} \left[\left(\sum_l \theta_l^{(e)} \phi_l^{(e)}(x_i, x_{i+1}) \right) \right]$$

and that this cost function is *linear* in θ. The distance term $d(\mathbf{x}^{(*,u)}, \mathbf{x}^{(u)})$ doesn't depend on θ, so doesn't contribute to the gradient. So if we *know* $\mathbf{x}^{*,i}$, the gradient is straightforward because C is linear in θ.

To be more explicit, we have

$$\frac{\partial C}{\partial \theta_j^{(v)}} = \sum_{i=1}^{N} \left[\phi_j^{(v)}\left(x_i^{(u)}, y_i^{(u)}\right) - \phi_j^{(v)}\left(x_i^{(*,u)}, y_i^{(u)}\right) \right]$$

and

$$\frac{\partial C}{\partial \theta_l^{(e)}} = \sum_{i=1}^{N-1} \left[\phi_l^{(e)}\left(x_i^{(u)}, x_{i+1}^{(u)}\right) - \phi_l^{(e)}\left(x_i^{(*,u)}, x_{i+1}^{(*,u)}\right) \right].$$

The problem is that we don't know $\mathbf{x}^{(*,u)}$ because it could change each time we change θ. Recall

$$\mathbf{x}^{(*,u)} = \underset{\mathbf{x}}{\operatorname{argmin}} \; C(\mathbf{x}; \mathbf{y}^{(u)}, \theta) - d(\mathbf{x}, \mathbf{x}^{(u)}).$$

So, to compute the gradient, we must first run an inference on the example to obtain $\mathbf{x}^{(*,u)}$. But this inference could be hard, depending on the form of

$$C(\mathbf{x}; \mathbf{y}^{(u)}, \theta) - d(\mathbf{x}, \mathbf{x}^{(u)})$$

(which is often known as the **loss augmented constraint violation**). We would like to choose $d(\mathbf{x}, \mathbf{x}^{(u)})$ so that we get a distance that doesn't make the inference harder. One good, widely used example is the **Hamming distance**.

The Hamming distance between two sequences is the number of locations in which they disagree. Write $\operatorname{diff}(m, n) = 1 - \mathbb{I}_{[m=n]}(m, n)$ for a function that returns zero if its arguments are the same, and one otherwise. Then we can express the Hamming distance as

$$d_h(\mathbf{x}, \mathbf{x}^{(u)}) = \sum_k \operatorname{diff}\left(x_k, x_k^{(u)}\right).$$

We could scale the Hamming distance, to express how quickly we expect the cost to grow. So we will choose a non-negative number ϵ, and write

$$d(\mathbf{x}, \mathbf{x}^{(u)}) = \epsilon d_h(\mathbf{x}, \mathbf{x}^{(u)}).$$

The expression for Hamming distance is useful, because it allows us to represent the distance term on a trellis. In particular, think about the trellis corresponding to the u'th example. Then to represent the cost

$$C(\mathbf{x}; \mathbf{y}^{(u)}, \theta) - d(\mathbf{x}, \mathbf{x}^{(u)})$$

we adjust the node costs on each column. For the k'th column, we subtract ϵ from each of the node costs *except* the one corresponding to the k'th term in $\mathbf{x}^{(u)}$. Then the sum of edge and node terms along any path will correspond to $C(\mathbf{x}; \mathbf{y}^{(u)}, \theta) - d(\mathbf{x}, \mathbf{x}^{(u)})$. In turn, this means we can construct $\mathbf{x}^{(*,u)}$ by dynamic programming to this offset trellis.

Now we can compute the gradient for any example, so learning is (conceptually) straightforward. In practice, computing the gradient at any example involves finding the best sequence predicted by the loss augmented constraint violation, then using this to compute the gradient. Every gradient evaluation involves a round of inference, making the method slow.

Remember This: *A key learning principle for CRFs is that inference on the model applied to training observations should produce the corresponding hidden states. This leads to an algorithm that repeatedly: (a) computes the current best inferred set of hidden states; (b) adjusts the cost functions so that the desired sequence scores better than the current best.*

14.4 You Should

14.4.1 Remember These Terms

14.4.2 Remember These Procedures

14.4.3 Be Able to

- Explain the label bias problem.
- Explain the difference between a CRF sequence model and an HMM.
- Use dynamic programming to infer a sequence of labels for a CRF sequence model.
- Set up and solve the learning problem for a CRF sequence model in your application domain.

Problems

14.1. You are presented with a set of training sequences for a model that must infer a binary sequence (of 0's and 1's) from a sequence of observed strings of two characters ("a"s and "b"s). You see the following set of pairs:

111111000000	aaaaaabbbbbb
000000111111	bbbbbbaaaaaa
0001111000	bbbaaaabbb
000000	aaaaaa
111111	bbbbbb

  (a) You decide to represent the model using an HMM, learning model parameters by maximum likelihood. What is $p(\text{'a'}|0)$?

  (b) Using this HMM model, what is $p(111111000000|aaaaaabbbbabb)$? Why?

  (c) You decide instead to represent the model using a CRF. Your model is represented by six parameters. Write c_{01}, etc. for the cost associated with a 1 following a 0, c_{1a} etc. for the cost of observing an "a" when the model is in state 1. Assume that all costs are finite. Now assume that $c_{11} < c_{10}$, $c_{00} < c_{01}$, $c_{1a} < c_{0a}$, and $c_{0b} < c_{1b}$. If $c_{01} + c_{10} + c_{1a} > c_{00} + c_{00} + c_{0a}$, show the model will infer 111111000000 when it observes "aaaaaabbbabb."

Programming Exercises

14.2. You determine, using your skill and judgement, that any word on a list of four letter words requires censorship. You implement a censoring procedure that chooses to replace either one or two letters in the word. The number of letters is chosen uniformly and at random. Your procedure then chooses the letters to replace by selecting either one or two distinct locations at random, then replacing the letter in that location with a letter randomly selected from the 26 lowercase letters. The list of words is *fair, lair, bear, tear, wear, tare, cart, tart, mart, marl, turl, hurl, duck, muck, luck, cant, want, aunt, hist, mist, silt, wilt, fall, ball, bell*

  (a) For each word in the list of 25, generate 20 censored examples using the model. Why is it hard to use this information to build a model of $p(\text{word}|\text{censored word})$ that is useful?

  (b) We will recover the censored word from the true word using a simplified version of discriminative learning of sequence models. We will work with 1 element sequences (i.e., each whole word is a sequence). The cost function for a pair of true/censored word is

      number of letters in common $+ \lambda$number of different letters

  so, for example, if the true word is "mist" and the censored word is "malt," the cost is $2 + 2\lambda$. Using your training dataset, choose the value of λ that produces the best performance (simple search should do this - you don't need stochastic gradient descent).

  (c) Now generate a test set of censored words, and infer their true form. How accurate is your model?

14.3. Warning: this exercise is fairly elaborate, though straightforward. We will correct text errors using a sequence model.

  (a) Obtain the text of a copyright-free book in plain characters. One natural source is Project Gutenberg, at https://www.gutenberg.org. Simplify this text by dropping all punctuation marks except spaces, mapping capital letters to lowercase, and mapping groups of many spaces to a single space. The result will have 27 symbols (26 lowercase letters and a space).

(b) Construct a corrupted version of the text by passing it through a process that, with probability p_c, replaces a character with a randomly chosen character, and otherwise reports the original character.

(c) Use the first half of the text to build a sequence model. You should look only at bigrams. You should use 27×27 $\psi_j(U, V)$; each of these is itself a 27×27 table containing zeros in all but one location. You should use 27×27 $\phi_j(U, V)$. Again, each of these is itself a 27×27 table containing zeros in all but one location. You should use the Hamming distance to augment the constraint violation loss, and train the method as in the text. Use $\epsilon = 1e - 2$.

(d) Now corrupt the second half of the text using the error process above. Using the model you fitted in the previous subexercise, denoise this text. How accurate is your method?

(e) Now try to fit your model with a much larger value of ϵ. What happens?

C H A P T E R   15

Mean Field Inference

Graphical models are important and useful, but come with a serious practical problem. For many models, we cannot compute either the normalizing constant or the maximum a posteriori state. It will help to have some notation. Write X for a set of observed values, $H_1, \ldots, H_N$ for the unknown (hidden) values of interest. We will assume that these are discrete. We seek the values of $H_1, \ldots, H_N$ that maximizes $P(H_1, \ldots, H_N | X)$. There is an exponential number of such possible values, so we must exploit some kind of structure in the problem to find the maximum. In the case of a model that could be drawn as a forest, this structure was easily found; for models which can't, mostly that structure isn't there. This means the model is formally intractable—there is no practical prospect of an efficient algorithm for finding the maximum.

There are two reasons not to use this problem as a reason to simply ignore graphical models. First, graphical models that quite naturally describe interesting application problems are intractable. This chapter will work with one such model for denoising images. Second, there are quite good approximation procedures for extracting information from intractable models. This chapter will describe one such procedure.

15.1   Useful but Intractable Models

Here is a formal model we can use. A **Boltzmann machine** is a distribution model for a set of binary random variables. Assume we have N binary random variables U_i, which take the values 1 or -1. The values of these random variables are not observed (the true values of the pixels). These binary random variables are not independent. Instead, we will assume that some (but not all) pairs are coupled. We could draw this situation as a graph (Fig. 15.1), where each node represents a U_i and each edge represents a coupling. The edges are weighted, so the coupling strengths vary from edge to edge.

Write $\mathcal{N}(i)$ for the set of random variables whose values are coupled to that of i—these are the neighbors of i in the graph. The joint probability model is

$$\log P(U|\theta) = \left[\sum_i \sum_{j \in \mathcal{N}(i)} \theta_{ij} U_i U_j \right] - \log Z(\theta) = -E(U|\theta) - \log Z(\theta).$$

Now $U_i U_j$ is 1 when U_i and U_j agree, and -1 otherwise (this is why we chose U_i to take values 1 or -1). The θ_{ij} are the edge weights; notice if $\theta_{ij} > 0$, the model generally prefers U_i and U_j to agree (as in, it will assign higher probability to states where they agree, unless other variables intervene), and if $\theta_{ij} < 0$, the model prefers they disagree.

© Springer Nature Switzerland AG 2019
D. Forsyth, *Applied Machine Learning*,
https://doi.org/10.1007/978-3-030-18114-7_15

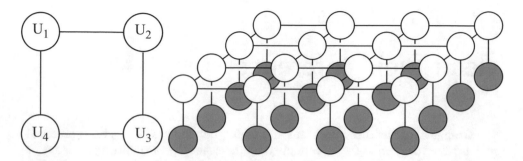

Figure 15.1: On the **left**, a simple Boltzmann machine. Each U_i has two possible states, so the whole thing has 16 states. Different choices of the constants coupling the U's along each edge lead to different probability distributions. On the **right**, this Boltzmann machine adapted to denoising binary images. The shaded nodes represent the known pixel values (X_i in the text) and the open nodes represent the (unknown, and to be inferred) true pixel values H_i. Notice that pixels depend on their neighbors in the grid. There are 2^{16} states for X in this simple example

Here $E(U|\theta)$ is sometimes referred to as the **energy** (notice the sign—higher energy corresponds to lower probability) and $Z(\theta)$ ensures that the model normalizes to 1, so that

$$Z(\theta) = \underset{\text{all values of U}}{\Sigma} \left[\exp\left(-E(U|\theta)\right)\right].$$

15.1.1 Denoising Binary Images with Boltzmann Machines

Here is a simple model for a binary image that has been corrupted by noise. At each pixel, we observe the corrupted value, which is binary. Hidden from us are the true values of each pixel. The observed value at each pixel is random, but depends only on the true value. This means that, for example, the value at a pixel can change, but the noise doesn't cause blocks of pixels to, say, shift left. This is a fairly good model for many kinds of transmission noise, scanning noise, and so on. The true value at each pixel is affected by the true value at each of its neighbors—a reasonable model, as image pixels tend to agree with their neighbors.

We can apply a Boltzmann machine. We split the U into two groups. One group represents the observed value at each pixel (I will use X_i, and the convention that i chooses the pixel), and the other represents the hidden value at each pixel (I will use H_i). Each observation is either 1 or -1. We arrange the graph so that the edges between the H_i form a grid, and there is a link between each X_i and its corresponding H_i (but no other—see Fig. 15.1).

Assume we know good values for θ. We have

$$P(H|X,\theta) = \frac{\exp(-E(H,X|\theta))/Z(\theta)}{\Sigma_H\left[\exp(-E(H,X|\theta))/Z(\theta)\right]} = \frac{\exp\left(-E(H,X|\theta)\right)}{\Sigma_H \exp\left(-E(H,X|\theta)\right)},$$

so posterior inference doesn't require evaluating the normalizing constant. This isn't really good news. Posterior inference still requires a sum over an exponential number of values. Unless the underlying graph is special (a tree or a forest) or very small, posterior inference is intractable.

You might think that focusing on MAP inference will solve this problem. Recall that MAP inference seeks the values of H to maximize $P(H|X,\theta)$ or equivalently, maximizing the log of this function. We seek

$$\underset{H}{\text{argmax}}\ \log P(H|X,\theta) = (-E(H,X|\theta)) - \log\left[\Sigma_H \exp\left(-E(H,X|\theta)\right)\right]$$

but the second term is not a function of H, so we could avoid the intractable sum. This doesn't mean the problem is tractable. Some pencil and paper work will establish that there is some set of constants a_{ij} and b_j so that the solution is obtained by solving

$$\underset{H}{\text{argmax}}\ \left(\textstyle\sum_{ij} a_{ij}h_ih_j\right) + \sum_j b_j h_j\ .$$
$$\text{subject to } h_i \in \{-1,1\}$$

This is a combinatorial optimization problem with considerable potential for unpleasantness. How nasty it is depends on some details of the a_{ij}, but with the right choice of weights a_{ij}, the problem is **max-cut**, which is NP-hard.

Remember This: *A natural model for denoising a binary image is to assume that there are unknown, true pixel values that tend to agree with the observed noisy pixel values and with one another. This model is intractable—you can't compute the normalizing constant, and you can't find the best set of true pixel values.*

15.1.2 A Discrete Markov Random Field

Boltzmann machines are a simple version of a much more complex device widely used in computer vision and other applications. In a Boltzmann machine, we took a graph and associated a binary random variable with each node and a coupling weight with each edge. This produced a probability distribution. We obtain a **Markov random field** by placing a random variable (doesn't have to be binary, or even discrete) at each node, and a coupling function (almost anything works) at each edge. Write U_i for the random variable at the i'th node, and $\theta(U_i, U_j)$ for the coupling function associated with the edge from i to j (the arguments tell you which function; you can have different functions on different edges).

We will ignore the possibility that the random variables are continuous. A **discrete Markov random field** has all U_i discrete random variables with a finite set of possible values. Write U_i for the random variable at each node, and $\theta(U_i, U_j)$ for the coupling function associated with the edge from i to j (the arguments tell you which function; you can have different functions on different edges). For a discrete Markov random field, we have

$$\log P(U|\theta) = \left[\sum_i \sum_{j \in \mathcal{N}(i)} \theta(U_i, U_j)\right] - \log Z(\theta).$$

It is usual—and a good idea—to think about the random variables as indicator functions, rather than values. So, for example, if there were three possible values at node i, we represent U_i with a 3D vector containing one indicator function for each value. One of the components must be one, and the other two must be zero. Vectors like this are sometimes known as **one-hot vectors**. The advantage of this representation is that it helps to keep track of the fact that the *values* that each random variable can take are not really to the point; it's the *interaction* between assignments that matters. Another advantage is that we can easily keep track of the parameters that matter. I will adopt this convention in what follows.

I will write $\mathbf{u}_i$ for the random variable at location i represented as a vector. All but one of the components of this vector are zero, and the remaining component is 1. If there are $\#(U_i)$ possible values for U_i and $\#(U_j)$ possible values for U_j, we can represent $\theta(U_i, U_j)$ as a $\#(U_i) \times \#(U_j)$ table of values. I will write $\Theta^{(ij)}$ for the table representing $\theta(U_i, U_j)$, and $\theta_{mn}^{(ij)}$ for the m, n'th entry of that table. This entry is the value of $\theta(U_i, U_j)$ when U_i takes its m'th value and U_j takes its n'th value. I write $\Theta^{(ij)}$ for a matrix whose m, n'th component is $\theta_{mn}^{(ij)}$. In this notation, I write

$$\theta(U_i, U_j) = \mathbf{u}_i^T \Theta^{(ij)} \mathbf{u}_j.$$

All this does not simplify computation of the normalizing constant. We have

$$Z(\theta) = \sum_{\text{all values of } \mathbf{u}} \left[\exp\left(\sum_i \sum_{j \in \mathcal{N}(i)} \mathbf{u}_i^T \Theta^{(ij)} \mathbf{u}_j\right)\right].$$

Note that the collection of all values of $\mathbf{u}$ has rather nasty structure, and is very big—it consists of all possible one-hot vectors representing each U.

15.1.3 Denoising and Segmenting with Discrete MRFs

A simple denoising model for images that aren't binary is just like the binary denoising model. We now use a discrete MRF. We split the U into two groups, H and X. We observe a noisy image (the X values) and we wish to reconstruct the true pixel values (the H). For example, if we are dealing with grey level images with 256 different possible grey values at each pixel, then each H has 256 possible values. The graph is a grid for the H and one link from an X to the corresponding H (like Fig. 15.1). Now we think about $P(H|X, \theta)$. As you would expect, the model is intractable—the normalizing constant can't be computed.

Worked Example 15.1   *A Simple Discrete MRF for Image Denoising*

Set up an MRF for grey level image denoising.

Solution: Construct a graph that is a grid. The grid represents the true value of each pixel, which we expect to be unknown. Now add an extra node for each grid element, and connect that node to the grid element. These nodes represent the observed value at each pixel. As before, we will separate the variables U into two sets, X for observed values and H for hidden values (Fig. 15.1). In most grey level images, pixels take one of 256 ($= 2^8$) values. For the moment, we work with a grey level image, so each variable takes one of 256 values. There is no reason to believe that any one pixel behaves differently from any other pixel, so we expect the $\theta(H_i, H_j)$ not to depend on the pixel location; there'll be one copy of the same function at each grid edge. By far the most usual case has

$$\theta(H_i, H_j) = \left[\begin{array}{ll} 0 & \text{if } H_i = H_j \\ c & \text{otherwise,} \end{array} \right.$$

where $c > 0$. Representing this function using one-hot vectors is straightforward. There is no reason to believe that the relationship between observed and hidden values depends on the pixel location. However, large differences between observed and hidden values should be more expensive than small differences. Write X_j for the observed value at node j, where j is the observed value node corresponding to H_i. We usually have

$$\theta(H_i, X_j) = (H_i - X_j)^2.$$

If we think of H_i as an indicator function, then this function can be represented as a vector of values; one of these values is picked out by the indicator. Notice there is a different vector at each H_i node (because there may be a different X_i at each).

Now write $\mathbf{h}_i$ for the hidden variable at location i represented as a vector, etc. Remember, all but one of the components of this vector are zero, and the remaining component is 1. The one-hot vector representing an observed value at location i is $\mathbf{x}_i$. I write $\Theta^{(ij)}$ for a matrix who's m, n'th component is $\theta^{(ij)}_{mn}$. In this notation, I write

$$\theta(H_i, H_j) = \mathbf{h}_i^T \Theta^{(ij)} \mathbf{h}_j$$

and

$$\theta(H_i, X_j) = \mathbf{h}_i^T \Theta^{(ij)} \mathbf{x}_j = \mathbf{h}_i^T \beta_i.$$

In turn, we have

$$\log p(H|X) = \left[\left(\sum_{ij} \mathbf{h}_i^T \Theta^{(ij)} \mathbf{h}_j \right) + \sum_i \mathbf{h}_i^T \beta_i \right] + \log Z.$$

Worked Example 15.2 *Denoising MRF—II*

Write out $\Theta^{(ij)}$ for the $\theta(H_i, H_j)$ with the form given in example 15.1 using the one-hot vector notation.

Solution: This is more a check you have the notation. $c\mathcal{I}$ is the answer.

Worked Example 15.3 *Denoising MRF—III*

Assume that we have $X_1 = 128$ and $\theta(H_i, X_j) = (H_i - X_j)^2$. What is β_1 using the one-hot vector notation? Assume pixels take values in the range $[0, 255]$.

Solution: Again, a check you have the notation. We have

$$\beta_1 = \begin{pmatrix} 128^2 & \text{first component} \\ \cdots & \\ (i - 128)^2 & i\text{'th component} \\ \cdots & \\ 127^2 & \end{pmatrix}.$$

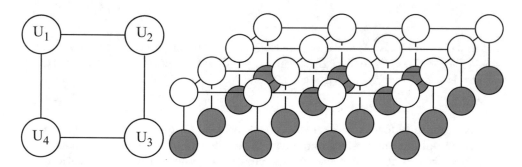

Figure 15.2: For problems like image segmentation, hidden labels may be linked to many observed labels. So, for example, the segment label at one pixel might depend on the values of many pixels. This is a sketch of such a graph. The shaded nodes represent the known pixel values (X_i in the text) and the open nodes represent the (unknown, and to be inferred) labels H_i. A particular hidden node may depend on many pixels, because we will use all these pixel values to compute the cost of labelling that node in a particular way

Segmentation is another application that fits this recipe. We now want to break the image into a set of regions. Each region will have a label (e.g., "grass," "sky," "tree," etc.). The X_i are the observed values of each pixel value, and the

H_i are the labels. In this case, the graph may have quite complex structure (e.g., Fig. 15.2). We must come up with a process that computes the cost of labelling a given pixel location in the image with a given label. Notice this process could look at many other pixel values in the image to come up with the label, but not at other labels. There are many possibilities. For example, we could build a logistic regression classifier that predicts the label at a pixel from image features around that pixel (if you don't know any image feature constructions, assume we use the pixel color; if you do, you can use anything that pleases you). We then model the cost of having a particular label at a particular point as the negative log probability of the label under that model. We obtain the $\theta(H_i, H_j)$ by assuming that labels on neighboring pixels should agree with one another, as in the case of denoising.

15.1.4 MAP Inference in Discrete MRFs Can Be Hard

As you should suspect, focusing on MAP inference doesn't make the difficulty go away for discrete Markov random fields.

Worked Example 15.4   *Useful Facts About MRFs*

Show that, using the notation of the text, we have: (a) for any i, $\mathbf{1}^T \mathbf{h}_i = 1$; (b) the MAP inference problem can be expressed as a quadratic program, with linear constraints, on discrete variables.

Solution: For (a) the equation is true because exactly one entry in $\mathbf{h}_i$ is 1, the others are zero. But (b) is more interesting. MAP inference is equivalent to maximizing $\log p(H|X)$. Recall $\log Z$ does not depend on the $\mathbf{h}$. We seek

$$\max_{\mathbf{h}_1, \ldots, \mathbf{h}_N} \left[\left(\sum_{ij} \mathbf{h}_i^T \Theta^{(ij)} \mathbf{h}_j \right) + \sum_i \mathbf{h}_i^T \beta_i \right] + \log Z$$

subject to very important constraints. We must have $\mathbf{1}^T \mathbf{h}_i = 1$ for all i. Furthermore, any component of any $\mathbf{h}_i$ must be either 0 or 1. So we have a quadratic program (because the cost function is quadratic in the variables), with linear constraints, on discrete variables.

Example 15.4 is a bit alarming, because it implies (correctly) that MAP inference in MRFs can be very hard. You should remember this. Gradient descent is no use here because the idea is meaningless. You can't take a gradient with respect to discrete variables. If you have the background, it's quite easy to prove by producing (e.g., from Example 15.4) an MRF where inference is equivalent to max-cut, which is NP-hard.

Worked Example 15.5 *MAP Inference for MRFs Is a Linear Program*

Show that, using the notation of the text, the MAP inference for an MRF problem can be expressed as a linear program, with linear constraints, on discrete variables.

Solution: If you have two binary variables z_i and z_j both in $\{0, 1\}$, then write $q_{ij} = z_i z_j$. We have that $q_{ij} \le z_i$, $q_{ij} \le z_j$, $q_{ij} \in \{0, 1\}$, and $q_{ij} \ge z_i + z_j - 1$. You should check (a) these inequalities and (b) that q_{ij} is uniquely identified by these inequalities. Now notice that each $\mathbf{h}_i$ is just a bunch of binary variables, and the quadratic term $\mathbf{h}_i^T \Theta^{(ij)} \mathbf{h}_j$ is linear in q_{ij}.

Example 15.5 is the start of an extremely rich vein of approximation mathematics, which we shall not mine. If you are of a deep mathematical bent, you can phrase everything in what follows in terms of approximate solutions of linear programs. For example, this makes it possible to identify MRFs for which MAP inference can be done in polynomial time; the family is more than just trees. We won't go there.

Remember This: *A natural model for denoising general images follows the line of the binary image model. One assumes that there are unknown, true pixel values that tend to agree with the observed noisy pixel values and with one another. This model is intractable—you can't compute the normalizing constant, and you can't find the best set of true pixel values. This is also a natural model of image segmentation, where the unknown values are segment identities.*

15.2 Variational Inference

We could just ignore intractable models, and stick to tractable models. This isn't a good idea, because intractable models are often quite natural. The discrete Markov random field model of an image is a fairly natural model. Image labels *should* depend on pixel values, and on neighboring labels. It is better to try and deal with the intractable model. One really successful strategy for doing so is to choose a tractable parametric family of probability models $Q(H; \theta)$, then adjust θ to find parameter values $\hat{\theta}$ that represent a distribution that is "close" in the right sense to $P(H|X)$. One then extracts information from $Q(H; \hat{\theta})$. This process is known as **variational inference**. What is remarkable is that (a) it is possible to find a $Q(H; \hat{\theta})$ without too much fuss and (b) information extracted from this distribution is often accurate and useful.

> **Remember This:** *Variational inference tries to find a tractable distribution $Q(H;\hat{\theta})$ that is "close" to an intractable $P(H|X)$. One then extracts information from $Q(H;\hat{\theta})$.*

15.2.1 The KL Divergence

Assume we have two probability distributions $P(X)$ and $Q(X)$. A measure of their similarity is the **KL divergence** (or sometimes **Kullback–Leibler divergence**) written as

$$\mathbb{D}(P \,\|\, Q) = \int P(X) \log \frac{P(X)}{Q(X)} dX$$

(you've clearly got to be careful about zeros in P and Q here). This likely strikes you as an odd measure of similarity, because it isn't symmetric. It is not the case that $\mathbb{D}(P \,\|\, Q)$ is the same as $\mathbb{D}(Q \,\|\, P)$, which means you have to watch your P's and Q's. Furthermore, some work will demonstrate that it does not satisfy the triangle inequality, so KL divergence lacks two of the three important properties of a metric.

KL divergence has some nice properties, however. First, we have

$$\mathbb{D}(P \,\|\, Q) \geq 0$$

with equality only if P and Q are equal almost everywhere (i.e., except on a set of measure zero).

> **Remember This:** *The KL divergence measures the similarity of two probability distributions. It is always non-negative, and is only zero if the two distributions are the same. However, it is not symmetric.*

Second, there is a suggestive relationship between KL divergence and maximum likelihood. Assume that X_i are IID samples from some *unknown* $P(X)$, and we wish to fit a parametric model $Q(X|\theta)$ to these samples. This is the usual situation we deal with when we fit a model. Now write $H(P)$ for the entropy of $P(X)$, defined by

$$H(P) = -\int P(X) \log P(X) dx = -\mathbb{E}_P[\log P].$$

The distribution P is unknown, and so is its entropy, but it is a constant. Now we can write

$$\mathbb{D}(P \,\|\, Q) = \mathbb{E}_P[\log P] - \mathbb{E}_P[\log Q].$$

Then

$$\mathcal{L}(\theta) = \sum_i \log Q(X_i|\theta) \approx \int P(X) \log Q(X|\theta) dX \quad = \quad \mathbb{E}_{P(X)}[\log Q(X|\theta)]$$

$$= \quad -H(P) - \mathbb{D}(P \| Q)(\theta).$$

Equivalently, we can write

$$\mathcal{L}(\theta) + \mathbb{D}(P \| Q)(\theta) = -H(P).$$

Recall P doesn't change (though it's unknown), so $H(P)$ is also constant (though unknown). This means that when $\mathcal{L}(\theta)$ goes up, $\mathbb{D}(P \| Q)(\theta)$ must go down. When $\mathcal{L}(\theta)$ is at a maximum, $\mathbb{D}(P \| Q)(\theta)$ must be at a minimum. All this means that, when you choose θ to maximize the likelihood of some dataset given θ for a parametric family of models, you are choosing the model in that family with smallest KL divergence from the (unknown) $P(X)$.

Remember This: *Maximum likelihood estimation recovers the parameters of a distribution in the chosen family that is closest in KL divergence to the data distribution.*

15.2.2 The Variational Free Energy

We have a $P(H|X)$ that is hard to work with (usually because we can't evaluate $P(X)$) and we want to obtain a $Q(H)$ that is "close to" $P(H|X)$. A good choice of "close to" is to require that

$$\mathbb{D}(Q(H) \| P(H|X))$$

is small. Expand the expression for KL divergence, to get

$$\begin{aligned} \mathbb{D}(Q(H) \| P(H|X)) &= \mathbb{E}_Q[\log Q] - \mathbb{E}_Q[\log P(H|X)] \\ &= \mathbb{E}_Q[\log Q] - \mathbb{E}_Q[\log P(H, X)] + \mathbb{E}_Q[\log P(X)] \\ &= \mathbb{E}_Q[\log Q] - \mathbb{E}_Q[\log P(H, X)] + \log P(X) \end{aligned}$$

which at first glance may look unpromising, because we can't evaluate $P(X)$. But $\log P(X)$ is fixed (although unknown). Now rearrange to get

$$\begin{aligned} \log P(X) &= \mathbb{D}(Q(H) \| P(H|X)) - (\mathbb{E}_Q[\log Q] - \mathbb{E}_Q[\log P(H, X)]) \\ &= \mathbb{D}(Q(H) \| P(H|X)) - \mathsf{E}_Q. \end{aligned}$$

Here

$$\mathsf{E}_Q = (\mathbb{E}_Q[\log Q] - \mathbb{E}_Q[\log P(H, X)])$$

is referred to as the **variational free energy**. We can't evaluate $\mathbb{D}(Q(H) \| P(H|X))$. But, because $\log P(X)$ is fixed, when E_Q goes down, $\mathbb{D}(Q(H) \| P(H|X))$ must go down too. Furthermore, a minimum of E_Q will correspond to a minimum of $\mathbb{D}(Q(H) \,\|\, P(H|X))$. And we can evaluate E_Q.

We now have a strategy for building approximate $Q(H)$. We choose a family of approximating distributions. From that family, we obtain the $Q(H)$ that minimizes E_Q (which will take some work). The result is the $Q(H)$ in the family that minimizes $\mathbb{D}(Q(H) \,\|\, P(H|X))$. We use that $Q(H)$ as our approximation to $P(H|X)$, and extract whatever information we want from $Q(H)$.

Remember This: *The variational free energy of Q gives a bound on the KL divergence $\mathbb{D}(Q(H) \,\|\, P(H|X))$, and is tractable if Q was chosen sensibly. We select the element of the family of Q with the smallest variational free energy.*

15.3 Example: Variational Inference for Boltzmann Machines

We want to construct a $Q(H)$ that approximates the posterior for a Boltzmann machine. We will choose $Q(H)$ to have one factor for each hidden variable, so $Q(H) = q_1(H_1)q_2(H_2)\dots q_N(H_N)$. We will then assume that all but one of the terms in Q are known, and adjust the remaining term. We will sweep through the terms doing this until nothing changes.

The i'th factor in Q is a probability distribution over the two possible values of H_i, which are 1 and -1. There is only one possible choice of distribution. Each q_i has one parameter $\pi_i = P(\{H_i = 1\})$. We have

$$q_i(H_i) = (\pi_i)^{\frac{(1+H_i)}{2}} (1 - \pi_i)^{\frac{(1-H_i)}{2}} .$$

Notice the trick that the power each term is raised to is either 1 or 0, and I have used this trick as a switch to turn on or off each term, depending on whether H_i is 1 or -1. So $q_i(1) = \pi_i$ and $q_i(-1) = (1 - \pi_i)$. This is a standard, and quite useful, trick. We wish to minimize the variational free energy, which is

$$\mathsf{E}_Q = (\mathbb{E}_Q[\log Q] - \mathbb{E}_Q[\log P(H, X)]).$$

We look at the $\mathbb{E}_Q[\log Q]$ term first. We have

$$
\begin{aligned}
\mathbb{E}_Q[\log Q] &= \mathbb{E}_{q_1(H_1)\dots q_N(H_N)}[\log q_1(H_1) + \dots \log q_N(H_N)] \\
&= \mathbb{E}_{q_1(H_1)}[\log q_1(H_1)] + \dots \mathbb{E}_{q_N(H_N)}[\log q_N(H_N)],
\end{aligned}
$$

where we get the second step by noticing that

$$\mathbb{E}_{q_1(H_1)\dots q_N(H_N)}[\log q_1(H_1)] = \mathbb{E}_{q_1(H_1)}[\log q_1(H_1)]$$

(write out the expectations and check this if you're uncertain).

Now we need to deal with $\mathbb{E}_Q[\log P(H, X)]$. We have

$$
\begin{aligned}
\log p(H, X) &= -E(H, X) - \log Z \\
&= \sum_{i \in H} \sum_{j \in \mathcal{N}(i) \cap H} \theta_{ij} H_i H_j + \sum_{i \in H} \sum_{j \in \mathcal{N}(i) \cap X} \theta_{ij} H_i X_j + K
\end{aligned}
$$

(where K doesn't depend on any H and is so of no interest). Assume all the q's are known except the i'th term. Write $Q_{\hat{i}}$ for the distribution obtained by omitting q_i from the product, so $Q_{\hat{1}} = q_2(H_2) q_3(H_3) \dots q_N(H_N)$, etc. Notice that

$$
\mathbb{E}_Q[\log P(H, X)] = \left(\begin{array}{l} q_i(-1) \mathbb{E}_{Q_{\hat{i}}}[\log P(H_1, \dots, H_i = -1, \dots, H_N, X)] + \\ q_i(1) \mathbb{E}_{Q_{\hat{i}}}[\log P(H_1, \dots, H_i = 1, \dots, H_N, X)] \end{array} \right).
$$

This means that if we fix all the q terms *except* $q_i(H_i)$, we must choose q_i to minimize

$$
\begin{aligned}
q_i(-1) \log q_i(-1) + q_i(1) \log q_i(1)\quad &- \\
q_i(-1) \mathbb{E}_{Q_{\hat{i}}}[\log P(H_1, \dots, H_i = -1, \dots, H_N, X)]\quad &+ \\
q_i(1) \mathbb{E}_{Q_{\hat{i}}}[\log P(H_1, \dots, H_i = 1, \dots, H_N, X)] &
\end{aligned}
$$

subject to the constraint that $q_i(1) + q_i(-1) = 1$. Introduce a Lagrange multiplier to deal with the constraint, differentiate and set to zero, and get

$$
\begin{aligned}
q_i(1) &= \frac{1}{c} \exp \left(\mathbb{E}_{Q_{\hat{i}}}[\log P(H_1, \dots, H_i = 1, \dots, H_N, X)] \right) \\
q_i(-1) &= \frac{1}{c} \exp \left(\mathbb{E}_{Q_{\hat{i}}}[\log P(H_1, \dots, H_i = -1, \dots, H_N, X)] \right) \\
\text{where } c &= \exp \left(\mathbb{E}_{Q_{\hat{i}}}[\log P(H_1, \dots, H_i = -1, \dots, H_N, X)] \right) + \\
&\quad \exp \left(\mathbb{E}_{Q_{\hat{i}}}[\log P(H_1, \dots, H_i = 1, \dots, H_N, X)] \right).
\end{aligned}
$$

In turn, this means we need to know $\mathbb{E}_{Q_{\hat{i}}}[\log P(H_1, \dots, H_i = -1, \dots, H_N, X)]$, etc., only up to a constant. Equivalently, we need to compute only $\log q_i(H_i) + K$ for K some unknown constant (because $q_i(1) + q_i(-1) = 1$). Now we compute

$$
\mathbb{E}_{Q_{\hat{i}}}[\log P(H_1, \dots, H_i = -1, \dots, H_N, X)].
$$

This is equal to

$$
\mathbb{E}_{Q_{\hat{i}}} \left[\sum_{j \in \mathcal{N}(i) \cap H} \theta_{ij}(-1) H_j + \sum_{j \in \mathcal{N}(i) \cap X} \theta_{ij}(-1) X_j + \text{terms not containing } H_i \right]
$$

which is the same as

$$
\sum_{j \in \mathcal{N}(i) \cap H} \theta_{ij}(-1) \mathbb{E}_{Q_{\hat{i}}}[H_j] + \sum_{j \in \mathcal{N}(i) \cap X} \theta_{ij}(-1) X_j + K
$$

and this is the same as

$$
\sum_{j \in \mathcal{N}(i) \cap H} \theta_{ij}(-1)((\pi_j)(1) + (1 - \pi_j)(-1)) + \sum_{j \in \mathcal{N}(i) \cap X} \theta_{ij}(-1) X_j + K
$$

and this is

$$\sum_{j \in \mathcal{N}(i) \cap H} \theta_{ij}(-1)(2\pi_j - 1) + \sum_{j \in \mathcal{N}(i) \cap X} \theta_{ij}(-1)X_j + K.$$

If you thrash through the case for

$$\mathbb{E}_{Q_{\hat{i}}}[\log P(H_1, \ldots, H_i = 1, \ldots, H_N, X)]$$

(which works the same) you will get

$$\log q_i(1) = \mathbb{E}_{Q_{\hat{i}}}[\log P(H_1, \ldots, H_i = 1, \ldots, H_N, X)] + K$$
$$= \sum_{j \in \mathcal{N}(i) \cap H} [\theta_{ij}(2\pi_j - 1)] + \sum_{j \in \mathcal{N}(i) \cap X} [\theta_{ij}X_j] + K$$

and

$$\log q_i(-1) = \mathbb{E}_{Q_{\hat{i}}}[\log P(H_1, \ldots, H_i = -1, \ldots, H_N, X)] + K$$
$$= \sum_{j \in \mathcal{N}(i) \cap H} [-\theta_{ij}(2\pi_j - 1)] + \sum_{j \in \mathcal{N}(i) \cap X} [-\theta_{ij}X_j] + K.$$

All this means that

$$\pi_i = \frac{e^a}{e^a + e^b},$$

where

$$a = e^{\left(\sum_{j \in \mathcal{N}(i) \cap H} [\theta_{ij}(2\pi_j - 1)] + \sum_{j \in \mathcal{N}(i) \cap X} [\theta_{ij}X_j] \right)}$$
$$b = e^{\left(\sum_{j \in \mathcal{N}(i) \cap H} [-\theta_{ij}(2\pi_j - 1)] + \sum_{j \in \mathcal{N}(i) \cap X} [-\theta_{ij}X_j] \right)}.$$

After this blizzard of calculation, our inference algorithm is straightforward. We visit each hidden node in turn, set the associated π_i to the value of the expression above *assuming all the other π_j are fixed at their current values*, and repeat until convergence. We can test convergence by checking the size of the change in each π_j.

We can now do anything to $Q(H)$ that we would have done to $P(H|X)$. For example, we might compute the values of H that maximize $Q(H)$ for MAP inference. It is wise to limit ones ambition here, because $Q(H)$ is an approximation. It's straightforward to set up and describe, but it isn't particularly good. The main problem is that the variational distribution is unimodal. Furthermore, we chose a variational distribution by assuming that each H_i was independent of all others. This means that computing, say, covariances will likely lead to the wrong numbers (although it's easy—almost all are zero, and the remainder are easy). Obtaining an approximation by assuming that H_i is independent of all others is often called a **mean field method**.

> **Remember This:** *A long, but easy, derivation yields a way to recover an approximation to the best denoised binary image. The inference algorithm is a straightforward optimization procedure.*

15.4   You Should

15.4.1   Remember These Terms

15.4.2   Remember These Facts

15.4.3   Be Able to

- Set up and solve a mean field model for denoising binary images.

PART SIX

Deep Networks

CHAPTER 16

Simple Neural Networks

All the classification and regression procedures we have seen till now assume that a reasonable set of features is available. If the procedure didn't work well, we needed to use domain knowledge, problem insight, or sheer luck to obtain more features. A neural network offers an alternative option: learn to make good features from the original signal. A neural network is made up of units. Each accepts a set of inputs and a set of parameters, and produces a number which is a non-linear function of the inputs and the parameters. It is straightforward to produce a k way classifier out of k units.

More interesting is to build a set of layers, by connecting the outputs of one layer of units to the inputs of the next. The first layer accepts inputs in whatever form is available. The final layer operates as a classifier. The intermediate layers each map their input into features that the next layer finds useful. As a result, the network learns features that the final classification layer finds useful. A stack of layers like this can be trained with stochastic gradient descent. The big advantage of this approach is that we don't need to worry about what features to use—the network is trained to form strong features. This is very attractive for applications like classifying images, and the best performing image classification systems known are built using neural networks.

This chapter describes units, how to make layers out of units, and how to train a neural network. Most of the tricks that are needed to build a good image classification system with a neural network are described in the next chapter.

16.1 Units and Classification

We will build complex classification systems out of simple units. A **unit** takes a vector $\mathbf{x}$ of inputs and uses a vector $\mathbf{w}$ of parameters (known as the **weights**), a scalar b (known as the **bias**), and a non-linear function F to form its output, which is

$$F(\mathbf{w}^T\mathbf{x} + b).$$

Over the years, a wide variety of non-linear functions have been tried. Current best practice is to use the **ReLU** (for rectified linear unit), where

$$F(u) = \max{(0, u)}.$$

For example, if $\mathbf{x}$ was a point on the plane, then a single unit would represent a line, chosen by the choice of $\mathbf{w}$ and b. The output for all points on one side of the line would be zero. The output for points on the other side would be a positive number that is larger for points that are further from the line.

Units are sometimes referred to as **neurons**, and there is a large and rather misty body of vague speculative analogy linking devices built out of units to neuroscience. I deprecate this practice; what we are doing here is quite useful and

D. Forsyth, *Applied Machine Learning*,
https://doi.org/10.1007/978-3-030-18114-7_16

interesting enough to stand on its own without invoking biological authority. Also, if you want to see a real neuroscientist laugh, explain to them how your neural network really models some gobbet of brain tissue or other.

16.1.1 Building a Classifier out of Units: The Cost Function

We will build a multiclass classifier out of units by modelling the class posterior probabilities using the outputs of the units. Each class will get the output of a single unit. We will organize these units into a vector $\mathbf{o}$, whose i'th component is o_i, which is the output of the i'th unit. This unit has parameters $\mathbf{w}^{(i)}$ and $b^{(i)}$. We want to use the i'th unit to model the probability that the input is of class i, which I will write $p(\text{class} = i | \mathbf{x}, \mathbf{w}^{(i)}, b^{(i)})$.

To build this model, I will use the **softmax function**. This is a function that takes a C dimensional vector and returns a C dimensional vector that is non-negative, and whose components sum to one. We have

$$\text{softmax}(\mathbf{u}) = \mathbf{s}(\mathbf{u}) = \left(\frac{1}{\sum_k e^{u_k}} \right) \begin{bmatrix} e^{u_1} \\ e^{u_2} \\ \dots \\ e^{u_C} \end{bmatrix}$$

(recall u_j is the j'th component of $\mathbf{u}$). We then use the model

$$p(\text{class} = j | \mathbf{x}, \mathbf{w}^{(i)}, b^{(i)}) = s_j(\mathbf{o}(\mathbf{x}, \mathbf{w}^{(i)}, b^{(i)})).$$

Notice that this expression passes important tests for a probability model. Each value is between 0 and 1, and the sum over classes is 1.

In this form, the classifier is not super interesting. For example, imagine that the features $\mathbf{x}$ are points on the plane, and we have two classes. Then we have two units, one for each class. There is a line corresponding to each unit; on one side of the line, the unit produces a zero, and on the other side, the unit produces a positive number that increases with perpendicular distance from the line. We can get a sense of what the decision boundary will be like from this. When a point is on the 0 side of both lines, the class probabilities will be equal (and so both $\frac{1}{2}$—two classes, remember). When a point is on the positive side of the j'th line, but the zero side of the other, the class probability for class j will be

$$\frac{e^{o_j(\mathbf{x}, \mathbf{w}^{(j)}, b^{(j)})}}{1 + e^{o_j(\mathbf{x}, \mathbf{w}^{(j)}, b^{(j)})}},$$

and the point will always be classified in the j'th class (remember, $o_j \geq 0$ because of the ReLU). Finally, when a point is on the positive side of both lines, the classifier boils down to choosing the class j that has the largest value of $o_j(\mathbf{x}, \mathbf{w}^{(j)}, b^{(j)})$. All this leads to the decision boundary shown in Fig. 16.1. Notice that this is piecewise linear, and somewhat more complex than the boundary of an SVM.

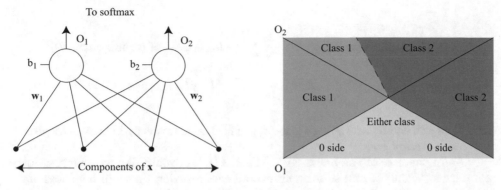

Figure 16.1: On the **left**, a simple classifier consisting of two units observing an input vector (in this case 4D), and providing outputs. On the **right**, the decision boundary for two units classifying a point in 2D into one of the two classes. As described in the text, the boundary consists of a set of rays. There are three regions: in one, the point is class 1; in the second, the point is class 2; and in the third, each class has the same probability. The angle of the dashed line depends on the magnitudes of $\mathbf{w}^{(1)}$ and $\mathbf{w}^{(2)}$

> **Remember This:** *Neural networks can make excellent classifiers. Outputs are passed through a softmax layer to produce estimates of the posterior an input belongs to a class. A very simple k class classifier can be made out of k units. This should be compared to logistic regression.*

16.1.2 Building a Classifier out of Units: Strategy

The essential difficulty here is to choose the $\mathbf{w}$'s and b's that result in the best behavior. We will do so by writing a cost function that estimates the error rate of the classification, then searching for a value that makes that function small. We have a set of N examples $\mathbf{x}_i$ and for each example we know the class. There are a total of C classes. We encode the class of an example using a **one hot** vector $\mathbf{y}_i$, which is C dimensional. If the i'th example is from class j, then the j'th component of $\mathbf{y}_i$ is 1, and all other components in the vector are 0. I will write y_{ij} for the j'th component of $\mathbf{y}_i$.

A natural cost function looks at the log likelihood of the data under the probability model produced from the outputs of the units. Stack all of the coefficients into a vector θ. If the i'th example is from class j, we would like

$$-\log p(\text{class} = j | \mathbf{x}_i, \theta)$$

to be small (notice the sign here; it's usual to minimize negative log likelihood). The components of $\mathbf{y}_i$ can be used as switches, as in the discussion of EM to obtain a loss function

$$\frac{1}{N}\sum_i L_{\log}(y_i, \mathbf{s}(\mathbf{o}(\mathbf{x}_i, \theta))) = \frac{1}{N}\sum_i [-\log p(\text{class of example } i | \mathbf{x}_i, \theta)]$$

$$= \frac{1}{N}\sum_{i \in \text{data}} [\{-\mathbf{y}_i^T \log \mathbf{s}(\mathbf{o}(\mathbf{x}_i, \theta))\}]$$

(recall the j'th component of $\log \mathbf{s}$ is $\log s_j$). This loss is variously known as **log-loss** or **cross-entropy loss**.

As in the case of the linear SVM (Sect. 2.1), we would like to achieve a low cost with a "small" θ, and so form an overall cost function that will have loss and penalty terms. It isn't essential to divide by N (the minimum is in the same place either way) but doing so means the loss of a model does not grow with the size of the training set, which is often convenient.

We will penalize large sets of weights, as in the linear SVM. Remember, we have C units (one per class) and so there are C distinct sets of weights. Write the weights for the k'th unit $\mathbf{w}^{(k)}$. Our penalty becomes

$$\frac{1}{2}\sum_{k \in \text{units}} \left(\mathbf{w}^{(k)}\right)^T \mathbf{w}^{(k)}.$$

As in the case of the linear SVM (Sect. 2.1), we write λ for a weight applied to the penalty. Our cost function is then

$$S(\theta, \mathbf{x}; \lambda) = \underbrace{\frac{1}{N}\sum_{i \in \text{data}} [\{-\mathbf{y}_i^T \log \mathbf{s}(\mathbf{o}(\mathbf{x}_i, \theta))\}]}_{\text{(misclassification loss)}} + \underbrace{\frac{\lambda}{2}\sum_{k \in \text{units}} \left(\mathbf{w}^{(k)}\right)^T \mathbf{w}^{(k)}}_{\text{(penalty)}}.$$

Remember This: *Networks are trained by descent on a loss. The usual loss for a classifier is a negative log-posterior, with the posterior modelled using a softmax function. It is usual to regularize with the magnitude of weights.*

16.1.3 Building a Classifier out of Units: Training

I have described a simple classifier built out of units. We must now train this classifier, by choosing a value of θ that results in a small loss. It may be quite hard to get the true minimum, and we may need to settle for a small value. We use stochastic gradient descent, because we have seen it before; because it is effective; and because it is the algorithm of choice when training more complex classifiers built out of units.

For the SVM, we selected one example at random, computed the gradient at that example, updated the parameters, and went again. For neural nets, it is more usual to use **minibatch training**, where we select a subset of the data uniformly and at random, compute a gradient using that subset, update, and go again. This is because in the best implementations many operations are vectorized, and using a minibatch can provide a gradient estimate that is clearly better than that obtained using only one example, but doesn't take longer to compute. The size of the minibatch is usually determined by memory or architectural considerations. It is often a power of two, for this reason.

Now imagine we have chosen a minibatch of M examples. We must compute the gradient of the cost function. The penalty term is easily dealt with, but the loss term is something of an exercise in the chain rule. We drop the index for the example, and so have to handle the gradient of

$$-\mathbf{y} \log \mathbf{s}(\mathbf{o}(\mathbf{x}, \theta)) = \sum_u y_u \log s_u (\mathbf{o}(\mathbf{x}, \theta))$$

with respect to the elements of θ. Applying the chain rule, we have

$$\frac{\partial}{\partial w_a^{(j)}} [y_u \log s_u (\mathbf{o}(\mathbf{x}, \theta))] = y_u \left[\sum_v \frac{\partial \log s_u}{\partial o_v} \frac{\partial o_v}{\partial w_a^{(j)}} \right]$$

and

$$\frac{\partial}{\partial b^{(j)}} [y_u \log s_u (\mathbf{o}(\mathbf{x}, \theta))] = y_u \left[\sum_v \frac{\partial \log s_u}{\partial o_v} \frac{\partial o_v}{\partial b^{(j)}} \right].$$

The relevant partial derivatives are straightforward. Write $\mathbb{I}_{[u=v]}(u, v)$ for the indicator function that is 1 when $u = v$ and zero otherwise. We have

$$
\begin{aligned}
\frac{\partial \log s_u}{\partial o_v} &= \mathbb{I}_{[u=v]} - \frac{e^{o_v}}{\sum_k e^{o_k}} \\
&= \mathbb{I}_{[u=v]} - s_v.
\end{aligned}
$$

To get the other partial derivatives, we need yet more notation (but this isn't new, it's a reminder). I will write $\mathbb{I}_{[o_u>0]}(o_u)$ for the indicator function that is 1 if its argument is greater than zero. Notice that if $j \neq u$,

$$\frac{\partial o_u}{\partial w_a^{(j)}} = 0 \text{ and } \frac{\partial o_u}{\partial b^{(j)}} = 0.$$

Then

$$\frac{\partial o_v}{\partial w_a^{(j)}} = x_a \left[\mathbb{I}_{[o_v>0]}(o_v) \right] \left[\mathbb{I}_{[u=j]}(u, j) \right]$$

and

$$\frac{\partial o_u}{\partial b^{(j)}} = \left[\mathbb{I}_{[o_u>0]}(o_u) \right] \left[\mathbb{I}_{[u=j]}(u, j) \right].$$

Once you have the gradient, you need to use it. Each step will look like $\theta^{(n+1)} = \theta^{(n)} - \eta_n \nabla_\theta \text{cost}$. You need to choose η_n for each step. This is widely known as the **learning rate**; an older term is **stepsize** (neither term is a super-accurate description). It is not usual for the stepsize to be the same throughout learning. We would like to take "large" steps early, and "small" steps late, in learning, so we would like η_n to be "large" for small n, and "small" for large n. It is tough to be precise about a good choice. As in stochastic gradient descent for a linear SVM, breaking learning into epochs ($e(n)$ is the epoch of the n'th iteration), then choosing two constants a and b to obtain

$$\eta_n = \frac{1}{a + be(n)}$$

is quite a good choice. Another rule that is quite widely used is to form

$$\eta_n = \eta(1/\gamma)^{e(n)},$$

where γ is larger than one. The constants, and the epoch size, will need to be chosen by experiment. As we build more complex collections of units, the need for a better process will become pressing.

Choosing the regularization constant follows the recipe we saw for a linear SVM. Hold out a validation dataset. Train for several different values of λ. Evaluate each system on the validation dataset, and choose the best. Notice this involves many rounds of training, which could make things slow. Evaluating the classifier is like evaluating any other classifier. You evaluate the error on a held-out dataset that *wasn't* used to choose the regularization constant, or during training.

16.2 Example: Classifying Credit Card Accounts

The UC Irvine Machine Learning Repository hosts a dataset on Taiwanese credit card accounts. The dataset records a variety of properties of credit card users, together with whether they defaulted on their payment or not. The dataset was donated by I-Cheng Yeh. The task is to predict whether an account will default or not. You can find the dataset at https://archive.ics.uci.edu/ml/datasets/default+of+credit+card+clients.

Straightforward methods work relatively poorly on this classification problem. About 22% of accounts default, so by just predicting that no account will default (the prior), you can get an error rate of 0.22. While this would likely produce happy customers, shareholders might be unhappy. A reasonable baseline is L1 regularized logistic regression (I used `glmnet` as in Sect. 11.4). An appropriate choice of regularization constant gets a cross-validated error rate with a mean of 0.19 and a standard deviation of about 0.002. This is better than the prior, but not much.

I trained a simple two-unit network using stochastic gradient descent as in the previous section. I normalized each feature separately to have zero mean and unit standard deviation. The figures show plots of the loss and error for a variety of different configurations and training options. As the figures show, this network can be made to behave slightly better than logistic regression, likely because the decision boundary is slightly more complicated. The dataset contains 30,000 examples. I used a 25,000 to train, and the rest for test. I used the gradient at a single example to update (rather than using a batch of examples). Every 100 updates, I computed

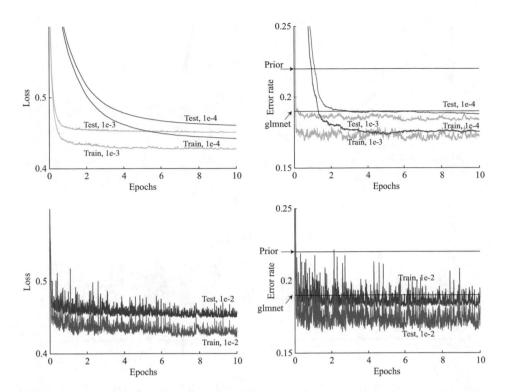

Figure 16.2: On the **left**, a plot of loss and on the **right**, a plot of error. The **top** row is for stepsizes 1e−4 and 1e−3, and the **bottom** row is for stepsize 1e−2. Each plot shows loss (resp. error) as training progresses for a two-unit classifier on the credit card account data of the text. These plots are for a regularization constant of 0.001. The test error for a stepsize of 1e−3 is significantly better than the cross-validated error produced by `glmnet`. Smaller stepsizes produce rather worse results. In the bottom row, the stepsize is rather large, resulting in quite noisy curves which are hard to compare. As the bottom row shows, a larger stepsize results in a rather less effective method

and recorded the value of both loss and error for both train and test sets. An epoch is 25,000 updates. The figures show plots of both loss and error for both train and test set as training proceeds. These plots are an important and usual part of keeping track of training.

The figures show the stepsize 1 used in each case. Generally, a curve for a large stepsize will be noisier than a curve for a small stepsize. Small stepsizes produce smoother curves, but small stepsizes usually result in poor exploration of the space of models and often produce rather higher training error (resp. loss). On the other hand, too large a stepsize means the method never settles into a good solution. In fact, a sufficiently large stepsize can cause training to diverge. I have separated the largest stepsize from the others, because the very noisy curve made the others hard to resolve. In each case, I adjusted the stepsize at the end of an epoch by multiplying by 1/1.1 (so in the last epoch, the stepsize is about 0.38 the stepsize

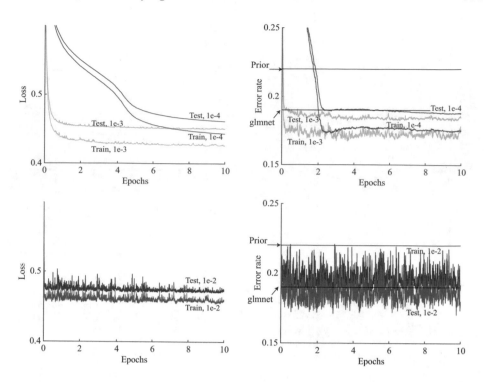

Figure 16.3: On the **left**, a plot of loss and on the **right**, a plot of error. The **top** row is for stepsizes 1e−4 and 1e−3, and the **bottom** row is for stepsize 1e−2. Each plot is for a two-unit classifier on the credit card account data. These plots are for a regularization constant of 0. You might expect a larger test–train gap than for Fig. 16.2, but the difference isn't large or reliable

at the start). You might notice that the curves are a little less noisy at the end of training than at the start. All of Figs. 16.2, 16.3, and 16.4 are on the same set of axes, and I have plotted the error rates for the two baselines as horizontal lines for reference. You should look at these figures in some detail.

For most problems, there is no natural stopping criterion (in some cases, you can stop training when error on a validation set reaches some value). I stopped training after ten epochs. As you can see from the curves, this choice seems reasonable. Each choice of stepsize and of regularization constant yields a different model. These models can be compared by looking at the curves, but it takes some search to find a good set of stepsizes and of regularization constant.

Figure 16.2 shows a model with a regularization constant of 0.001, which seems to be quite good. Reducing the regularization constant to 0 produces a rather worse set of models (Fig. 16.3), as does increasing the constant to 0.1 (Fig. 16.4). You should notice that the regularization constant has quite complicated effects. A larger regularization constant tends to yield (but doesn't guarantee) a smaller gap between test and train curves.

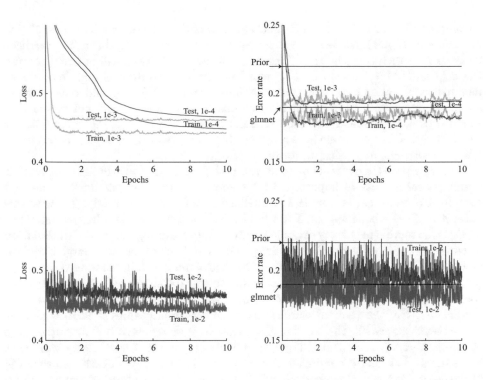

Figure 16.4: On the **left**, a plot of loss and on the **right**, a plot of error. The **top** row is for stepsizes 1e−4 and 1e−3, and the **bottom** row is for stepsize 1e−2. Each plot shows loss (resp. error) as training progresses for a two-unit classifier on the credit card account data of the text. These plots are for a regularization constant of 0.1. You might expect a smaller test–train gap than for Fig. 16.2 or for Fig. 16.3, but the difference isn't large or reliable. Test errors are notably worse than the cross-validated error produced by `glmnet`. In the bottom row, the stepsize is rather large, resulting in quite noisy curves which are hard to compare. As the bottom row shows, a larger stepsize results in a rather less effective method

Comparing Figs. 16.3 and 16.4 yields another point to remember. A model can produce lower loss *and* higher error rates. The training loss for 1e−3 and 1e−4 is better in Fig. 16.3 than in Fig. 16.4, but the error rates are higher in Fig. 16.3 than in Fig. 16.4. To explain this, remember the loss is *not* the same as the error rate. A method can improve the loss without changing the error rate by improving the probability of the right class, even if doing so doesn't change the ranking of the classes.

You should build a simple classifier like this for some choice of dataset. If you try, you will notice that the performance quite strongly depends both on choice of stepsize and of regularization constant, and there can be some form of minor interactions between them. Another important consideration is **initialization**. You need to choose the initial values of all of the parameters. There are many parameters; in our case, with a d dimensional $\mathbf{x}$ and C classes, we have $(d+1) \times C$ parameters. If you initialize each parameter to zero, you will find that the gradient is also zero, which is not helpful. This occurs because all the o_u will be zero (because the $w_{u,i}$ and the b_u are zero). It is usual to initialize to draw a sample of a zero mean normal random variable for each initial value. Each should have a small standard deviation. If you replicate my experiments above, you will notice that the size of this standard deviation matters quite a lot (I used the value 0.01 for the figures). Generally, one gets a system like this to work by trying a variety of different settings, and using an awful lot of data.

If you try this experiment (and you really should), you will also notice that **preprocessing** data is important to get good results. You should have noticed that, for the credit card example, I normalized each feature separately. Try with this dataset and see what happens if you fail to do this, but use the other parameters I used (I was unable to get any sensible progress from training). A partial explanation of why there might be problems is easy. The first feature has very large standard deviation (of the order of $1e5$) and the others have much smaller standard deviation (of the order of 1). If you fit a good classifier to this data, you will discover that the coefficient of the first feature is quite large, but some other features have quite small coefficients. But the gradient search starts with very small values of each coefficient. If the stepsize is small, the search must spend very many steps building up a large value of the coefficient; if it is large, it is likely to get small coefficients wrong. It is a general fact that a good choice of preprocessing can significantly improve the performance of a neural network classifier. However, I know of no general procedure to establish the right preprocessing for a particular problem other than trying a lot of different approaches.

Here is another significant obstacle that occurs in training. Imagine the system gets into a state where for some unit u, $o_u = 0$ for every training data item. This could happen, for example, if the learning rate was too large, or you chose an unlucky initialization. Then the system can't get out of this state, because the gradient for that unit will be zero for every training data item, too. Such units are referred to as **dead units**. For a very simple classifier like our two-unit model, this problem can be contained by keeping the learning rate small enough. In more complex architectures (below), it is also contained by having a large number of units. Figure 16.5 shows this effect occurring in an extreme version in the two-unit example for credit cards—here both units are dead.

Plots of loss can be very informative, and skilled practitioners are good at diagnosing training problems from these plots. Here are some basics. If the learning rate is small, the system will make very slow progress but may (eventually) end up in a good state. If the learning rate is large, the system will make fast progress initially, but will then stop improving, because the state will change too quickly to find a good solution. If the learning rate is very large, the system might even diverge. If the learning rate is just right, you should get fast descent to a good value, and then slow but fairly steady improvement. Of course, just as in the case of SVMs, the plot of loss against step isn't a smooth curve, but rather noisy. Complicated models can display quite startling phenomena, not all of which are understood or explained. I found an amusing collection of examples of training problems at lossfunctions.tumblr.com—you should look at this.

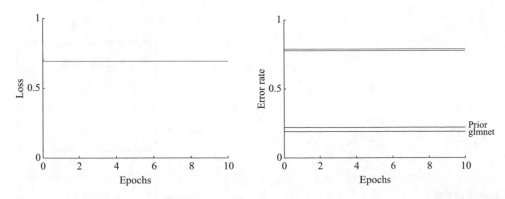

Figure 16.5:   On the **left**, a plot of loss and on the **right**, a plot of error. Each plot shows loss (resp. error) as training progresses for a two-unit classifier on the credit card account data of the text. This is an extreme case (notice the axes are different from those in the other figures) where both units died very early in training. The stepsize is 1e−3, but this has no effect because the gradient is zero. The error rate is very high, because each unit produces a zero and so the probabilities are 0.5. The tiebreaking mechanism is "choose default" which gives a particularly poor performance here

16.3   Layers and Networks

We have built a multiclass classifier out of units by using one unit per class, then interpreting the outputs of the units as probabilities using a softmax function. This classifier is at best only mildly interesting. The way to get something really interesting is to ask what the features for this classifier should be. To date, we have not looked closely at features. Instead, we've assumed that they come with the dataset or should be constructed from domain knowledge. Remember that, in the case of regression, we could improve predictions by forming non-linear functions of features. We can do better than that; we could *learn* what non-linear functions to apply, by using the output of one set of units to form the inputs of the next set.

16.3.1   Stacking Layers

We will focus on systems built by organizing the units into **layers**; these layers form a **neural network** (a term I dislike, for the reasons above, but use because everybody else does). There is an input layer, consisting of the units that receive feature inputs from outside the network. There is an output layer, consisting of units whose outputs are passed outside the network. These two might be the same, as they were in the previous section. The most interesting cases occur when they are not the same. There may be **hidden layers**, whose inputs come from other layers and whose outputs go to other layers. In our case, the layers are ordered, and outputs of a given layer act as inputs to the next layer only (as in Fig. 16.6—we don't allow connections to wander all over the network). For the moment, assume that each unit in a layer receives an input from every unit in the previous layer; this means that our network is **fully connected**. Other architectures are possible,

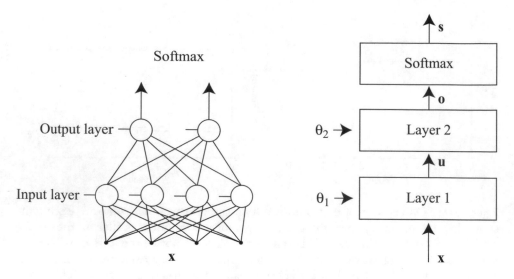

Figure 16.6: On the **left**, an input layer connected to an output layer. The units in the input layer take the inputs and compute features; the output layer turns these features into output values that will be turned into class probabilities with the softmax function. On the **right**, an abstraction of these layers. I have illustrated the softmax as a layer of its own. Each layer of units accepts a vector and some parameters, and produces another vector. The softmax layer takes no parameters

but right now the most important question is how to train the resulting object.

Figure 16.6 shows a simple generalization of our classifier, with two layers of units connected to one another. The best way to think about the systems we are building is in terms of multiple layers of functions that accept vector inputs (and often parameters) and produce vector outputs. A layer of units is one such function. But a softmax layer is another. The figure shows this abstraction applied to the collection of units. Here the first layer accepts an input $\mathbf{x}$ and some parameters $\theta^{(1)}$, and produces an output $\mathbf{u}$. The second layer accepts this $\mathbf{u}$ as input, together with some parameters, then makes an output $\mathbf{o}$. The third layer is softmax—it accepts inputs and makes outputs, but has no parameters. The whole stack of layers produces a model of the conditional probability of each class, conditioned on the input $\mathbf{x}$ and the parameters θ.

We will train objects like this with stochastic gradient descent. The important question is how to compute the gradient. The issue here is that $\mathbf{s}$ depends on $\theta^{(1)}$ only by way of $\mathbf{u}$—changing $\theta^{(1)}$ causes $\mathbf{s}$ to change only by changing $\mathbf{u}$.

I could write the object pictured in Fig. 16.6 as

$$\mathbf{s}(\mathbf{o}(\mathbf{u}(\mathbf{x}, \theta^{(1)}), \theta^{(2)}))$$

(which is a clumsy notation for something made quite clear by the picture). This is more cleanly written as

$$\mathbf{s},$$

where

$$
\begin{aligned}
\mathbf{s} &= \mathbf{s}(o) \\
\mathbf{o} &= \mathbf{o}(\mathbf{u}, \theta^{(2)}) \\
\mathbf{u} &= \mathbf{u}(\mathbf{x}, \theta^{(1)}).
\end{aligned}
$$

You should see these equations as a form of map for a computation. You feed in $\mathbf{x}$; this gives $\mathbf{u}$; which gives $\mathbf{o}$; which gives $\mathbf{s}$; from which you compute your loss.

16.3.2 Jacobians and the Gradient

Now to compute the gradient. I will do the logic in two ways, because one might work better for you than the other. For graphical thinkers, look at Fig. 16.6. The loss L will change only if $\mathbf{s}$ changes. If the input is fixed, to determine the effects of a change in $\theta^{(2)}$, you must compute the effect of that change on $\mathbf{o}$; then the effect of the change in $\mathbf{o}$ on $\mathbf{s}$; and finally, the effect of the that change in $\mathbf{s}$ on the loss. And if the input is fixed, determining the effects of a change in $\theta^{(1)}$ is more complicated. You must compute the effect of that change on $\mathbf{u}$; then the effect of the change in $\mathbf{u}$ on $\mathbf{o}$; then the effect of the change in $\mathbf{o}$ on $\mathbf{s}$; and finally, the effect of that change in $\mathbf{s}$ on the loss. It may be helpful to trace these changes along the figure with a pencil.

If you find equations easier to reason about, the last paragraph is just the chain rule, but in words. The loss is $L(\mathbf{s})$. Then you can apply the chain rule to the set of equations

$$\mathbf{s},$$

where

$$
\begin{aligned}
\mathbf{s} &= \mathbf{s}(o) \\
\mathbf{o} &= \mathbf{o}(\mathbf{u}, \theta^{(2)}) \\
\mathbf{u} &= \mathbf{u}(\mathbf{x}, \theta^{(1)}).
\end{aligned}
$$

It will help to represent the relevant derivatives cleanly with a general notation. Assume we have a vector function $\mathbf{f}$ of a vector variable $\mathbf{x}$. I will write $\#(\mathbf{x})$ to mean the number of components of $\mathbf{x}$, and x_i for the i'th component. I will write $\mathcal{J}_{\mathbf{f};\mathbf{x}}$ to mean

$$
\begin{pmatrix}
\frac{\partial f_1}{\partial x_1} & \cdots & \frac{\partial f_1}{\partial x_{\#(\mathbf{x})}} \\
\cdots & \cdots & \cdots \\
\frac{\partial f_{\#(\mathbf{f})}}{\partial x_1} & \cdots & \frac{\partial f_{\#(\mathbf{s})}}{\partial f_{\#(\mathbf{o})}}
\end{pmatrix}
$$

and refer to such a matrix of first partial derivatives as a **Jacobian** (in some circles, it is called the derivative of $\mathbf{f}$, but this convention can become confusing). The Jacobian simplifies writing out the chain rule. You should check (using whatever form of the chain rule you recall) that

$$
\nabla_{\theta^{(2)}} L = (\nabla_{\mathbf{s}} L) \times J_{\mathbf{s};\mathbf{o}} \times J_{\mathbf{o};\theta^{(2)}}.
$$

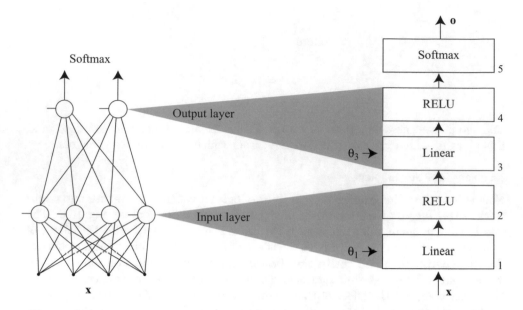

Figure 16.7: On the **left**, an input layer connected to an output layer as in Fig. 16.6. On the **right**, a further abstraction of these layers, which is useful in computing the gradient. This sees a unit as a composite of two layers. The first computes the linear function, then the second computes the ReLU

Now to get the derivative of the loss with respect to $\theta^{(1)}$ is more interesting. The loss depends on $\theta^{(1)}$ only via $\mathbf{u}$. So we must have

$$\nabla_{\theta^{(1)}} L = (\nabla_{\mathbf{s}} L) \times J_{\mathbf{s};\mathbf{o}} \times J_{\mathbf{o};\mathbf{u}} \times J_{\mathbf{u};\theta^{(1)}}.$$

The reasoning of the last two paragraphs can be extended to cover any number of layers, parameters, interconnections, and so on, as long as we have notation to keep track of each layer's inputs and outputs.

> **Remember This:** *Stacking multiple layers of units into a neural network results in learned features that should cause the final classifier to perform well. There are a variety of conventions as to what makes a layer. The chain rule yields the gradient of a network loss with respect to parameters.*

16.3.3 Setting up Multiple Layers

To go further, we will need to keep track of some details with new notation. We will see a neural network as a collection of D layers of functions. Each function accepts a vector and produces a vector. These layers are *not* layers of units. Instead, it is convenient to see a unit as a composite of two layers. The first uses a set of parameters to compute a linear function of its inputs. The second computes the ReLU from the result of the linear function. As Fig. 16.7 shows, this means that a network of two layers of units followed by a softmax has $D = 5$.

This notation allows us to abstract away from what each function does, and so compute the gradient. I will write the r'th function $\mathbf{o}^{(r)}$. Not every function will accept parameters (the ReLU and softmax functions don't; the linear functions do). I will use the convention that the r'th layer receives parameters $\theta^{(r)}$; this parameter vector will be empty if the r'th layer does not accept parameters. In this notation, the output of a network applied to $\mathbf{x}$ could be written as

$$\mathbf{o}^{(D)}(\mathbf{o}^{(D-1)}(\ldots(\mathbf{o}^1(\mathbf{x},\theta^{(1)}),\theta^{(2)}),\ldots),\theta^{(D)})$$

which is messy. More clean is to write

$$\mathbf{o}^{(D)},$$

where

$$
\begin{aligned}
\mathbf{o}^{(D)} &= \mathbf{o}^{(D)}(\mathbf{u}^{(D)},\theta^{(D)}) \\
\mathbf{u}^{(D)} &= \mathbf{o}^{(D-1)}(\mathbf{u}^{(D-1)},\theta^{(D-1)}) \\
\ldots &= \ldots \\
\mathbf{u}^{(2)} &= \mathbf{o}^{(1)}(\mathbf{u}^{(1)},\theta^1) \\
\mathbf{u}^{(1)} &= \mathbf{x}.
\end{aligned}
$$

These equations really are a map for a computation. You feed in $\mathbf{x}$; this gives $\mathbf{u}^{(1)}$; which gives $\mathbf{u}^{(2)}$; and so on, up to $\mathbf{o}^{(D)}$. This is important, because it allows us to write an expression for the gradient fairly cleanly (Fig. 16.8 captures some of this).

Our losses will usually consist of two terms. The first is an average of per-item losses, and so takes the form

$$\frac{1}{N}\sum_i L(\mathbf{y}_i,\mathbf{o}^{(D)}(\mathbf{x}_i,\theta)).$$

The second (which we won't always use) is a term that depends on the parameters, which will be used to regularize as with SVMs and regression as above. The gradient of the regularization term is easy.

16.3.4 Gradients and Backpropagation

To get the gradient of the loss term, drop the index of the example and focus on

$$L(\mathbf{y},\mathbf{o}^{(D)}),$$

where

$$
\begin{aligned}
\mathbf{o}^{(D)} &= \mathbf{o}^{(D)}(\mathbf{u}^{(D)},\theta^{(D)}) \\
\mathbf{u}^{(D)} &= \mathbf{o}^{(D-1)}(\mathbf{u}^{(D-1)},\theta^{(L-1)}) \\
\ldots &= \ldots \\
\mathbf{u}^{(2)} &= \mathbf{o}^{(1)}(\mathbf{u}^{(1)},\theta^1) \\
\mathbf{u}^{(1)} &= \mathbf{x}.
\end{aligned}
$$

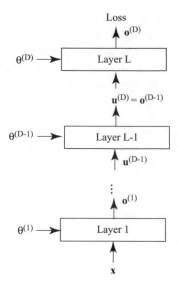

Figure 16.8: Notation for layers, inputs, and parameters, for reference

Again, think of these equations as a map for a computation. Now consider $\nabla_\theta L$ and extend our use of the chain rule from Sect. 16.1.3, very aggressively. We have

$$\nabla_{\theta^{(D)}} L(\mathbf{y}, \mathbf{o}^{(D)}) = (\nabla_{\mathbf{o}^{(D)}} L) \times J_{\mathbf{o}^{(D)};\theta^{(D)}}$$

(I ignored $\theta^{(D)}$ in Sect. 16.1.3, because layer D was a softmax layer and didn't have any parameters).

Now think about $\nabla_{\theta^{(D-1)}} L$. The loss depends on $\theta^{(D-1)}$ in a somewhat round-about way; layer $D-1$ uses $\theta^{(D-1)}$ to produce its outputs, and these are fed into layer D *as that layer's inputs*. So we must have

$$\nabla_{\theta^{(D-1)}} L(\mathbf{y}_i, \mathbf{o}^{(D)}(\mathbf{x}_i, \theta)) = (\nabla_{\mathbf{o}^{(D)}} L) \times J_{\mathbf{o}^{(D)};\mathbf{u}^{(D)}} \times J_{\mathbf{o}^{(D-1)};\theta^{(D-1)}}$$

(look carefully at the subscripts on the Jacobians, and remember that $\mathbf{u}^{(D)} = \mathbf{o}^{(D-1)}$). And $\mathbf{o}^{(D)}$ depends on $\theta^{(D-2)}$ through $\mathbf{u}^{(D)}$ which is a function of $\mathbf{u}^{(D-1)}$ which is a function of $\theta^{(D-2)}$, so that

$$\nabla_{\theta^{(D-2)}} L(\mathbf{y}_i, \mathbf{o}^{(D)}(\mathbf{x}_i, \theta)) = (\nabla_{\mathbf{o}^{(D)}} L) \times J_{\mathbf{o}^{(D)};\mathbf{u}^{(D)}} \times J_{\mathbf{o}^{(D-1)};\mathbf{u}^{(D-1)}} \times J_{\mathbf{o}^{(D-2)};\theta^{(D-2)}}$$

(again, look carefully at the subscripts on each of the Jacobians, and remember that $\mathbf{u}^{(D)} = \mathbf{o}^{(D-1)}$ *and* $\mathbf{u}^{(D-1)} = \mathbf{o}^{(D-2)}$).

We can now get to the point. We have a recursion:

$$\begin{aligned}
\mathbf{v}^{(D)} &= (\nabla_{\mathbf{o}^{(D)}} L) \\
\nabla_{\theta^{(D)}} L &= \mathbf{v}^{(D)} \mathcal{J}_{\mathbf{o}^{(D)};\theta^{(D)}} \\
\nabla_{\theta^{(D-1)}} L &= \mathbf{v}^{(D)} \mathcal{J}_{\mathbf{o}^{(D)};\mathbf{u}^{(D)}} \mathcal{J}_{\mathbf{o}^{(D-1)};\theta^{(D-1)}} \\
&\cdots \\
\nabla_{\theta^{(i-1)}} L &= \mathbf{v}^{(D)} \mathcal{J}_{\mathbf{o}^{(D)};\mathbf{u}^{(D)}} \cdots \mathcal{J}_{\mathbf{o}^{(i)};\mathbf{u}^{(i)}} \mathcal{J}_{\mathbf{o}^{(i-1)};\theta^{(i-1)}} \\
&\cdots
\end{aligned}$$

But look at the form of the products of the matrices. We don't need to remultiply all those matrices; instead, we are attaching a new term to a product we've already computed. All this is more cleanly written as:

$$
\begin{aligned}
\mathbf{v}^{(D)} &= \left(\nabla_{\mathbf{o}}^{(D)} L \right) \\
\nabla_{\theta^{(D)}} L &= \mathbf{v}^{(D)} \mathcal{J}_{\mathbf{o}^{(D)};\theta^{(D)}} \\
\mathbf{v}^{(D-1)} &= \mathbf{v}^{(D)} \mathcal{J}_{\mathbf{o}^{(D)};\mathbf{u}^{(D)}} \\
\nabla_{\theta^{(D-1)}} L &= \mathbf{v}^{(D-1)} \mathcal{J}_{\mathbf{o}^{(D-1)};\theta^{(D-1)}} \\
& \ldots \\
\mathbf{v}^{(i-1)} &= \mathbf{v}^{(i)} \mathcal{J}_{\mathbf{o}^{(i)};\mathbf{u}^{(i)}} \\
\nabla_{\theta^{(i-1)}} L &= \mathbf{v}^{(i-1)} \mathcal{J}_{\mathbf{o}^{(i-1)};\theta^{(i-1)}} \\
& \ldots
\end{aligned}
$$

I have not added notation to keep track of the point at which the partial derivative is evaluated (it should be obvious, and we have quite enough notation already). When you look at this recursion, you should see that, to evaluate $\mathbf{v}^{(i-1)}$, you will need to know $\mathbf{u}^{(k)}$ for $k \geq i - 1$. This suggests the following strategy. We compute the $\mathbf{u}$'s (and, equivalently, $\mathbf{o}$'s) with a "forward pass," moving from the input layer to the output layer. Then, in a "backward pass" from the output to the input, we compute the gradient. Doing this is often referred to as **backpropagation**.

Remember This: *The gradient of a multilayer network follows from the chain rule. A straightforward recursion known as backpropagation yields an efficient algorithm for evaluating the gradient. Information flows up the network to compute outputs, then back down to get gradients.*

16.4 Training Multilayer Networks

A multilayer network represents an extremely complex, highly non-linear function, with an immense number of parameters. Such an architecture has been known for a long time, but hasn't been particularly successful until recently. Hindsight suggests the problem is that networks are hard to train successfully. There is quite good evidence that having many layers can improve practical performance *if* one can train the resulting network. For some kinds of problem, multilayer networks with a very large number of layers (sometimes called **deep networks**) easily outperform all other known methods. Neural networks seem to behave best when one is solving a classification problem and has an awful lot of data. We will concentrate on this case.

Getting a multilayer neural network to behave well faces a number of important structural obstacles. There isn't (as of writing) any kind of clear theoretical guide to what will and won't work. What this means is that building really useful applications involves mastering a set of tricks, and building intuition as to when each trick will be helpful. There is now a valuable community of people on the internet who share tricks, code implementing the tricks, and their general experience with those tricks.

Datasets: Fully connected layers have many parameters, meaning that multiple layers will need a lot of data to train, and will take many training batches. There is some reason to believe that multilayer neural networks were discounted in application areas for quite a long time because people underestimated just how much data and how much training was required to make them perform well. Very large datasets seem to be essential for successful applications, but there isn't any body of theory that will give a helpful guide to when one has enough data, etc.

Computing Friction: Evaluating multilayer neural networks and their gradients can be slow. Modern practice stresses the use of GPUs, which significantly improve training time. There are fierce software obstacles, too. If you're uncertain on this point, use the description above to build and train a three-layer network from scratch in a reasonable programming environment. You'll find that you spend a lot of time and effort on housekeeping code (connecting layers to one another; evaluating gradients; and so on). Having to do this every time you try a new network is a huge obstacle. Modern practice stresses the use of customized software environments, which accept a network description and do all the housekeeping for you. I describe some current environments in Sect. 16.4.1, but by the time these words appear in print, new ones might have popped up.

Redundant Units: In our current layer architectures, the units have a kind of symmetry. For example, one could swap the first and the second unit and their weights, swap a bunch of connections to the next layer, and have exactly the same classifier, *but with major changes to the weight vector.* This means that many units might be producing about the same output for the same input, and we would be unable to diagnose this. One problem that results is units in later layers might choose only one of the equivalent units, and rely on that one. This is a poor strategy, because that particular unit might behave badly—it would be better to look at all the redundant units. You might expect that a random initialization resolves this problem, but there is some evidence that advanced tricks that force units to look at most of their inputs can be quite helpful (Sect. 16.4.2).

Gradient Obfuscation: One obstacle that remains technically important has to do with the gradient. Look at the recursion I described for backpropagation. The gradient update at the L'th (top) layer depends pretty directly on the parameters in that layer. But now consider a layer close to the input end of the network. The gradient update has been multiplied by several Jacobian matrices. The update may be very small (if these Jacobians shrink their input vectors) or unhelpful (if layers close to the output have poor parameter estimates). For the gradient update to be really helpful, we'd like the layers higher up the network to be right, but we can't achieve this with lower layers that are confused, because they pass their outputs up. If a layer low in the network is in a nonsensical state, it may be very hard to get it out of that state. In turn, this means that adding layers to a network might improve performance, but also might make it worse because the training turns out poorly.

There are a variety of strategies for dealing with gradient problems. We might try to initialize each layer with a good estimate, as in Sect. 16.4.3. Poor gradient estimates can sometimes be brought under control with gradient rescaling tricks (Sect. 16.4.4). Changing the structure of the layers can help. The most useful variant is the convolutional layer, a special form of layer that has fewer parameters and applies very well to image and sound signals. These are described in the next chapter. Changing the way that layers connect can help too, and I describe some tricks in the next chapter.

Remember This: *Multilayer neural networks are trained with multi-batch stochastic gradient descent, often using variants of the gradient to update. Training can be difficult, but a number of tricks can help.*

16.4.1 Software Environments

In Sect. 16.3.3, I wrote a multilayer network as

$$\mathbf{o}^{(D)},$$

where

$$
\begin{aligned}
\mathbf{o}^{(D)} &= \mathbf{o}^{(D)}(\mathbf{u}^{(D)}, \theta^{(D)}) \\
\mathbf{u}^{(D)} &= \mathbf{o}^{(D-1)}(\mathbf{u}^{(D-1)}, \theta^{(L-1)}) \\
\ldots &= \ldots \\
\mathbf{u}^{(2)} &= \mathbf{o}^{(1)}(\mathbf{u}^{(1)}, \theta^1) \\
\mathbf{u}^{(1)} &= \mathbf{x}.
\end{aligned}
$$

I then used the equations as a map for a computation. This map showed how to use the chain rule to compute gradients. To use this map, we need to know: (a) the form of the computation in the layer; (b) the derivative of the layer with respect to its inputs; and (c) the derivative of the layer with respect to its parameters. There are a variety of types of layer that are particularly useful (we've seen a softmax layer and a fully connected layer; in the next chapter, we'll see a convolutional layer).

There are now several software environments that can accept a description of a network as a map of a computation, like the one above, and automatically construct a code that implements that network. In essence, the user writes a map, provides inputs, and decides what to do with gradients to get descent. These environments support the necessary housekeeping to map a network onto a GPU, evaluate the network and its gradients on the GPU, train the network by updating parameters, and so on. The easy availability of these environments has been an important factor in the widespread adoption of neural networks.

At the time of writing, the main environments available are:

- **Darknet:** This is an open source environment developed by Joe Redmon. You can find it at https://pjreddie.com/darknet/. There is some tutorial material there.
- **MatConvNet:** This is an environment for MATLAB users, originally written by Andrea Vedaldi and supported by a community of developers. You can find it at http://www.vlfeat.org/matconvnet. There is a tutorial at that URL.
- **MXNet:** This is a software framework from Apache that is supported on a number of public cloud providers, including Amazon Web Services and Microsoft Azure. It can be invoked from a number of environments, including R and MATLAB. You can find it at https://mxnet.apache.org.
- **PaddlePaddle:** This is an environment developed at Baidu Research. You can find it at http://www.paddlepaddle.org. There is tutorial material on that page; I understand there is a lot of tutorial material in Chinese, but I can't read Chinese and so can't find it or offer URLs. You should search the web for more details.
- **PyTorch:** This is an environment developed at Facebook's AI research. You can find it at https://pytorch.org. There video tutorials at https://pytorch.org/tutorials/.
- **TensorFlow:** This is an environment developed at Google. You can find it at https://www.tensorflow.org. There is extensive tutorial material at https://www.tensorflow.org/tutorials/.
- **Keras:** This is an environment developed by François Chollet, intended to offer high-level abstractions independent of what underlying computational framework is used. It is supported by the TensorFlow core library. You can find it at https://keras.io. There is tutorial material at that URL.

Each of these environments has their own community of developers. It is now common in the research community to publish code, networks, and datasets openly. This means that, for much cutting edge research, you can easily find a code base that implements a network; and all the parameter values that the developers used to train a network; and a trained version of the network; and the dataset they used for training and evaluation. But these aren't the only environments. You can find a useful comparison at https://en.wikipedia.org/wiki/Comparison_of_deep-learning_software that describes many other environments.

Remember This: *Training even a simple network involves a fair amount of housekeeping code. There are a number of software environments that simplify setting up and training complicated neural networks.*

16.4.2 Dropout and Redundant Units

Regularizing by the square of the weights is all very well, but doesn't ensure that units don't just choose one of their redundant inputs. A very useful regularization strategy is to try and ensure that no unit relies too much on the output of any other unit. One can do this as follows. At each training step, randomly select some units, set their outputs to zero (and reweight the inputs of the units receiving input from them), and then take the step. Now units are trained to produce reasonable outputs even if some of their inputs are randomly set to zero—units can't rely too much on one input, because it might be turned off. Notice that this sounds sensible, but it isn't quite a proof that the approach is sound that comes from experiment. The approach is known as **dropout**.

There are some important details we can't go into. Output units are not subject to dropout, but one can also turn off inputs randomly. At test time, there is no dropout. Every unit computes its usual output in the usual way. This creates an important training issue. Write p for the probability that a unit is dropped out, which will be the same for all units subject to dropout. You should think of the expected output of the i'th unit at *training* time as $(1-p)o_i$ (because with probability p, it is zero). But at test time, the next unit will see o_i; so at training time, you should reweight the inputs by $1/(1-p)$. In exercises, we will use packages that arrange all the details for us.

> **Remember This:** *Dropout can force units to look at inputs from all of a set of redundant units, and so regularize a network.*

16.4.3 Example: Credit Card Accounts Revisited

The Taiwan credit card account data of Sect. 16.2 has 30,000 examples. I split this dataset into 25,000 training examples and 5000 validation examples. I will use results on these validation examples to illustrate performance of various different network architectures, but we won't choose an architecture, so we don't need a test set to evaluate the chosen architecture.

I will compare four different architectures (Fig. 16.9). The simplest—I from now on—has one layer of 100 units with ReLU non-linearities. These units accept the input features and produce a 100 dimensional feature vector. A linear function maps the resulting 100 dimensional feature space to a two-dimensional feature space, and then a softmax produces class-conditional probabilities. You should think of this network as a single layer that produces a 100 dimensional feature vector, followed by a logistic regression classifier that classifies the feature vector. Notice this is different from the classifier of Sect. 16.2 because that used just two units, one per class; here the units produce a large feature vector. Architecture II has a second layer of units (again, ReLU non-linearities) that maps the first 100 dimensional feature vector to a second 100 dimensional feature vector; this is followed by the linear function and softmax as with I. Architecture III has a third layer of units (again, ReLU non-linearities) that maps the second 100 dimensional feature vector to a second 100 dimensional feature vector; this is followed by the linear function and softmax as with I. Finally, architecture IV has a third layer of units (again, ReLU non-linearities) that maps the third 100 dimensional feature

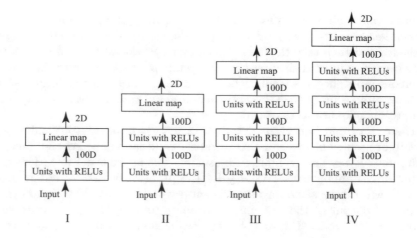

Figure 16.9: The four architectures I used to classify the Taiwan credit card data

vector to a second 100 dimensional feature vector; this is followed by the linear function and softmax as with *I*. For each of the architectures, I used dropout of 50% of the units for each ReLU layer. I used a regularization constant of 0.01, but did not find any particular improvement resulting from changes of this constant.

I used minibatch stochastic gradient descent to minimize, with a minibatch size of 500. I chose this number for convenience, rather than through detailed experiment. Every step, I computed the loss and error rate for each of the training and validation sets. Every 400 steps, I reduced the learning rate by computing

$$\eta = \frac{\eta}{1.77}$$

(so after 1600 steps, the learning rate is reduced by a factor of ten).

Initialization presented serious problems. For architecture *I*, I initialized each parameter with a sample of a zero mean normal random variable with small variance for each initial value, and learning proceeded without difficulty. This strategy did not work for the other architectures (or, at least, I couldn't get it to work for them). For each of these architectures, I found that a random initialization produced a system that very quickly classified all examples with one class. This is likely the result of dead units caused by poor gradient steps in the early part of the process.

To train architectures *II–IV* properly, I found it necessary to initialize with the previous architectures earlier layers. So I initialized architecture *II*'s first ReLU layer with a trained version of architecture *I*'s first ReLU layer (and its second layer randomly); architecture *III*'s first two ReLU layers with a trained version of architecture *II*'s first two ReLU layers; and so on. This trick is undignified, but I found it to work. We will see it again (Sect. 18.1.5).

Figure 16.10 compares the four architectures. Generally, having more layers improved the error rate, but not as much as one might hope for. You should notice

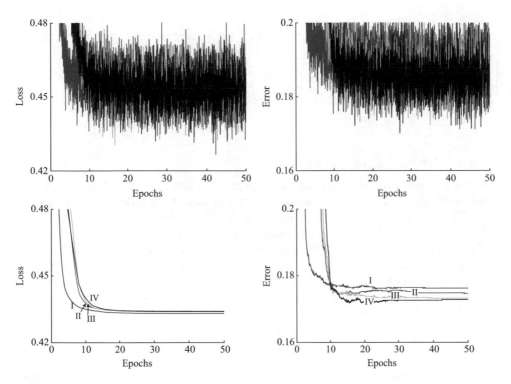

Figure 16.10:   On the **left** loss and on the **right** error plotted for training (**top**, noisy, hard to resolve) and validation (**bottom**, smooth) datasets for various neural networks applied to classifying the Taiwan credit card data. I used the four architectures described in the text. The training curves are hard to resolve (but if you are viewing this figure in color, the colors are keyed, so red is I, blue II, green III, and grey IV). Note that improvements in loss don't necessarily yield improvements in error. Note also that adding layers results in a more accurate classifier, but the improvement in accuracy for this dataset is not large

one important curiosity. The validation loss is actually slightly larger for networks with more layers, but the validation error rate is smaller. The loss is not the same as the error rate, and so it is quite common to see improvements in loss that don't result in reduction in error. One important mechanism occurs when a network improves its loss by improving the class-conditional posteriors for examples *that it already classifies correctly*. If there are many such examples, you might see a significant improvement in loss with no improvement in error. This is a significant nuisance that can't be fixed by training to minimize error rate, because error rate isn't differentiable in parameters (or rather, the derivative of error at each example is zero for almost every parameter setting, which isn't helpful).

> **Remember This:** *Stacking multiple layers of units results in learned features that should cause the final classifier to perform well. Such multilayer neural networks make excellent classifiers, but can be hard to train without enough data.*

16.4.4 Advanced Tricks: Gradient Scaling

Everyone is surprised the first time they learn that the best direction to travel in when you want to minimize a function is not, in fact, backwards down the gradient. The gradient *is* uphill, but repeated downhill steps are often not particularly efficient. An example can help, and we will look at this point several ways because different people have different ways of understanding this point.

We can look at the problem with algebra. Consider $f(x, y) - (1/2)(\epsilon x^2 + y^2)$, where ϵ is a small positive number. The gradient at (x, y) is $(\epsilon x, y)$. For simplicity, use a fixed learning rate η, so we have

$$\begin{bmatrix} x^{(r)} \\ y^{(r)} \end{bmatrix} = \begin{bmatrix} (1 - \epsilon\eta)x^{(r-1)} \\ (1 - \eta)y^{(r-1)} \end{bmatrix}.$$

If you start at, say $(x^{(0)}, y^{(0)})$ and repeatedly go downhill along the gradient, you will travel very slowly to your destination. You can show that

$$\begin{bmatrix} x^{(r)} \\ y^{(r)} \end{bmatrix} = \begin{bmatrix} (1 - \epsilon\eta)^r x^{(0)} \\ (1 - \eta)^r y^{(0)} \end{bmatrix}.$$

The problem is that the gradient in y is quite large (so y must change quickly) and the gradient in x is small (so x changes slowly). In turn, for steps in y to converge we must have $|1 - \eta| < 1$; but for steps in x to converge, we require only the much weaker constraint $|1 - \epsilon\eta| < 1$. Imagine we choose the largest η we dare for the y constraint. The y value will very quickly have small magnitude, though its sign will change with each step. But the x steps will move you closer to the right spot only extremely slowly.

Another way to see this problem is to reason geometrically. Figure 16.11 shows this effect for this function. The gradient is at right angles to the level curves of the function. But when the level curves form a narrow valley, the gradient points across the valley rather than down it. The effect isn't changed by rotating and translating the function (Fig. 16.12).

You may have learned that Newton's method resolves this problem. This is all very well, but to apply Newton's method we would need to know the matrix of second partial derivatives. A network can easily have thousands to millions of parameters, and we simply can't form, store, or work with matrices of these dimensions. Instead, we will need to think more qualitatively about what is causing trouble.

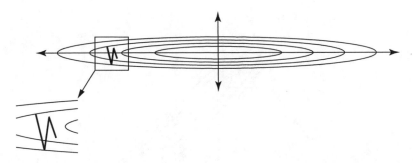

Figure 16.11: A plot of the level curves (curves of constant value) of the function $f(x, y) = (1/2)(\epsilon x^2 + y^2)$. Notice that the value changes slowly with large changes in x, and quickly with small changes in y. The gradient points mostly toward the x-axis; this means that gradient descent is a slow zig-zag across the "valley" of the function, as illustrated. We might be able to fix this problem by changing coordinates, *if* we knew what change of coordinates to use

One useful insight into the problem is that fast changes in the gradient vector are worrying. For example, consider $f(x) = (1/2)(x^2 + y^2)$. Imagine you start far away from the origin. The gradient won't change much along reasonably sized steps. But now imagine yourself on one side of a valley like the function $f(x) = (1/2)(x^2 + \epsilon y^2)$; as you move along the gradient, the gradient in the x direction gets smaller very quickly, then points back in the direction you came from. You are not justified in taking a large step in this direction, because if you do you will end up at a point with a very different gradient. Similarly, the gradient in the y direction is small, and stays small for quite large changes in y value. You would like to take a small step in the x direction and a large step in the y direction.

You can see that this is the impact of the second derivative of the function (which is what Newton's method is all about). But we can't do Newton's method. We would like to travel further in directions where the gradient doesn't change much, and less far in directions where it changes a lot. There are several methods for doing so.

Momentum: We should like to discourage parameters from "zig-zagging" as in the example above. In these examples, the problem is caused by components of the gradient changing sign from step to step. It is natural to try and smooth the gradient. We could do so by forming a moving average of the gradient. Construct a vector $\mathbf{v}$, the same size as the gradient, and initialize this to zero. Choose a positive number $\mu < 1$. Then we iterate

$$\begin{aligned} \mathbf{v}^{(r+1)} &= \mu \mathbf{v}^{(r)} + \eta \nabla_\theta E \\ \theta^{(r+1)} &= \theta^{(r)} - \mathbf{v}^{(r+1)}. \end{aligned}$$

Notice that, in this case, the update is an average of all past gradients, each weighted by a power of μ. If μ is small, then only relatively recent gradients will participate in the average, and there will be less smoothing. Larger μ leads to more smoothing.

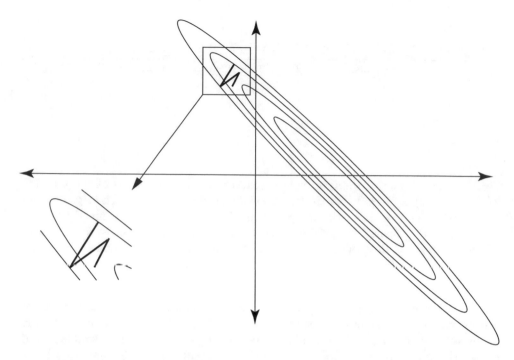

Figure 16.12:   Rotating and translating a function rotates and translates the gradient; this is a picture of the function of Fig. 16.11, but now rotated and translated. The problem of zig-zagging remains. This is important, because it means that we may have serious difficulty choosing a good change of coordinates

A typical value is $\mu = 0.9$. It is reasonable to make the learning rate go down with epoch when you use momentum, but keep in mind that a very large μ will mean you need to take several steps before the effect of a change in learning rate shows.

Adagrad: We will keep track of the size of each component of the gradient. In particular, we have a running cache **c** which is initialized at zero. We choose a small number α (typically 1e−6), and a fixed η. Write $g_i^{(r)}$ for the i'th component of the gradient $\nabla_\theta E$ computed at the r'th iteration. Then we iterate

$$
\begin{aligned}
c_i^{(r+1)} &= c_i^{(r)} + (g_i^{(r)})^2 \\
\theta_i^{(r+1)} &= \theta_i^{(r)} - \eta \frac{g_i^{(r)}}{(c_i^{(r+1)})^{\frac{1}{2}} + \alpha}
\end{aligned}
$$

Notice that each component of the gradient has its own learning rate, set by the history of previous gradients.

RMSprop: This is a modification of Adagrad, to allow it to "forget" large gradients that occurred far in the past. Again, write $g_i^{(r)}$ for the i'th component of the gradient $\nabla_\theta E$ computed at the r'th iteration. We choose another number, Δ, (the **decay rate**; typical values might be 0.9, 0.99, or 0.999), and iterate

$$c_i^{(r+1)} = \Delta c_i^{(r)} + (1 - \Delta)(g_i^{(r)})^2$$

$$\theta_i^{(r+1)} = \theta_i^{(r)} - \eta \frac{g_i^{(r)}}{(c_i^{(r+1)})^{\frac{1}{2}} + \alpha}.$$

Adam: This is a modification of momentum that rescales gradients, tries to forget large gradients, and adjusts early gradient estimates to correct for bias. Again, write $g_i^{(r)}$ for the i'th component of the gradient $\nabla_\theta E$ computed at the r'th iteration. We choose three numbers β_1, β_2, and ϵ (typical values are 0.9, 0.999, and 1e−8, respectively), and some stepsize or learning rate η. We then iterate

$$\mathbf{v}^{(r+1)} = \beta_1 * \mathbf{v}^{(r)} + (1 - \beta_1) * \nabla_\theta E$$

$$c_i^{(r+1)} = \beta_2 * c_i^{(r)} + (1 - \beta_2) * (g_i^r)^2$$

$$\hat{\mathbf{v}} = \frac{\mathbf{v}^{(r+1)}}{1 - \beta_1^t}$$

$$\hat{c}_i = \frac{\hat{c}_i^{(r+1)}}{1 - \beta_2^t}$$

$$\theta_i^{(r+1)} = \theta_i^{(r)} - \eta \frac{\hat{v}_i}{\sqrt{\hat{c}_i} + \epsilon}.$$

Remember This: *If you are not getting improvements during training, use a gradient scaling trick.*

16.5   You Should

16.5.1   Remember These Terms

16.5.2   Remember These Facts

16.5.3   Remember These Procedures

- **Backpropagation** is a catch-all term describing a recursion used to compute the state and gradients of a neural network. Inputs are presented to the network, and propagated forward along the layers to evaluate the output; once this is known, the gradient of the loss with respect to parameters can be computed by working backwards toward the input.

16.5.4   Be Able to

- Run at least the tutorials for your chosen neural network environment.
- Set up and train a simple multilayer neural network classifier for a straight-forward problem.
- Apply gradient scaling procedures to improve network training.

Problems

16.1. Draw the decision boundary for a classifier built with three units (one per class) that classifies a 2D point $\mathbf{x}$ into three classes.

16.2. Write the loss of a network classifying an example $\mathbf{x}$ as

$$L(\mathbf{y}, \mathbf{o}^{(D)}),$$

where

$$\mathbf{o}^{(D)} = \mathbf{o}^{(D)}(\mathbf{u}^{(D)}, \theta^{(D)})$$
$$\mathbf{u}^{(D)} = \mathbf{o}^{(D-1)}(\mathbf{u}^{(D-1)}, \theta^{(L-1)})$$
$$\ldots = \ldots$$
$$\mathbf{u}^{(2)} = \mathbf{o}^{(1)}(\mathbf{u}^{(1)}, \theta^1)$$
$$\mathbf{u}^{(1)} = \mathbf{x}.$$

Now consider $\nabla_\theta L$.

(a) Show

$$\nabla_{\theta^{(D)}} L(\mathbf{y}, \mathbf{o}^{(D)}) = (\nabla_{\mathbf{o}^{(D)}} L) \times J_{\mathbf{o}^{(D)}; \theta^{(D)}}.$$

(b) Show

$$\nabla_{\theta^{(D-1)}} L(\mathbf{y}_i, \mathbf{o}^{(D)}(\mathbf{x}_i, \theta)) = (\nabla_{\mathbf{o}^{(D-1)}} L) \times J_{\mathbf{o}^{(D)}; \mathbf{u}^{(D)}} \times J_{\mathbf{o}^{(D-1)}; \theta^{(D-1)}}.$$

(c) Show

$$\nabla_{\theta^{(D-2)}} L(\mathbf{y}_i, \mathbf{o}^{(D)}(\mathbf{x}_i, \theta)) = (\nabla_{\mathbf{o}^{(D)}} L) \times J_{\mathbf{o}^{(D)}; \mathbf{u}^{(D)}} \times J_{\mathbf{o}^{(D-1)}; \mathbf{u}^{(D-1)}}$$
$$\times J_{\mathbf{o}^{(D-2)}; \theta^{(D-2)}}.$$

16.3. Confirm that the gradient computed by the recursion

$$\mathbf{v}^{(D)} = \left(\nabla_{\mathbf{o}}^{(D)} L\right)$$
$$\nabla_{\theta^{(D)}} L = \mathbf{v}^{(D)} J_{\mathbf{o}^{(D)}; \theta^{(D)}}$$
$$\mathbf{v}^{(D-1)} = \mathbf{v}^{(D)} J_{\mathbf{o}^{(D)}; \mathbf{u}^{(D)}}$$
$$\nabla_{\theta^{(D-1)}} L = \mathbf{v}^{(D-1)} J_{\mathbf{o}^{(D-1)}; \theta^{(D-1)}}$$
$$\ldots$$
$$\mathbf{v}^{(i-1)} = \mathbf{v}^{(i)} J_{\mathbf{o}^{(i)}; \mathbf{u}^{(i)}}$$
$$\nabla_{\theta^{(i-1)}} L = \mathbf{v}^{(i-1)} J_{\mathbf{o}^{(i-1)}; \theta^{(i-1)}}$$
$$\ldots$$

is correct. What is $\nabla_{\theta^{(1)}} L$?

16.4. Write the loss of a network classifying an example $\mathbf{x}$ as

$$L(\mathbf{y}, \mathbf{o}^{(D)}),$$

where

$$\mathbf{o}^{(D)} = \mathbf{o}^{(D)}(\mathbf{u}^{(D)}, \theta^{(D)})$$
$$\mathbf{u}^{(D)} = \mathbf{o}^{(D-1)}(\mathbf{u}^{(D-1)}, \theta^{(L-1)})$$
$$\ldots = \ldots$$
$$\mathbf{u}^{(2)} = \mathbf{o}^{(1)}(\mathbf{u}^{(1)}, \theta^1)$$
$$\mathbf{u}^{(1)} = \mathbf{x}.$$

Explain how to use backpropagation to compute

$$\nabla_{\mathbf{x}} L.$$

Programming Exercises

16.5. Reproduce the example of Sect. 16.2, using the constants in that example, the initialization in that example, and the preprocessing example. *Hint:* use one of the packages sketched in Sect. 16.4.1, or this will be tricky. It's a good warm-up problem to get used to an environment.

   (a) What happens if you do *not* preprocess the features?

   (b) By adjusting initialization, learning rate, preprocessing, and so on, what is the best test error you can get?

16.6. Reproduce the example of Sect. 16.4.3, using the constants in that example, the initialization in that example, and the preprocessing example. *Hint:* use one of the packages sketched in Sect. 16.4.1, or this will be very tricky indeed. Use the previous exercise as a warm-up problem.

   (a) What happens if you do *not* preprocess the features?

   (b) By adjusting initialization, learning rate, preprocessing, and so on, what is the best test error you can get?

   (c) Can you get multilayer networks to train without using the trick in that section?

   (d) Does dropout help or hurt the accuracy?

16.7. The UC Irvine machine learning data repository hosts a collection of data on the whether p53 expression is active or inactive. You can find out what this means, and more information about the dataset, by reading: Danziger, S.A., Baronio, R., Ho, L., Hall, L., Salmon, K., Hatfield, G.W., Kaiser, P., and Lathrop, R.H. "Predicting Positive p53 Cancer Rescue Regions Using Most Informative Positive (MIP) Active Learning," *PLOS Computational Biology*, 5(9), 2009; Danziger, S.A., Zeng, J., Wang, Y., Brachmann, R.K. and Lathrop, R.H. "Choosing where to look next in a mutation sequence space: Active Learning of informative p53 cancer rescue mutants," *Bioinformatics*, 23(13), 104–114, 2007; and Danziger, S.A., Swamidass, S.J., Zeng, J., Dearth, L.R., Lu, Q., Chen, J.H., Cheng, J., Hoang, V.P., Saigo, H., Luo, R., Baldi, P., Brachmann, R.K. and Lathrop, R.H. "Functional census of mutation sequence spaces: the example of p53 cancer rescue mutants," *IEEE/ACM transactions on computational biology and bioinformatics*, 3, 114–125, 2006.

You can find this data at https://archive.ics.uci.edu/ml/datasets/p53+Mutants. There are a total of 16,772 instances, with 5409 attributes per instance. Attribute 5409 is the class attribute, which is either active or inactive. There are several versions of this dataset. You should use the version K8.data.

Train a multilayer neural network to classify this data, using stochastic gradient descent. You will need to drop data items with missing values. You should estimate a regularization constant using cross-validation, trying at least three values. Your training method should touch at least 50% of the training set data. You should produce an estimate of the accuracy of this classifier on held-out data consisting of 10% of the dataset, chosen at random. Preprocess the features as in Sect. 16.4.3.

   (a) What happens if you do *not* preprocess the features?

   (b) By adjusting initialization, learning rate, preprocessing, and so on, what is the best test error you can get?

(c) Can you get multilayer networks to train without using the trick in that section?

(d) Does dropout help or hurt the accuracy?

(e) Do gradient scaling tricks help or hurt the training process?

CHAPTER 17

Simple Image Classifiers

There are two problems that lie at the core of image understanding. The first is **image classification**, where we decide what class an image of a fixed size belongs to. It's usual to work with a collection of images of objects. These objects will be largely centered in the image, and largely isolated. Each image will have an associated object name, using a taxonomy of classes provided in advance. You should think of catalog images of clothing or furniture. Another possible example is mugshot photos or pictures of people on websites (where the taxonomy is names). Judging by the amount of industry money pouring into image classification research, there are valuable applications for solutions.

The second problem is **object detection**, where we try to find the locations of objects of a set of classes in the image. So we might try to mark all cars, all cats, all camels, and so on. As far as anyone knows, the right way to think about object detection is that we search a collection of windows in an image, apply an image classification method to each window, then resolve disputes between overlapping windows. How windows are to be chosen for this purpose is an active and quickly changing area of research. Object detection is another problem receiving tremendous attention from industry.

Neural networks have enabled spectacular progress in both problems. We now have very accurate methods for large scale image classification and quite effective and fast methods for object detection. This chapter describes the main methods used in building these methods, and finishes with two fairly detailed examples of simple image classification. The next chapter covers modern methods for image classification and object detection.

17.1 Image Classification

An instructive image classification dataset is the MNIST dataset of handwritten digits. This dataset is very widely used to check simple methods. It was originally constructed by Yann Lecun, Corinna Cortes, and Christopher J.C. Burges. You can find this dataset in several places. The original dataset is at http://yann.lecun.com/exdb/mnist/. The version I used was prepared for a Kaggle competition (so I didn't have to decompress Lecun's original format). I found it at http://www.kaggle.com/c/digit-recognizer.

Images have important, quite general, properties (Fig. 17.1). Images of "the same thing"—in the case of MNIST, the same handwritten digit—can look fairly different. Small shifts and small rotations do not change the class of an image. Making the image somewhat brighter of somewhat darker does not change the class of the image either. Making the image somewhat larger, or making it somewhat smaller (then cropping or filling in pixels as required) does not change the class either. This means that individual pixel values are not particularly informative—

D. Forsyth, *Applied Machine Learning*,
https://doi.org/10.1007/978-3-030-18114-7_17

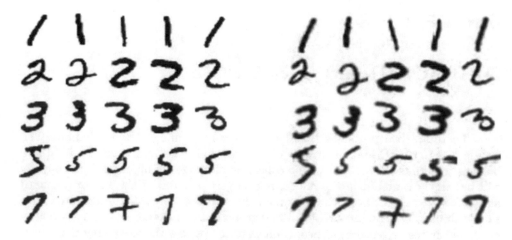

Figure 17.1: On the **left**, a selection of digits from the MNIST dataset. Notice how images of the same digit can vary, which makes classifying the image demanding. It is quite usual that pictures of "the same thing" look quite different. On the **right**, digit images from MNIST that have been somewhat rotated and somewhat scaled, then cropped fit the standard size. Small rotations, small scales, and cropping really doesn't affect the identity of the digit

you can't tell whether a digit image is, for example, a zero by looking at a given pixel, because the ink might slide to the left or to the right of the pixel without changing the digit. In turn, you should not expect applying logistic regression directly to the pixel values to be particularly helpful. For MNIST, this approach yields an error rate that is quite poor compared to better methods (try it—glmnet can handle this).

Another important property of images is that they have many pixels. Building a fully connected layer where every unit sees every pixel is impractical—each unit might have millions of inputs, none of which is particularly useful. But if you think of a unit as a device for constructing features, this construction is odd, because it suggests that one needs to use every pixel in an image to construct a useful feature. This isn't consistent with experience. For example, if you look at the images in Fig. 17.2, you will notice another important property of images. Local patterns can be quite informative. Digits like 0 and 8 have loops. Digits like 4 and 8 have crossings. Digits like 1, 2, 3, 5, and 7 have line endings, but no loops or crossings. Digits like 6 and 9 have loops and line endings. Furthermore, spatial relations between local patterns are informative. A 1 has two line endings above one another; a 3 has three line endings above one another. These observations suggest a strategy that is a central tenet of modern computer vision: you construct features that respond to patterns in small, localized neighborhoods; then other features look at patterns of *those* features; then others look at patterns of those, and so on.

17.1.1 Pattern Detection by Convolution

For the moment, think of an image as a two-dimensional array of intensities. Write $\mathcal{I}_{uv}$ for the pixel at position u, v. We will construct a small array (a **mask** or **kernel**) $\mathcal{W}$, and compute a new image $\mathcal{N}$ from the image and the mask, using the rule

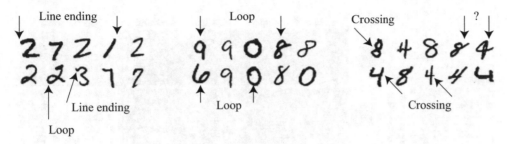

Figure 17.2: Local patterns in images are quite informative. MNIST images, shown here, are simple images, so a small set of patterns is quite helpful. The relative location of patterns is also informative. So, for example, an eight has two loops, one above the other. All this suggests a key strategy: construct features that respond to patterns in small, localized neighborhoods; then other features that look at patterns of *those* features; then others that look at patterns of those, and so on. Each pattern (here line endings, crossings, and loops) has a range of appearances. For example, a line ending sometimes has a little wiggle as in the three. Loops can be big and open, or quite squashed. The list of patterns isn't comprehensive. The "?" shows patterns that I haven't named, but which appear to be useful. In turn, this suggests learning the patterns (and patterns of patterns; and so on) that are most useful for classification

$$\mathcal{N}_{ij} = \sum_{kl} \mathcal{I}_{i-k,j-l} \mathcal{W}_{kl}.$$

Here we sum over all k and l that apply to $\mathcal{W}$; for the moment, do not worry about what happens when an index goes out of the range of $\mathcal{I}$. This operation is known as **convolution**. The form of the operation is important in signal processing mathematics, but makes it quite hard to understand what convolution is good for. We will generalize the idea.

Notice that if we flip $\mathcal{W}$ in both directions to obtain $\mathcal{M}$, we can write the new image as

$$\mathcal{N} = \text{conv}(\mathcal{I}, \mathcal{M})$$
$$\text{where}$$
$$\mathcal{N}_{ij} = \sum_{kl} \mathcal{I}_{kl} \mathcal{M}_{k-i,l-j}.$$

In what follows, I will always apply this flip, and use the term "convolution" to refer to the operator `conv` defined above. This isn't consistent with the signal processing literature, but is quite usual in the machine learning literature. Now reindex yet again, by substituting $u = k - i$, $v = l - j$, and noticing that if u runs over the range 0 to ∞, so does $u - i$ to get

$$\mathcal{N}_{ij} = \sum_{uv} \mathcal{I}_{i+u, j+v} \mathcal{M}_{uv}.$$

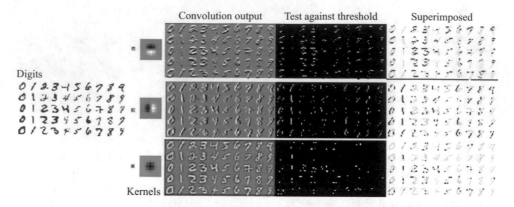

Figure 17.3: On the **far left**, some images from the MNIST dataset. Three kernels appear on the **center left**; the small blocks show the kernels scaled to the size of the image, so you can see the size of the piece of image the kernel is applied to. The larger blocks show the kernels (mid grey is zero; light is positive; dark is negative). The kernel in the top row responds most strongly to a dark bar above a light bar; that in the middle row responds most strongly to a dark bar to the left of a light bar; and the bottom kernel responds most strongly to a spot. **Center** shows the results of applying these kernels to the images. You will need to look closely to see the difference between a medium response and a strong response. **Center right** shows pixels where the response exceeds a threshold. You should notice that this gives (from top to bottom): a horizontal bar detector; a vertical bar detector; and a line ending detector. These detectors are moderately effective, but not perfect. **Far right** shows detector responses (in black) superimposed on the original image (grey) so you can see the alignment between detections and the image

This operation is linear. You should check that:

- if $\mathcal{I}$ is zero, then $\text{conv}(\mathcal{I}, \mathcal{M})$ is zero;
- $\text{conv}(k\mathcal{I}, \mathcal{M}) = k\text{conv}(\mathcal{I}, \mathcal{M})$; and
- $\text{conv}(\mathcal{I} + \mathcal{J}, \mathcal{M}) = \text{conv}(\mathcal{I}, \mathcal{M}) + \text{conv}(\mathcal{J}, \mathcal{M})$.

The value of $\mathcal{N}_{ij}$ is a dot-product, as you can see by reindexing $\mathcal{M}$ and the piece of image that lies under $\mathcal{M}$ to be vectors. This view explains why a convolution is interesting: it is a very simple pattern detector. Assume that $\mathbf{u}$ and $\mathbf{v}$ are unit vectors. Then $\mathbf{u} \cdot \mathbf{v}$ is largest when $\mathbf{u} = \mathbf{v}$, and smallest when $\mathbf{u} = -\mathbf{v}$. Using the dot-product analogy, for $\mathcal{N}_{ij}$ to have a large and positive value, the piece of image that lies under $\mathcal{M}$ must "look like" $\mathcal{M}$. Figure 17.3 gives some examples.

The proper model for `conv` is this. To compute the value of $\mathcal{N}$ at some location, you take the window $\mathcal{W}$ of $\mathcal{I}$ at that location that is the same size as $\mathcal{N}$; you multiply together the elements of $\mathcal{M}$ and $\mathcal{W}$ that lie on top of one another; and you sum the results (Fig. 17.4). Thinking of this as an operation on windows allows us to generalize in very useful ways.

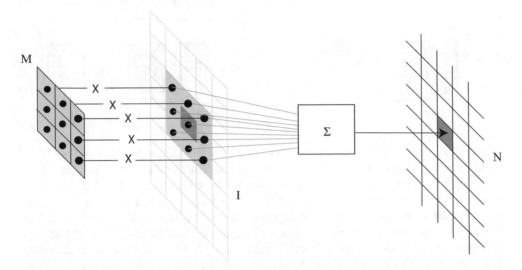

Figure 17.4: To compute the value of $\mathcal{N}$ at some location, you shift a copy of $\mathcal{M}$ to lie over that location in $\mathcal{I}$; you multiply together the non-zero elements of $\mathcal{M}$ and $\mathcal{I}$ that lie on top of one another; and you sum the results

In the original operation, we used a window at every location in $\mathcal{I}$, but we may prefer to look at (say) a window at every second location. The centers of the windows we wish to look at lie on a grid of locations in $\mathcal{I}$. The number of pixels skipped between points on the grid is known as its **stride**. A grid with stride 1 consists of each spatial location. A grid with stride 2 consists of every second spatial location in $\mathcal{I}$, and so on. You can interpret a stride of 2 as either performing `conv` then keeping the value at every second pixel in each direction. Better is to think of the kernel striding across the image—perform the `conv` operation as above, but now move the window by two pixels before multiplying and adding.

The description of the original operation avoided saying what would happen if the window at a location went outside $\mathcal{I}$. We adopt the convention that $\mathcal{N}$ contains entries only for windows that lie inside $\mathcal{I}$. But we can apply **padding** to $\mathcal{I}$ to ensure that $\mathcal{N}$ has the size we want. Padding attaches a set of rows (resp. columns) to the top and bottom (resp. left and right) of $\mathcal{I}$ to make it a convenient size. Usually, but not always, the new rows or columns contain zeros. By far the most common case uses $\mathcal{M}$ that are square with odd dimension (making it much easier to talk about the center). Assume $\mathcal{I}$ is $n_x \times n_y$ and $\mathcal{M}$ is $(2k+1) \times (2k+1)$; if we pad $\mathcal{I}$ with k rows on top and bottom and k columns on each side, $\texttt{conv}(\mathcal{I}, \mathcal{M})$ will be $n_x \times n_y$ (Fig. 17.5).

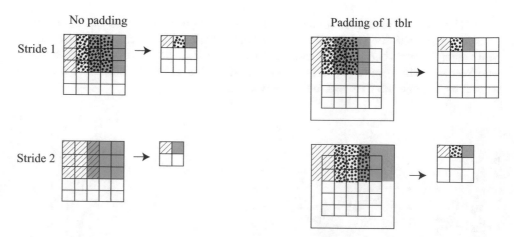

Figure 17.5: The effects of stride and padding on `conv`. On the **left**, `conv` without padding accepts an $\mathcal{I}$, places a 3×3 $\mathcal{M}$ on grid locations determined by the stride, then reports values for valid windows. When the stride is 1, a 5×5 $\mathcal{I}$ becomes a 3×3 $\mathcal{N}$. When the stride is 2, a 5×5 $\mathcal{I}$ becomes a 2×2 $\mathcal{N}$. The hatching and shading show the window used to compute the corresponding value in $\mathcal{N}$. On the **right**, `conv` with padding accepts an $\mathcal{I}$, pads it (in this case, by one row top and bottom, and one column left and right), places a 3×3 $\mathcal{M}$ on grid locations in the padded result determined by the stride, then reports values for valid windows. When the stride is 1, a 5×5 $\mathcal{I}$ becomes a 5×5 $\mathcal{N}$. When the stride is 2, a 5×5 $\mathcal{I}$ becomes a 3×3 $\mathcal{N}$. The hatching and shading show the window used to compute the corresponding value in $\mathcal{N}$

Images are naturally 3D objects with two spatial dimensions (up–down, left–right) and a third dimension that chooses a slice (R, G, or B for a color image). This structure is natural for representations of image patterns, too—two dimensions that tell you where the pattern is and one that tells you what it is. The results in Fig. 17.3 show a block consisting of three such slices. These slices are the response of a pattern detector *for a fixed pattern*, where there is one response for each spatial location in the block, and so are often called **feature maps**.

We will generalize `conv` and apply it to 3D blocks of data (which I will call **blocks**). Write $\mathcal{I}$ for an input block of data, which is now $x \times y \times d$. Two dimensions—usually the first two, but this can depend on your software environment—are spatial and the third chooses a slice. Write $\mathcal{M}$ for a 3D kernel, which is $k_x \times k_y \times d$. Now choose padding and a stride. This determines a grid of locations in the spatial dimensions of $\mathcal{I}$. At each location, we must compute the value of $\mathcal{N}$. To do so, take the 3D window $\mathcal{W}$ of $\mathcal{I}$ at that location that is the same size as $\mathcal{N}$; you multiply together the elements of $\mathcal{M}$ and $\mathcal{W}$ that lie on top of one another; and you sum the results (Fig. 17.4). This sum now goes over the third dimension as well. This produces a two-dimensional $\mathcal{N}$.

To make this operation produce a block of data, use a 4D block of kernels. This **kernel block** consists of D kernels, each of which is a $k_x \times k_y \times d$ dimensional kernel. If you apply each kernel as in the previous paragraph to an $x \times y \times d$ dimensional

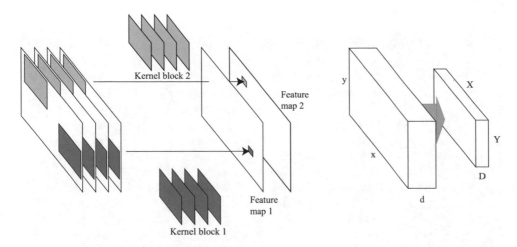

Figure 17.6: On the **left**, two kernels (now 3D, as in the text) applied to a set of feature maps produce one new feature map per kernel, using the procedure of the text (the bias term isn't shown). Abstract this as a process that takes an $x \times y \times d$ block to an $X \times Y \times D$ block (as on the **right**)

$\mathcal{I}$, you obtain an $X \times Y \times D$ dimensional block $\mathcal{N}$, as in Fig. 17.6. What X and Y are depends on k_x, k_y, the stride and the padding. A **convolutional layer** takes a kernel block and a bias vector of D bias terms. The layer applies the kernel block to an input block (as above), then adds the corresponding bias value to each slice.

A convolutional layer is a very general and very useful idea. A fully connected layer is one form of convolutional layer. You can build a simple pattern detector out of a convolutional layer followed by a ReLU layer. You can build a linear map that reduces dimension can be built out of a convolutional layer.

Useful Fact: 17.1 *Definition: Convolutional Layer*

A convolutional layer makes 3D blocks of data from 3D blocks of data, using a stride, padding, a block of kernels, and a vector of bias terms. The details are in the text.

Remember This: *A fully connected layer can be thought of as a convolutional layer followed by a ReLU layer. Assume you have an $x \times y \times d$ block of data. Reshape this to be a $(xyd) \times 1 \times 1$ block. Apply a convolutional layer whose kernel block has size $(xyd) \times 1 \times D$, and then a ReLU. This pair of layers produces the same result as a fully connected layer of D units.*

Remember This: *Take the output of a convolutional layer and apply a ReLU. First, think about what happens to one particular piece of image the size of one particular kernel. If that piece is "sufficiently similar" to the kernel, we will see a positive response at the relevant location. If the piece is too different, we will see a zero. This is a pattern detector as in Fig. 17.3. What "sufficiently similar" means is tuned by changing the bias for that kernel. For example, a bias term that is negative with large magnitude means the image block will need to be very like the kernel to get a non-zero response. This pattern detector is (basically) a unit—apply a ReLU to a linear function of the piece of image, plus a constant. Now it should be clear what happens when all kernels are applied to the whole image. Each pixel in a slice represents the result of a pattern detector applied to the piece of image corresponding to the pixel. Each slice of the resulting block represents the result of a different pattern detector. The elements of the output block are often thought of as* **features***.*

Remember This: *There isn't a standard meaning for the term convolutional layer. I'm using one of the two that are widely used. Software implementations tend to use my definition. Very often, research papers use the alternative, which is my definition followed by a non-linearity (almost always a ReLU). This is because convolutional layers mostly are followed by ReLUs in research papers, but it is more efficient in software implementations to separate the two.*

Different software packages use different defaults about padding. One default assumes that no padding is applied. This means that a kernel block of size $k_x \times k_y \times d \times D$ applied to a block of size $x \times y \times d$ with stride 1 yields a block of size $(n_x - k_x + 1) \times (n_y - k_y + 1) \times D$ (check this with a pencil and paper). Another assumes that the input block is padded with zeros so that the output block is $n_x \times n_y \times D$.

Remember This: *In Fig. 17.3, most values in the output block are zero (black pixels in that figure). This is typical of pattern detectors produced in this way. This is an experimental fact that seems to be related to deep properties of images.*

> **Remember This:**   *A kernel block that is $1 \times 1 \times n_z \times D$ is known as a 1×1 **convolution**. This is a linear map in an interesting way. Think of the input and output blocks as sets of column vectors. So the input block is a set of $n_x \times n_y$ column vectors, each of which has dimension $n_z \times 1$ (i.e., there is a column vector at each location of the input block). Write $\mathbf{i}_{uv}$ for the vector at location u, v in the input block, and $\mathbf{o}_{uv}$ for the vector at location u, v in the output block. Then there is a $D \times n_z$ matrix $\mathcal{M}$ so that the 1×1 convolution maps $\mathbf{i}_{uv}$ to*
>
> $$\mathbf{o}_{uv} = \mathcal{M}\mathbf{i}_{uv}.$$
>
> *This can be extremely useful when the input has very high dimension, because $\mathcal{M}$ can be used to reduce dimension and is learned from data.*

17.1.2   Convolutional Layers upon Convolutional Layers

Convolutional layers take blocks of data and make blocks of data, as do ReLU layers. This suggests the output of a convolutional layer could be passed through a ReLU, then connected to another convolutional layer, and so on. Doing this turns out to be an excellent idea.

Think about the output of the first convolutional layer. Each location receives inputs from pixels in a window about that location. The output of the ReLU, as we have seen, forms a simple pattern detector. Now if we put a second layer on top of this, each location in the second layer receives inputs from first layer values in a window about that location. This means that locations in the second layer are affected by a larger window of pixels than those in the first layer. You should think of these as representing "patterns of patterns." If we place a third layer on top of the second layer, locations in that third layer will depend on an even larger window of pixels. A fourth layer will depend on a yet larger window, and so on. The key point here is that we can choose the patterns by *learning* what kernels will be applied at each layer.

The **receptive field** of a location in a data block (or, equivalently, a unit) is the set of image pixels that affect the value of the location. Usually, all that matters is the size of the receptive field. The receptive field of a location in the first convolutional layer will be given by the kernel of that layer. Determining the receptive field for later layers requires some bookkeeping (among other things, you must account for any stride or pooling effects).

If you have several convolutional layers with stride 1, then each block of data has the same spatial dimensions. This tends to be a problem, because the pixels that feed a unit in the top layer will tend to have a large overlap with the pixels that feed the unit next to it. In turn, the values that the units take will be similar, and so there will be redundant information in the output block. It is usual to try and

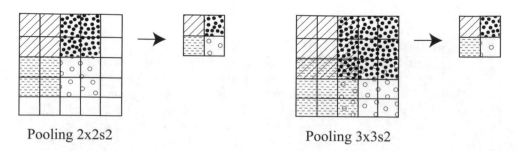

Pooling 2x2s2   Pooling 3x3s2

Figure 17.7: In a pooling layer, pooling units compute a summary of their inputs, then pass it on. The most common case is 2×2, illustrated here on the **left**. We tile each feature map with 2×2 windows that do not overlap (so have stride 2). Pooling units compute a summary of the inputs (usually either the max or the average), then pass that on to the corresponding location in the corresponding feature map of the output block. As a result, the spatial dimensions of the output block will be about half those of the input block. On the **right**, the common alternative of pooling in overlapping 3×3 windows with stride 2

deal with this by making blocks get smaller. One natural strategy is to occasionally have a layer that has stride 2.

An alternative strategy is to use **pooling**. A pooling unit reports a summary of its inputs. In the most usual arrangement, a pooling layer halves each spatial dimension of a block. For the moment, ignore the entirely minor problems presented by a fractional dimension. The new block is obtained by pooling units that pool a window at each feature map of the input block to form each feature map of the output block. If these units pool a 2×2 window with stride 2 (i.e., they don't overlap), the output block is half the size of the input block. We adopt the convention that the output reports only valid input windows, so that this takes an $x \times y \times d$ block to an $\text{floor}(x/2) \times \text{floor}(y/2) \times d$ block. So, as Fig. 17.7 shows, a $5 \times 5 \times 1$ block becomes a $2 \times 2 \times 1$ block, but one row and one column are ignored. A common alternative is pooling a 3×3 window with a stride of 2; in this case, a $5 \times 5 \times 1$ block becomes a $2 \times 2 \times 1$ block without ignoring rows or columns. Each unit reports either the largest of the inputs (yielding a **max pooling** layer) or the average of its inputs (yielding an **average pooling** layer).

17.2 Two Practical Image Classifiers

We can now put together image classifiers using the following rough architecture. A convolutional layer receives image pixel values as input. The output is fed to a stack of convolutional layers, each feeding the next, possibly with ReLU layers intervening. There are occasional max-pooling layers, or convolutional layers with stride 2, to ensure that the data block gets smaller and the receptive field gets bigger as the data moves through the network. The output of the final layer is fed to one or more fully connected layers, with one output per class. Softmax takes

Positive

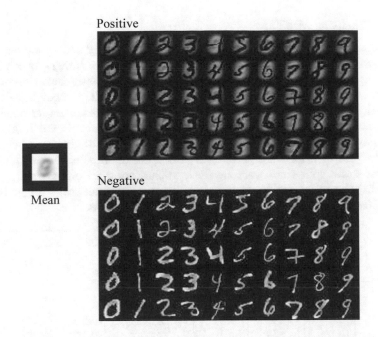

Mean

Negative

Figure 17.8: The mean of MNIST training images is shown on the **left**, surrounded by a black frame so that you can resolve it against the background. On the **right**, the positive (**top**) and negative (**bottom**) components of the difference between mean and image for some training images. Lighter pixels have larger magnitude. Notice the blob of small positive values where there tends to be ink, and the strong negative values where this particular image has ink. This gives the network some information about where ink is expected to lie in general images, which seems to help training in practice

these outputs and turns them into class probabilities. The whole is trained by batch gradient descent, or a variant, as above, using a log-loss.

Notice that different image classification networks differ by relatively straightforward changes in architectural parameters. Mostly, the same thing will happen to these networks (variants of batch gradient descent on a variety of costs; dropout; evaluation). In turn, this means that we should use some form of specification language to put together a description of the architecture of interest. Ideally, in such an environment, we describe the network architecture, choose an optimization algorithm, and choose some parameters (dropout probability, etc.). Then the environment assembles the net, trains it (ideally, producing log files we can look at), and runs an evaluation. The tutorials mentioned in Sect. 16.4.1 each contain examples of image classifiers for the relevant environments. In the examples shown here, I used MatConvNet, because I am most familiar with Matlab.

17.2.1   Example: Classifying MNIST

MNIST images have some very nice features that mean they are a good case to start with. Our relatively simple network architecture accepts images of a fixed size. This property is quite common, and applies to most classification architectures. This isn't a problem for MNIST, because all the MNIST images have the same size. Another nice feature is that pixels are either ink pixels or paper pixels—there are few intermediate values, and none of them are meaningful or helpful. In more general images, $\mathcal{I}$ and $0.9 \times \mathcal{I}$ show the same thing, just at different brightnesses. This doesn't happen for MNIST images. Yet another nice feature is that there is a fixed test–train split that everyone uses, so that comparisons are easy. Without a fixed split, the difference in performance between two networks might be due to random effects, because the networks see different test sets.

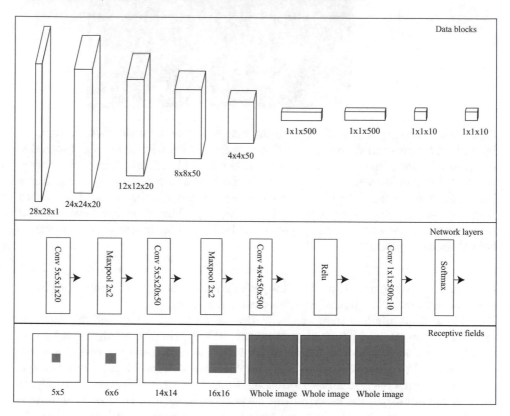

Figure 17.9:   Three different representations of the simple network used to classify MNIST digits for this example. Details in the text

Much of the information in an MNIST image is redundant. Many pixels are paper pixels for every (or almost every) image. These pixels should likely be ignored by every classifier, because they contain little or no information. For

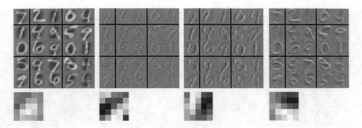

Figure 17.10: Four of the 20 kernels in the first layer of my trained version of the MNIST network. The kernels are small (5 × 5) and have been blown up so you can see them. The outputs for each kernel on a set of images are shown above the kernel. The output images are scaled so that the largest value over all outputs is light, the smallest is dark, and zero is mid grey. This means that the images can be compared by eye. Notice that (rather roughly) the **far left** kernel looks for contrast; **center left** seems to respond to diagonal bars; **center right** to vertical bars; and **far right** to horizontal bars

other pixels, the value of the pixel is less important than how different the pixel is from the expected value at that location. Experience shows that it is surprisingly hard for neural networks to learn from heavily redundant image data. It is usual to preprocess images to remove some redundancies. For MNIST, the usual is to subtract the mean of the training images from the input image. Figure 17.8 shows how doing so seems to enhance the information content of the image.

Figure 17.9 shows the network used for this example. This network is a standard simple classification network for MNIST, distributed with MatConvNet. There are three different representations of the network here. The network layers representation, in the center of the figure, records the type of each layer and the size of the relevant convolution kernels. The first layer accepts the image which is a $28 \times 28 \times 1$ block of data (the data block representation), and applies a convolution. By convention, "conv $5 \times 5 \times 1 \times 20$" means a convolution layer, with a 20 different kernels each $5 \times 5 \times 1$. The effects of some of the learned kernels in this layer are visualized in Fig. 17.10.

In the implementation I used, the convolution was not padded so that the resulting data block was $24 \times 24 \times 20$ (check that you know why this is correct). A value in this data block is computed from a 5×5 window of pixels, so the receptive field is 5×5. Again, by convention, every convolutional layer has a bias term, so the total number of parameters in the first layer is $(5 \times 5 \times 1) \times 20 + 20$ (check this statement, too). The next layer is a 2×2 max-pooling layer, which again is not padded. This takes a $24 \times 24 \times 20$ block and produces a $12 \times 12 \times 20$ block. The receptive field for values in this block is 6×6 (you should check this with a pencil and paper drawing; it's right).

Another convolutional layer and another max-pooling layer follow, reducing the data to a $4 \times 4 \times 50$ block. Every value in this block is potentially affected by every image pixel, and this is true for all following blocks. Yet another convolutional

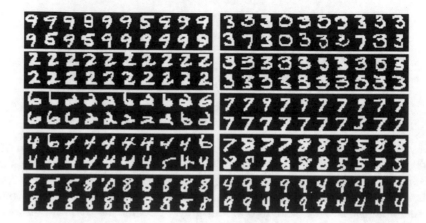

Figure 17.11: Visualizing the patterns that the final stage ReLUs respond to for the simple CIFAR example. Each block of images shows the images that get the largest output for each of 10 ReLUs (the ReLUs were chosen at random from the 500 available). Notice that these ReLU outputs don't correspond to class— these outputs go through a fully connected layer before classification—but each ReLU clearly responds to a pattern, and different ReLUs respond more strongly to different patterns

layer reduces this to a $1 \times 1 \times 500$ block (again, where every value is potentially affected by every pixel in the image). That goes through a ReLU (outputs visualized in Fig. 17.11). You should think of the result as a 500 dimensional feature vector describing the image, and the convolutional layer and softmax that follow are logistic regression applied to that feature vector.

I trained this network for 20 epochs using tutorial code circulated with MatConvNet. Minibatches are pre-selected so that each training data item is touched once per epoch, so an epoch represents a single pass through the data. It is common in image classification to report loss, top-1 error, and top-5 error. Top-1 error is the frequency that the correct class has the highest posterior. Top-5 error is the frequency that the correct class appears in the five classes with largest posterior. This can be useful when the top-1 error is large, because you may observe improvements in top-5 error even when the top-1 error doesn't change. Figure 17.12 shows the loss, top-1 error, and top-5 error for training and validation sets plotted as a function of epoch. This network has a low error rate, so of the 10,000 test examples there are only 89 errors, which are shown in Fig. 17.13.

17.2.2 Example: Classifying CIFAR-10

CIFAR-10 is a dataset of 32×32 color images in ten categories, collected by Alex Krizhevsky, Vinod Nair, and Geoffrey Hinton. It is often used to evaluate image classification algorithms. There are 50,000 training images and 10,000 test images, and the test–train split is standard. Images are evenly split between the classes. Figure 17.14 shows the categories, and examples from each category. There is no overlap between the categories (so "automobile" consists of sedans, etc., and "truck" consists of big trucks). You can download this dataset from https://www. cs.toronto.edu/~kriz/cifar.html.

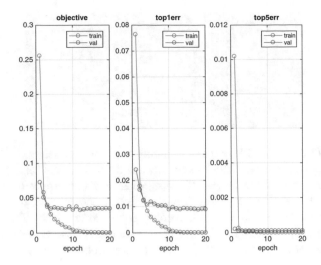

Figure 17.12: This figure shows the results of training the network of Fig. 17.9 on the MNIST training set. Loss, top-1 error, and top-5 error for training and validation sets plotted as a function of epoch for the network of the text. The loss (recorded here as "objective") is the log-loss. Note: the low validation error; the gap between train and validation error; and the very low top-5 error. The validation error is actually quite high for this dataset—you can find a league table at http://rodrigob. github.io/are_we_there_yet/build/classification_datasets_results.html

Figure 17.15 shows the network used to classify CIFAR-10 images. This network is again a standard classification network for CIFAR-10, distributed with MatConvNet. Again, I have shown the network in three different representations. The network layer representation, in the center of the figure, records the type of each layer and the size of the relevant convolution kernels. The first layer accepts the image which is a $32 \times 32 \times 3$ block of data (the data block representation), and applies a convolution.

In this network, the convolution *was* padded so that the resulting data block

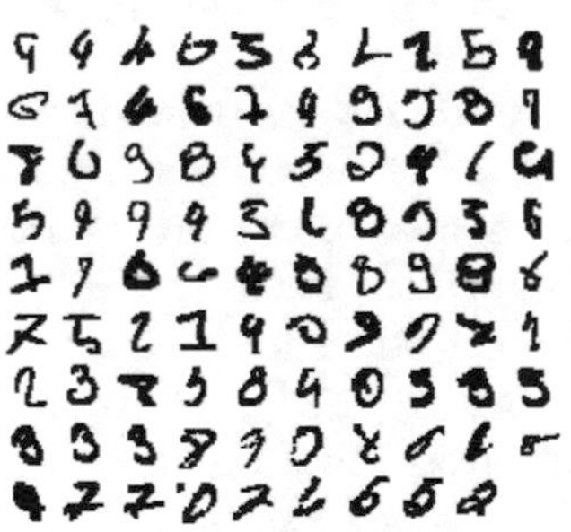

```
| 5 | 9 | 2 | 0 | 5 | 2 | 6 | 1 | 8 | 9 |
| 5 | 1 | 6 | 5 | 2 | 4 | 5 | 7 | 3 | 9 |
| 7 | 6 | 5 | 0 | 4 | 3 | 0 | 9 | 1 | 4 |
| 3 | 4 | 4 | 9 | 3 | 1 | 0 | 7 | 5 | 8 |
| 2 | 9 | 0 | 4 | 8 | 0 | 5 | 3 | 9 | 8 |
| 2 | 2 | 8 | 3 | 4 | 2 | 3 | 0 | 7 | 7 |
| 1 | 2 | 7 | 5 | 0 | 4 | 6 | 3 | 8 | 3 |
| 8 | 9 | 9 | 9 | 7 | 7 | 2 | 8 | 6 | 5 |
| 9 | 2 | 2 | 8 | 7 | 3 | 6 | 0 | 8 |
```

Figure 17.13: **Left:** All 89 errors from the 10,000 test examples in MNIST and **right** the predicted labels for these examples. True labels are mostly fairly clear, though some of the misclassified digits take very odd shapes

Figure 17.14: The CIFAR-10 image classification dataset consists of 60,000 images, in a total of ten categories. The images are all 32×32 color images. This figure shows 20 images from each of the 10 categories and the labels of each category. On the **far right**, the mean of the images in each category. I have doubled the brightness of the means, so you can resolve color differences. The per category means are different, and suggest that some classes look like a blob on a background, and others (e.g., ship, truck) more like an outdoor scene

was $32 \times 32 \times 32$. You should check whether you agree with these figures, and you can tell by how much the image needed to be padded to achieve this (a drawing might help). A value in this data block is computed from a 5×5 window of pixels, so the receptive field is 5×5. Again, by convention, every convolutional layer has a bias term, so the total number of parameters in the first layer is $(5 \times 5 \times 3) \times 32 + 32$. The next layer is a 3×3 max pooling layer. The notation $3s2$ means that the pooling blocks have a stride of 2, so they overlap. The block is padded for this pooling layer, by attaching a single column at the right and a single row at the bottom to get a $33 \times 33 \times 32$ block. With this padding and stride, the pooling takes $33 \times 33 \times 32$ block and produces a $16 \times 16 \times 32$ block (you should check this with a pencil and paper drawing; it's right). The receptive field for values in this block is 7×7 (you should check this with a pencil and paper drawing; it's right, too).

The layer labelled "Apool 3s2" is an average pooling layer which computes an average in a 3×3 window, again with a stride of 2. The block is padded before this layer in the same way the block before the max-pooling layer was padded. Eventually, we wind up with a 64 dimensional feature vector describing the image, and the convolutional layer and softmax that follow are logistic regression applied to that feature vector.

Just like MNIST, much of the information in a CIFAR-10 image is redundant. It's now somewhat harder to see the redundancies, but Fig. 17.14 should make you suspect that some classes have different backgrounds than others. Figure 17.14 shows the class mean for each class. There are a variety of options for normalizing these images (more below). For this example, I whitened pixel values for each pixel in the image grid independently (Procedure 17.1, which is widely used). Whitened images tend to be very hard for humans to interpret. However, the normalization involved deals with changes in overall image brightness and moderate shifts in color rather well, and can significantly improve classification.

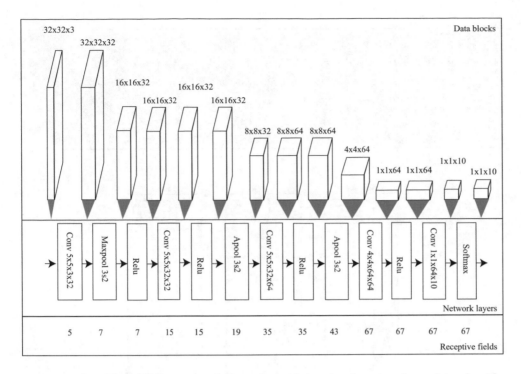

Figure 17.15: Three different representations of the simple network used to classify CIFAR-10 images for this example. Details in the text

---

Procedure: 17.1 *Simple Image Whitening*

At Training Time: Start with N training images $\mathcal{I}^{(i)}$. We assume that these are 3D blocks of data. Write $I_{uvw}^{(i)}$ for the u, v, w'th location in the i'th image. Compute $\mathcal{M}$ and $\mathcal{S}$, where the u, v, w'th location in each is given by

$$M_{uvw} = \frac{\sum_i I_{uvw}^{(i)}}{N}$$

$$S_{uvw} = \sqrt{\frac{\sum_i (I_{uvw}^{(i)} - M_{uvw}^{(i)})^2}{N}}.$$

Choose some small number ϵ to avoid dividing by zero. Now the i'th whitened image, $\mathcal{W}^{(i)}$, has for its u, v, w'th location

$$W_{uvw}^{(i)} = (I_{uvw}^{(i)} - M_{uvw})/(S_{uvw} + \epsilon).$$

Use these whitened images to train.
At Test Time: For a test image $\mathcal{T}$, compute $\mathcal{W}$ which has for its u, v, w'th location

$$W_{uvw} = (T_{uvw} - M_{uvw})/(S_{uvw} + \epsilon)$$

and classify that.

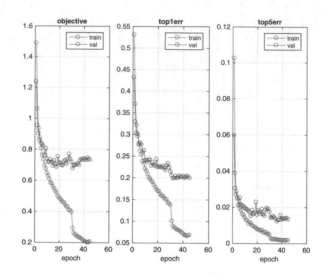

Figure 17.16: This figure shows the results of training the network of Fig. 17.15 on the CIFAR-10 training set. Loss, top-1 error, and top-5 error for training and validation sets, plotted as a function of epoch for the network of the text. The loss (recorded here as "objective") is the log-loss. Note: the low validation error; the gap between train and validation error; and the very low top-5 error. The validation error is actually quite high for this dataset—you can find a league table at http://rodrigob.github.io/are_we_there_yet/build/classification_datasets_results.html

Figure 17.17: Some of the approximately 2000 test examples misclassified by the network trained in the text. Each row corresponds to a category. The images in that row belong to that category, but are classified as belonging to some other category. At least some of these images look like "uncommon" views of the object or "strange" instances—it's plausible that the network misclassifies images when the view is uncommon or the object is a strange instance of the category

Figure 17.18: Some of the approximately 2000 test examples misclassified by the network trained in the text. Each row corresponds to a category. The images in that row are classified as belonging to that category, but actually belong to another. At least some of these images look like "confusing" views—for example, you can find birds that do look like aircraft, and aircraft that do look like birds

I trained this network for 20 epochs using tutorial code circulated with Mat-ConvNet. Minibatches are pre-selected so that each training data item is touched once per epoch, so an epoch represents a single pass through the data. It is common in image classification to report loss, top-1 error, and top-5 error. Top-1 error is the frequency that the correct class has the highest posterior. Top-5 error is the frequency that the correct class appears in the five classes with largest posterior. This can be useful when the top-1 error is large, because you may observe improvements in top-5 error even when the top-1 error doesn't change. Figure 17.16 shows the loss, top-1 error, and top-5 error for training and validation sets plotted as a function of epoch. This classifier misclassifies about 2000 of the test examples, so it is hard to show all errors. Figure 17.17 shows examples from each class that are misclassified as belonging to some other class. Figure 17.18 shows examples that are misclassified into each class.

The phenomenon that ReLUs are pattern detectors is quite reliable. Figure 17.19 shows the 20 images that give the strongest responses for each of 10 ReLUs in the final ReLU layer. These ReLUs clearly have a quite strong theory of a pattern, and different ReLUs respond most strongly to quite different patterns. More sophisticated visualizations search for images that get the strongest response from units at various stages of complex networks; it's quite reliable that these images show a form of order or structure.

17.2.3 Quirks: Adversarial Examples

Adversarial examples are a curious experimental property of neural network image classifiers. Here is what happens. Assume you have an image **x** that is correctly classified with label l. The network will produce a probability distribution over

Figure 17.19: Visualizing the patterns that the final stage ReLUs respond to for the simple CIFAR example. Each block of images shows the images that get the largest output for each of 10 ReLUs (the ReLUs were chosen at random from the 64 available in the top ReLU layer). Notice that these ReLU outputs don't correspond to class—these outputs go through a fully connected layer before classification— but each ReLU clearly responds to a pattern, and different ReLUs respond more strongly to different patterns

labels $P(L|\mathbf{x})$. Choose some label k that is not correct. It is possible to use modern optimization methods to search for a modification to the image $\delta\mathbf{x}$ such that

$$\delta\mathbf{x} \quad \text{is small}$$
$$\text{and}$$
$$P(k|\mathbf{x} + \delta\mathbf{x}) \quad \text{is large.}$$

You might expect that $\delta\mathbf{x}$ is "large"; what is surprising is that mostly it is so tiny as to be imperceptible to a human observer. The property of being an adversarial example seems to be robust to image smoothing, simple image processing, and printing and photographing. The existence of adversarial examples raises the following, rather alarming, prospect: You could make a template that you could hold over a stop sign, and with one pass of a spraypaint can, turn that sign into something that is interpreted as a minimum speed limit sign by current computer vision systems. I haven't seen this demonstration done yet, but it appears to be entirely within the reach of modern technology, and it and activities like it offer significant prospects for mayhem.

What is startling about this behavior is that it is exhibited by networks that are very good at image classification, *assuming* that no one has been fiddling with the images. So modern networks are very accurate on untampered pictures, but may behave very strangely in the presence of tampering. One can (rather vaguely) identify the source of the problem, which is that neural network image classifiers have far more degrees of freedom than can be pinned down by images. This observation doesn't really help, though, because it doesn't explain why they (mostly) work rather well, and it doesn't tell us what to do about adversarial examples. There have been a variety of efforts to produce networks that are robust to adversarial examples, but evidence right now is based only on experiment (some networks behave better than others) and we are missing clear theoretical guidance.

17.3 You Should

17.3.1 Remember These Definitions

17.3.2 Remember These Terms

17.3.3 Remember These Facts

17.3.4 Remember These Procedures

17.3.5 Be Able to

- Explain what convolutional layers do.
- Compute the size of a data block resulting from applying a convolutional layer with given size and stride to a block with given padding.
- Explain what a 1×1 convolution does and why it might be useful.
- Train and run a simple image classifier in your chosen framework.
- Explain why preprocessing data might help a neural network based classifier.
- Explain what an adversarial example is.

Programming Exercises

17.1. Download tutorial code for a simple MNIST classifier for your chosen programming framework, and train and run a classifier using that code. You should be able to do this exercise without access to a GPU.

17.2. Now reproduce the example of Sect. 17.2.1 in your chosen programming framework. The section contains enough detail about the structure of the network for you to build that network. This isn't a super good classifier; the point of the exercise is being able to translate a description of a network to an instance. Use the standard test–train split, and train with straightforward stochastic gradient descent. Choose a minibatch size that works for this example and your hardware. Again, you should be able to do this exercise without access to a GPU.

(a) Does using momentum improve training?

(b) Does using dropout in the first two layers result in a better performing network?

(c) Modify this network architecture to improve performance. Reading ahead will suggest some tricks. What works best?

17.3. Download tutorial code for a simple CIFAR-10 classifier for your chosen programming framework, and train and run a classifier using that code. You might very well be able to do this exercise without access to a GPU.

17.4. Now reproduce the example of Sect. 17.2.2 in your chosen programming framework. The section contains enough detail about the structure of the network for you to build that network. This isn't a super good classifier; the point of the exercise is being able to translate a description of a network to an instance. Use the standard test–train split, and train with straightforward stochastic gradient descent. Choose a minibatch size that works for this example and your hardware. Again, you might very well be able to do this exercise without access to a GPU.

(a) Does using momentum improve training?

(b) Does using dropout in the first two layers result in a better performing network?

(c) Modify this network architecture to improve performance. Reading ahead will suggest some tricks. What works best?

C H A P T E R   18

Classifying Images and Detecting Objects

Neural networks have gone from being one curiosity in lists of classification methods to being the prime engine of a huge and very successful industry. This has happened in a very short time, less than a decade. The main reason is that, with enough training data and enough training ingenuity, neural networks produce very successful classification systems, much better than anyone has been able to produce with other methods. They are particularly good at classifying images. As Fig. 18.1 shows, the top-5 error rate on one (very large and very hard) image classification dataset has collapsed in quite a short period. The primary reason seems to be that the features that are being used by the classifier are themselves learned from data. The learning process seems to ensure that the features are useful for classification. It's easy to see that it might do so; the news here is that it does.

There are two important trends that have advanced this area. One is the development of large, challenging (but not unreasonably hard) datasets that are publicly available and where accuracy is evaluated using conventions that are fair and open. The second is the widespread dissemination of successful models. If someone produces a really good image classifier, you can usually find an implementation on the internet fairly soon afterwards. This means that it's easy to fiddle with successful architectures and try to make them better. Very often, these implementations come with pretrained models.

This chapter will describe the main recent successes in image classification and object detection using neural networks. It's unlikely you would be able to build anything I describe here from the text alone, but you can likely find a trained version elsewhere. You should get a good enough grasp of what people do, what seems to work, and why to apply and use models that have been shared.

18.1 Image Classification

I will describe several important network architectures in the following subsections, but building any of these from scratch based only on this description would be a heroic (and likely unsuccessful) venture. What you should do is download a version for the environment you prefer, and play with that. You can find pretrained models at:

- https://pjreddie.com/darknet/imagenet/ (for darknet);
- http://www.vlfeat.org/matconvnet/pretrained/ (for MatConvNet);
- https://mxnet.apache.org/api/python/gluon/model_zoo.html (for MXNet);
- https://github.com/PaddlePaddle/models (for PaddlePaddle; it helps to be able to read Chinese);

© Springer Nature Switzerland AG 2019
D. Forsyth, *Applied Machine Learning*,
https://doi.org/10.1007/978-3-030-18114-7_18

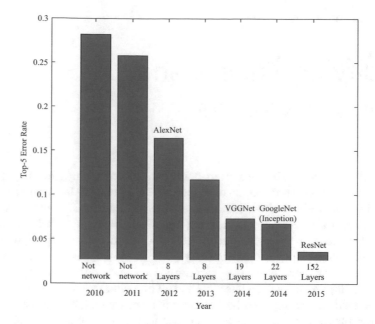

Figure 18.1: The top-5 error rate for image classification using the ImageNet dataset has collapsed from 28% to 3.6% from 2010 to 2015. There are two 2014 entries here, which makes the fall in error rate look slower. This is because each of these methods is significant, and discussed in the sections below. Notice how increasing network depth seems to have produced reduced error rates. This figure uses ideas from an earlier figure by Kaiming He. Each of the named networks is described briefly in a section below

- https://pytorch.org/docs/stable/torchvision/models.html (for PyTorch);
- https://github.com/tensorflow/models (for TensorFlow);
- https://keras.io (for Keras; look for "examples" in the sidebar).

18.1.1 Datasets for Classifying Images of Objects

MNIST and CIFAR-10 are no longer cutting edge image classification datasets. The networks I described are quite simple, but work rather well on these problems. The very best methods are now extremely good. Rodrigo Benenson maintains a website giving best performance to date on these datasets at http://rodrigob.github.io/are_we_there_yet/build/classification_datasets_results.html. The best error rate recorded there for MNIST is 0.21% (i.e., a total of 21 test examples wrong in the 10,000 example test set). For CIFAR-10, the best error rate is 3.47% (i.e., a total of 347 test examples wrong; much better than our 2000 odd). Mostly, methods work so well that improvements must be very small, and so it is difficult to see what is an important change and what is a lucky accident. These datasets are now mostly used for warming-up purposes—to check that an idea isn't awful, or that a method can work on an "easy" dataset.

> **Remember This:** *MNIST and CIFAR-10 are warm-up datasets. You can find MNIST at http://yann.lecun.com/exdb/mnist/ or at http://www. kaggle.com/c/digit-recognizer. You can find CIFAR-10 at https://www.cs. toronto.edu/~kriz/cifar.html.*

It is difficult to say precisely what makes a dataset hard. It is very likely that having more categories makes a dataset harder than having few categories. It is very likely that having a lot of training data per category makes a dataset easier. It is certain that labelling errors and differences between test images and training images will cause problems. Modern datasets tend to be built carefully using protocols that try to ensure that the label for each data item is right. For example, one can have images labelled independently, then check whether the labels agree. There isn't any way of checking to see that the training set is like the test set, but one can collect first, then split later.

MNIST and CIFAR-10 contain pictures of largely isolated objects. A harder dataset is CIFAR-100. This is very like CIFAR-10, but now with 100 categories. Images are 32×32 color images in 100 categories, collected by Alex Krizhevsky, Vinod Nair, and Geoffrey Hinton. There are 50,000 training images (so now 500 per category, rather than 5000) and 10,000 test images, and the test–train split is standard. Images are evenly split between the classes. The categories are grouped rather roughly into superclasses, so that there are several different insect categories, several different reptile categories, and so on.

> **Remember This:** *CIFAR-100 is a small hard image classification dataset. You can download this dataset from https://www.cs.toronto.edu/ ~kriz/cifar.html. CIFAR-100 accuracy is also recorded at http://rodrigob. github.io/are_we_there_yet/build/classification_datasets_results.html. The best error rate (24.28%) is a crude indicator that this dataset is harder than CIFAR-10 or MNIST.*

There are several important big image classification datasets. Datasets tend to develop over time, and should be followed by looking at a series of workshops. The **Pascal** visual object classes challenges are a set of workshops held from 2005 to 2012 to respond to challenges in image classification. The workshops, which were a community wide effort led by the late Mark Everingham, resulted in a number of tasks and datasets which are still used. There is more information, including leaderboards, best practice, organizers, etc., at http://host.robots.ox.ac.uk/pascal/ VOC/.

> **Remember This:**   *PASCAL VOC 2007 remains a standard image classification dataset. You can find this at http://host.robots.ox.ac.uk/pascal/ VOC/voc2007/index.html. The dataset uses a collection of 20 object classes that became a form of standard.*

There is very little point in classifying images of objects into classes that aren't useful, but it isn't obvious what classes should be used. One strategy is to organize classes in the same way that nouns for objects are organized. **WordNet** is a lexical database of the English language, organized hierarchically in a way that tries to represent the distinctions that people draw between objects. So, for example, a `cat` is a `felid` which is a `carnivore` which is a `placental mammal` which is a `vertebrate` which is a `chordate` which is an `animal` (and so on...). You can explore WordNet at https://wordnet.princeton.edu. **ImageNet** is a collection organized according to a semantic hierarchy taken from WordNet. ImageNet Large Scale Visual Recognition Challenge (ILSVRC) workshops were held from 2010 to 2017, organized around a variety of different challenges.

> **Remember This:**   *ImageNet is an extremely important large-scale image classification dataset. A very commonly used standard is the ILSVRC2012 dataset, with 1000 classes and 1.28 million training images. There's a standard validation set of 50,000 images (50 per category). You can find this at http://www.image-net.org/challenges/LSVRC/2012/ nonpub-downloads. The dataset uses a collection of 1000 object classes that became a form of standard.*

18.1.2   Datasets for Classifying Images of Scenes

Objects tend to appear together in quite structured ways, so if you see a giraffe you might also expect to see an acacia or a lion, but you wouldn't expect to see a submarine or a couch. Different contexts tend to result in different groups of objects. So in grassland you might see a giraffe or a lion, and in the living room you might see a couch, but you don't expect a giraffe in a living room. This suggests that environments are broken up into clusters that look different and tend to contain different objects. Such clusters are widely called **scene**s in the vision community. An important image classification challenge is to take an image of a scene and predict what the scene is.

One important scene classification dataset is the **SUN** dataset. This is widely used for training, and for various classification challenges. There is a benchmark dataset with 397 categories. The full dataset contains over 900 categories and many million images. Workshop challenges, including particular datasets used and leaderboards, appear at http://lsun.cs.princeton.edu/2016/ (LSUN 2016); and http://lsun.

cs.princeton.edu/2017/ (LSUN 2017). The challenges use a selected subset of the scene categories.

> **Remember This:** *SUN is a large-scale scene classification dataset that has been the core of several challenge workshops. The dataset appears at https://groups.csail.mit.edu/vision/SUN/.*

Another important dataset is the **Places-2** dataset. There are 10 million images in over 400 categories, with annotations of scene attributes and a variety of other materials.

> **Remember This:** *Places-2 is a large-scale scene classification dataset. You can find this at http://places2.csail.mit.edu.*

18.1.3 Augmentation and Ensembles

Three important practical issues need to be addressed to build very strong image classifiers.

- **Data sparsity:** Datasets of images are never big enough to show all effects accurately. This is because an image of a horse is still an image of a horse even if it has been through a small rotation, or has been resized to be a bit bigger or smaller, or has been cropped differently, and so on. There is no way to take account of these effects in the architecture of the network.
- **Data compliance:** We want each image fed into the network to be the same size.
- **Network variance:** The network we have is never the best network; training started at a random set of parameters, and has a strong component of randomness in it. For example, most minibatch selection algorithms select random minibatches. Training the same architecture on the same dataset twice will not yield the same network.

All three can be addressed by some care with training and test data.

Generally, the way to address data sparsity is **data augmentation**, by expanding the training dataset to include different rotations, scalings, and crops of images. Doing so is relatively straightforward. You take each training image, and generate a collection of extra training images from it. You can obtain this collection by: resizing and then cropping the training image; using different crops of the same training image (assuming that training images are a little bigger than the size of image you will work with); rotating the training image by a small amount, resizing and cropping; and so on.

There are some cautions. When you rotate then crop, you need to be sure that no "unknown" pixels find their way into the final crop. You can't crop too much, because you need to ensure that the modified images are still of the relevant class, and too aggressive a crop might cut out the horse (or whatever) entirely. This somewhat depends on the dataset. If each image consists of a tiny object on a large background, and the objects are widely scattered, crops need to be cautious; but if the object covers a large fraction of the image, the cropping can be quite aggressive.

Cropping is usually the right way to ensure that each image has the same size. Resizing images might cause some to stretch or squash, if they have the wrong aspect ratio. This likely isn't a great idea, because it will cause objects to stretch or squash, making them harder to recognize. It is usual to resize images to a convenient size without changing the aspect ratio, then crop to a fixed size.

There are two ways to think about network variance (at least!). If the network you train isn't the best network (because it can't be), then it's very likely that training multiple networks and combining the results in some way will improve classification. You could combine results by, for example, voting. Small improvements can be obtained reliably like this, but the strategy is often deprecated because it isn't particularly elegant or efficient. A more usual approach is to realize that the network might very well handle one crop of a test image rather better than others (because it isn't the best network, etc.). Small improvements in performance can be obtained very reliably by presenting multiple crops of a test image to a given network, and combining the results for those crops.

18.1.4 AlexNet

The first really successful neural network image classifier was **AlexNet**, described in "ImageNet Classification with Deep Convolutional Neural Networks," a NIPS 2012 paper by Alex Krizhevsky, Ilya Sutskever, and Geoffrey Hinton. AlexNet is quite like the simple networks we have seen—a sequence of convolutional layers that reduce the spatial dimensions of the data block, followed by some fully connected layers—but has a few special features. GPU memories in 2012 were much smaller than they are now, and the network architecture is constructed so that the data blocks can be split across two GPUs. There are new normalization layers, and there is a fully connected layer that reduces a data block in size in a new way.

The impact of splitting the data blocks is quite significant. As Fig. 18.2 shows, the image passes into a convolutional layer with 96 kernels followed by a ReLU, response normalization (which modifies values in a block, but doesn't change its size), and max pooling. This would normally result in a data block of dimension $55 \times 55 \times 96$, but here each GPU gets a block consisting of the output of a different half of the kernels (so there are two $55 \times 55 \times 48$ blocks). Each goes through another convolutional layer of 128 kernels (size $5 \times 5 \times 48$), with a total of 256 kernels. The blocks on GPU 1 and GPU 2 may contain quite different features; the block on GPU 1 at B in the figure does *not* see the block on GPU 2 at A. The block at C for each GPU is constructed using the block at B for both GPUs, but then blocks move through the network without interacting until the dense layer (which turns E into F). This means that features on one GPU could encode rather different properties, and this actually happens in practice.

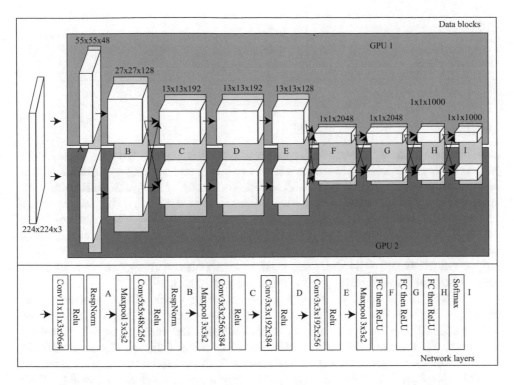

Figure 18.2: Two views of the architecture of AlexNet, the first convolutional neural network architecture to beat earlier feature constructions at image classification. There are five convolutional layers with ReLU, response normalization, and pooling layers interspersed. **Top** shows the data blocks at various stages through the network and **bottom** shows all the layers (capital letters key stages in the network to blocks of data). Horizontal and diagonal arrows in the top box indicate how data is split between GPUs, details in the main text. The response normalization layer is described in the text. I have compacted the final fully connected layers to fit the figure in

For each location in a block, response normalization layers then scale the value at that location using a summary of nearby values. Response normalization like this is no longer widely used, so I will omit details. This network was trained using substantial data augmentation, as above. Units in the first two layers are dropped out with a probability of 0.5. Training uses the usual stochastic gradient descent, but with momentum. AlexNet was a spectacular success, achieving top-1 and top-5 error rates of 37.5% and 17.0%, respectively, on the ImageNet ILSVRC-2010 challenge. These scores are significantly better than any other method had produced in the past, and stimulated widespread investigation into network architectures that might do better.

> **Remember This:** *AlexNet was a spectacular success at classifying ImageNet images.*

18.1.5 VGGNet

AlexNet has some odd features. It has relatively few layers. It splits data blocks across GPUs. The kernels in the first layer are large, and have a large stride. And it has response normalization layers. VGGNet is a family of networks built to investigate these and other issues. Using the best member of the family, the best practices in cropping, evaluation, data augmentation, and so on, VGGNet obtained top-1 and top-5 error rates of 23.7% and 6.8%, respectively, on the ImageNet ILSVRC-2014 challenge. This was a substantial improvement. Table 18.1 describes the five most important VGGNets (the sixth was used to establish that response normalization wasn't helpful for everything; this doesn't matter to us).

Table 18.1 is a more compact presentation of much of the information in Fig. 18.2, but for the five VGGNets. The table shows the flow of information downwards. The naming conventions work like this. The term "convX-Y" means a convolutional layer of Y $X \times X$ kernels followed by a ReLU layer. The term "FC-X" means a fully connected layer that produces an X dimensional vector. For example, in VGGNet-A, a $224 \times 224 \times 3$ image passes into a layer, labelled "conv3-64." This consists of a convolutional layer of 64 $3 \times 3 \times 3$ kernels, followed by a ReLU layer. The block then passes into a maxpool layer, pooling over 2×2 windows with stride 2. The result goes to a convolutional layer of 128 $3 \times 3 \times 3$ kernels, followed by a ReLU layer. Eventually, the block of data goes to a fully connected layer that produces a 4096 dimensional vector ("FC-4096"), passes through another of these to an FC-1000 layer, and then to a softmax layer.

Reading across the table gives the different versions of the network. Notice that there are significantly more layers with trainable weights than for AlexNet. The E version (widely known as **VGG-19**) is the most widely used; others were mainly used in training, and to establish that more layers give better performance. The networks have more layers as the version goes up. Terms in bold identify layers introduced when the network changes (reading right). So, for example, the B version has a conv3-64 term that the A version doesn't have, and the C, D, and E versions keep; the C version has a conv1-512 term that the A and B versions don't have, and the D and E versions replace with a conv3-512 term.

You should expect that training a network this deep is hard (recall Sect. 16.4.3). VGGNet training followed a more elaborate version of the procedure I used in Sect. 16.4.3. Notice that the B version is the A version together with two new terms, etc. Training proceeded by training the A version. Once the A version was trained, the new layers were inserted to make a B version (keeping the parameter values of the A version's layers), and the new network was trained from that initialization. All parameter values in the new network were updated. The C version was then trained from B, and so on. All training is by minibatch stochastic gradient

Network architecture				
A	B	C	D	E
Number of layers with learnable weights				
11	13	16	16	19
Input (224 × 224 × 3 image)				
conv3-64	conv3-64	conv3-64	conv3-64	conv3-64
	conv3-64	conv3-64	conv3-64	conv3-64
maxpool2 × 2s2				
conv3-64	conv3-64	conv3-64	conv3-64	conv3-64
	conv3-64	conv3-64	conv3-64	conv3-64
maxpool2 × 2s2				
conv3-128	conv3-128	conv3-128	conv3-128	conv3-128
	conv3-128	conv3-128	conv3-128	conv3-128
maxpool2 × 2s2				
conv3-256	conv3-256	conv3-256	conv3-256	conv3-256
conv3-256	conv3-256	conv3-256	conv3-256	conv3-256
		conv1-256	**conv3-256**	conv3-256
				conv3-256
maxpool2 × 2s2				
conv3-512	conv3-512	conv3-512	conv3-512	conv3-512
conv3-512	conv3-512	conv3-512	conv3-512	conv3-512
		conv1-512	**conv3-512**	conv3-512
				conv3-512
maxpool2 × 2s2				
conv3-512	conv3-512	conv3-512	conv3-512	conv3-512
conv3-512	conv3-512	conv3-512	conv3-512	conv3-512
		conv1-512	**conv3-512**	conv3-512
				conv3-512
maxpool2 × 2s2				
FC-4096				
FC-4096				
FC-1000				
softmax				

TABLE 18.1: This table summarizes the architecture of five VGGNets. Details in the text

descent with momentum. The first two layers were subject to dropout (probability of dropout 0.5). Data was aggressively augmented.

Experiment suggests that the features constructed by VGG-19 and networks like it are canonical in some way. If you have a task that involves computing something from an image, using VGG-19 features for that task very often works. Alternatively, you could use VGG-19 as an initialization for training a network for your task. VGG-19 is still widely used as a **feature stack**—a network that was trained for classification, but whose features are being used for something else.

> **Remember This:**   *VGGNet outperformed AlexNet at classifying Ima-*
> *geNet images. There are several versions. VGG-19 is still used to produce*
> *image features for other tasks.*

18.1.6   Batch Normalization

There is good experimental evidence that large values of inputs to any layer within a neural network lead to problems. One source of the problem could be this. Imagine some input to some unit has a large absolute value. If the corresponding weight is relatively small, then one gradient step could cause the weight to change sign. In turn, the output of the unit will swing from one side of the ReLU's non-linearity to the other. If this happens for too many units, there will be training problems because the gradient is then a poor prediction of what will actually happen to the output. So we should like to ensure that relatively few values at the input of any layer have large absolute values. We will build a new layer, sometimes called a **batch normalization layer**, which can be inserted between two existing layers.

Write $\mathbf{x}^b$ for the input of this layer, and $\mathbf{o}^b$ for its output. The output has the same dimension as the input, and I shall write this dimension d. The layer has two vectors of parameters, γ and β, each of dimension d. Write $\mathrm{diag}(\mathbf{v})$ for the matrix whose diagonal is $\mathbf{v}$, and with all other entries zero. Assume we know the mean ($\mathbf{m}$) and standard deviation ($\mathbf{s}$) of each component of $\mathbf{x}^b$, where the expectation is taken over all relevant data. The layer forms

$$\mathbf{x}^n = [\mathrm{diag}(\mathbf{s} + \epsilon)]^{-1} (\mathbf{x}^b - \mathbf{m})$$
$$\mathbf{o}^b = [\mathrm{diag}(\gamma)] \mathbf{x}^n + \beta.$$

Notice that the output of the layer is a differentiable function of γ and β. Notice also that this layer *could* implement the identity transform, if $\gamma = \mathrm{diag}(\mathbf{s} + \epsilon)$ and $\beta = \mathbf{m}$. We adjust the parameters in training to achieve the best performance. It can be helpful to think about this layer as follows. The layer rescales its input to have zero mean and unit standard deviation, then allows training to readjust the mean and standard deviation as required. In essence, we expect that large values encountered between layers are likely an accident of the difficulty training a network, rather than required for good performance.

The difficulty here is we don't know either $\mathbf{m}$ or $\mathbf{s}$, because we don't know the parameters used for previous layers. Current practice is as follows. First, start with $\mathbf{m} = \mathbf{0}$ and $\mathbf{s} = \mathbf{1}$ for each layer. Now choose a minibatch, and train the network using that minibatch. Once you have taken enough gradient steps and are ready to work on another minibatch, re-estimate $\mathbf{m}$ as the mean of values of the inputs to the layer, and $\mathbf{s}$ as the corresponding standard deviations. Now obtain another minibatch, and proceed. Remember, γ and β are parameters that are trained, just like the others (using gradient descent, momentum, AdaGrad, or whatever). Once the network has been trained, one then takes the mean (resp. standard deviation) of the layer inputs over the training data for $\mathbf{m}$ (resp. $\mathbf{s}$). Most neural network implementation environments will do all the work for you. It is quite usual to place a batch normalization layer between each layer within the network.

For some problems, minibatches are small, usually because one is using a large model or a large data item and its hard to cram many items into the GPU. If you have many GPUs, you can consider synchronizing the minibatches and then averaging over all the minibatches being presented to the GPU—this isn't for everybody. If the minibatch is small, then the estimate of **m** and **s** obtained using a minibatch will be noisy, and batch normalization typically performs poorly. For many problems involving images, you can reasonably expect that groups of features should share the same scale. This justifies using **group normalization**, where the feature channels are normalized in groups across a minibatch. The advantage of doing so is that you will have more values to use when estimating the parameters; the disadvantage is that you need to choose which channels form groups.

There is a general agreement that normalization improves training, but some disagreement about the details. Experiments comparing two networks, one with normalization the other without, suggest that the same number of steps tends to produce a lower error rate when batch normalized. Some authors suggest that convergence is faster (which isn't quite the same thing). Others suggest that larger learning rates can be used.

Remember This: *Batch normalization improves training by discouraging large numbers in datablocks that aren't required for accuracy. When minibatches are small, it can be better to use group normalization, where one normalizes over groups of features.*

18.1.7 Computation Graphs

In Sect. 113, I wrote a simple network in the following form:

$$\mathbf{o}^{(D)},$$

where

$$\mathbf{o}^{(D)} = \mathbf{o}^{(D)}(\mathbf{u}^{(D)}, \theta^{(D)})$$
$$\mathbf{u}^{(D)} = \mathbf{o}^{(D-1)}(\mathbf{u}^{(D-1)}, \theta^{(D-1)})$$
$$\dots = \dots$$
$$\mathbf{u}^{(2)} = \mathbf{o}^{(1)}(\mathbf{u}^{(1)}, \theta^{1})$$
$$\mathbf{u}^{(1)} = \mathbf{x}.$$

These equations really were a map for a computation. You feed in $\mathbf{x}$; this gives $\mathbf{u}^{(1)}$; which gives $\mathbf{u}^{(2)}$; and so on, up to $\mathbf{o}^{(D)}$. The gradient follows from passing information back down this map. These procedures don't require that any layer has only one input or that any layer has only one output. All we need is to connect the inputs and the outputs in a directed acyclic graph, so that at any node we

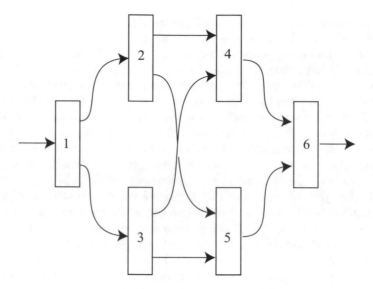

Figure 18.3:   A simple computation graph. You should reassure yourself that a straightforward adjustment to backpropagation will yield all gradients of interest for this network

know what it means for information to go forward (resp. backward). This graph is known as a **computation graph**. Figure 18.3 shows an example that you should use to check that you understand how gradients would be computed. A key feature of good software environments is that they support building complex computation graphs.

18.1.8 Inception Networks

Up to here, we have seen image classification networks as a sequence of layers, where each layer has one input and one output, and information passes from layer to layer in order, and in blocks. This isn't necessary for backpropagation to work. It's enough to have a set of blocks (equivalent to our layers), each with possibly more than one input and possibly more than one outputs. As long as you know how to differentiate each output with respect to each input, and as long as outputs are connected to inputs in a directed acyclic graph, backpropagation works.

This means that we can build structures that are far richer than a sequence of layers. A natural way to do this is to build layers of modules. Figure 18.4 shows two **inception modules** (of a fairly large vocabulary that you can find in the literature; there are some pointers at the end of the chapter). The base block passes its input to each output. A block labelled "AxB" is a convolution layer of $A \times B$ kernels followed by a layer of ReLUs; a stack block stacks each of the data blocks from its input to form its output.

Modules consist of a set of streams that operate independently on a data block; the resulting blocks are then stacked. Stacking means each stream must produce a block of the same spatial size, so all the streams must have consistent stride. Each

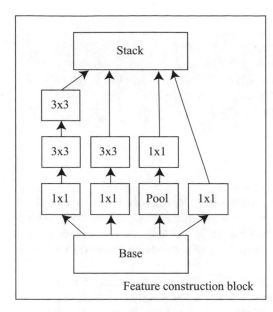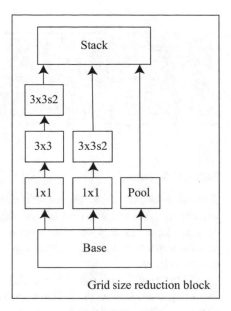

Figure 18.4: On the **left** an inception module for computing features. On the **right**, a module that reduces the size of the grid. The feature module features with: 5×5 support (far left stream); 3×3 support (left stream); 1×1 support after pooling (right stream); and 1×1 support without pooling. These are then stacked into a block. The grid size reduction module takes a block of features on a grid, and reduces the size of the grid. The stream on the left constructs a reduced size grid of features that have quite broad support (5×5 in the input stream); the one in the center constructs a reduced size grid of features that have medium support (3×3 in the input stream); and the one on the right just pools. The outputs of these streams are then stacked

of the streams has a 1×1 convolution in it, which is used for dimension reduction. This means that if you stack two modules, each stream in the top module can select from features which look at the incoming data over different spatial scales. This selection occurs because the network learns the linear map that achieves dimension reduction. In the network, some units can specialize in big (or small, or mixed size) patterns, and later units can choose to make their patterns out of big (or small, or mixed size) components.

There are many different inception modules, and a rich collection of possible networks built out of them. Networks built out of these modules are usually called inception networks. Inception networks tend to be somewhat smaller and faster than VGG-19. An inception network (with appropriate practices in cropping, evaluation, data augmentation, and so on) obtained top-1 and top-5 error rates of 21.2% and 5.6%, respectively, on the ImageNet ILSVRC-2012 classification challenge dataset. This was a substantial improvement. As you would expect, training can be tricky. It's usual to use RMSprop.

> **Remember This:** *Inception networks outperformed VGG-19 on Im-ageNet. Inception networks are built of modules. Feature modules select from incoming data using a 1×1 convolution, then construct features at different spatial scales, then stack them. Other modules reduce the size of the spatial grid. Training can be tricky.*

18.1.9 Residual Networks

A randomly initialized deep network can so severely mangle its inputs that only a wholly impractical amount of training will cause the latest layers to do anything useful. As a result, there have been practical limits on the number of layers that can be stacked. One recent strategy for avoiding this difficulty is to use **residual connections**.

Our usual process takes a data block $\mathcal{X}^{(l)}$, forms a function of that block $\mathcal{W}(\mathcal{X}^{(l)})$, then applies a ReLU to the result. To date, the function involves applying either a fully connected layer or a convolution, then adding bias terms. Writing $F(\cdot)$ for a ReLU, we have

$$\mathcal{X}^{(l+1)} = F(\mathcal{W}(\mathcal{X}^{(l)})).$$

Now assume the linear function does not change the size of the block. We replace this process with

$$\mathcal{X}^{(l+1)} = F(\mathcal{W}(\mathcal{X}^{(l)})) + \mathcal{X}^{(l)}$$

(where F, $\mathcal{W}$, etc. are as before). The usual way to think about this is that a layer now passes on its input, but adds a residual term to it. The point of all this is that, at least in principle, this residual layer can represent its output as a small offset on its input. If it is presented with large inputs, it can produce large outputs by passing on the input. Its output is also significantly less mangled by stacking layers, because its output is largely given by its input plus a non-linear function. These residual connections can be used to bypass multiple blocks. Networks that use residual connections are often known as **ResNets**.

There is good evidence that residual connections allow layers to be stacked very deeply indeed (for example, 1001 layers to get under 5% error on CIFAR-10; beat that if you can!). One reason is that there are useful components to the gradient for each layer that do not get mangled by previous layers. You can see this by considering the Jacobian of such a layer with respect to its inputs. You will see that this Jacobian will have the form

$$\mathcal{J}_{\mathbf{o}^{(l)};\mathbf{u}^l} = (\mathcal{I} + \mathcal{M}_l),$$

where $\mathcal{I}$ is the identity matrix and $\mathcal{M}_l$ is a set of terms that depend on the map $\mathcal{W}$. Now remember that, when we construct the gradient at the k'th layer, we evaluate by multiplying a set of Jacobians corresponding to the layers above. This product in turn must look like

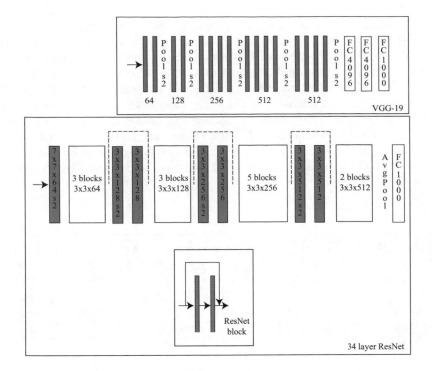

Figure 18.5: Comparing VGG-19 to a 34 layer ResNet requires an even more compact graphical representation. Each shaded box is a convolutional layer of $3 \times 3 \times D$ kernels followed by a ReLU. The number of kernels is given the notation below the box, and D follows by matching sizes. Every layer with learnable parameters is represented by a box, so VGG-19 has 19 such layers, together with pooling layers. The 34 layer ResNet has 34 such layers. There are a few specialized layers (text in the box), but most appear in the blocks (inset) which have two $3 \times 3 \times D$ layers with a residual connection that skips both. These blocks are stacked, as indicated in the figure. The dashed lines around greyed blocks represent a residual connection that causes the size of the data block to change

$$(\nabla_{\mathbf{o}^{(D)}} L) \, J_{\mathbf{o}^{(D)}; \mathbf{u}^{(D)}} \times J_{\mathbf{o}^{(D-1)}; \mathbf{u}^{(D-1)}} \times \cdots \times J_{\mathbf{o}^{k}; \theta^k}$$

which is

$$(\nabla_{\mathbf{o}^{(D)}} L) \, (\mathcal{I} + \mathcal{M}_D)(\mathcal{I} + \mathcal{M}_{D-1}) \ldots (\mathcal{I} + \mathcal{M}_{l+1}) \mathcal{J}_{\mathbf{x}^{k+1}; \theta^k}$$

which is

$$(\nabla_{\mathbf{o}^{(D)}} L) \, (\mathcal{I} + \mathcal{M}_D + \mathcal{M}_{D-1} \ldots + \mathcal{M}_{l+1} + \ldots) \mathcal{J}_{\mathbf{x}^{k+1}; \theta^k},$$

which means that some components of the gradient at that layer do not get mangled by being passed through a sequence of poorly estimated Jacobians.

For some choices of function, the size of the block changes. In this case, we cannot use the form $\mathcal{X}^{(l+1)} = F(\mathcal{W}(\mathcal{X}^{(l)}) + \mathcal{X}^{(l)}$, but instead use

$$\mathcal{X}^{(l+1)} = F(\mathcal{W}(\mathcal{X}^{(l)}) + \mathcal{G}(\mathcal{X}^{(l)}),$$

where $\mathcal{G}$ represents a learned linear projection of $\mathcal{X}^{(l)}$ to the right size block.

It is possible to train very deep networks with this structure very successfully. Figure 18.5 compares a 34 layer residual network with a VGG-19 network. A network with this structure (with appropriate practices in cropping, evaluation, data augmentation, and so on) obtained top-1 and top-5 error rates of 24.2% and 7.4%, respectively, on the ImageNet ILSVRC-2012 classification challenge validation dataset. This is somewhat worse than the inception network performance, but accuracy can be significantly improved by building deeper networks (hard to draw) and using ensembles, voting over different crops, and so on. A model using 152 layers (ResNet-152) obtained a top-5 error of 3.57% ImageNet ILSVRC-2015 challenge. ResNet-152 is widely used as a feature stack, and is usually more accurate than VGGNet.

> **Remember This:** *ResNets are the go-to for image classification. ResNets use a network block that adds a processed version of the input to the input. This means that helpful gradient values are available even for very deep networks. ResNet models can be built with extremely deep networks, and are widely used to make features for tasks other than image classification.*

18.2 Object Detection

An object detection program must mark the locations of each object from a known set of classes in test images. Object detection is hard for many reasons. First, objects can look different when you look at them from different directions. For example, a car seen from above can look very different from a car seen from the side. Second, objects can appear in images at a wide range of scales and locations. For example, a single image can contain large faces (from people standing close to the camera) and small faces (from people in the background). Third, many objects (like people) deform without changing their identity. Fourth, there are often nasty hierarchical structures to worry about. For example, chairs have legs, backs, bolts, washers, nuts, cushions, stitches (on the cushions), and so on. Finally, most scenes contain an awful lot of objects (think about the number of bolts in a picture of a lecture hall—each chair has many) and most are not worth mentioning.

18.2.1 How Object Detectors Work

Object detectors are built out of image classifiers. Here is the simplest way to build (say) a face detector. Build an image classifier that can tell whether a face is present in an image window of fixed size or not. This classifier produces a high score for faces, and a low score for non-faces. Take this face classifier, and search through a set of windows selected from the image. Use the resulting scores to decide which windows contain faces. This very simple model exposes the big questions to be addressed. We must:

- **Decide on a window shape:** This is easy. There are two possibilities: a box, or something else. Boxes are easy to represent, and are used for almost all practical detectors. The alternative—some form of mask that cuts the object out of the image—is hardly ever used, because it is hard to represent.
- **Build a classifier for windows:** This is easy—we've seen multiple constructions for image classifiers.
- **Decide which windows to look at:** This turns out to be an interesting problem. Searching all windows isn't efficient.
- **Choose which windows with high classifier scores to report:** This is interesting, too, because windows will overlap, and we don't want to report the same object multiple times in slightly different windows.
- **Report the precise locations of all faces using these windows:** This is also interesting. It turns out our window is likely not the best available, and we can improve it after deciding that it contains a face.

Which window to look at is hard, and most innovation has occurred here. Each window is a hypothesis about the **configuration** (position and size) of the object. The very simplest procedure for choosing windows is to use all windows on some grid (if you want to find larger faces, use the same grid on a smaller version of the image). No modern detector looks at a grid because it is inefficient. A detector that looks at closely spaced windows may be able to **localize** (estimate position and size of) the object more accurately. But more windows mean the classifier's false positive rate must be extremely small to produce a useful detector. Tiling the image tends to produce far too many windows, many of which are fairly obviously bad (for example, a box might cut an object in half).

Deciding which windows to report presents minor but important problems. Assume you look at 32×32 windows with a stride of 1. Then there will be many windows that overlap the object fairly tightly, and these should have quite similar scores. Just thresholding the value of the score will mean that we report many instances of the same object in about the same place, which is unhelpful. If the stride is large, no window may properly overlap the object and it might be missed. Instead, most methods adopt variants of a greedy algorithm usually called **non-maximum suppression**. First, build a sorted list of all windows whose score is over threshold. Now repeat until the list is empty: choose the window with highest score, and accept it as containing an object; now remove all windows with large enough overlap on the object window.

Deciding precisely where the object is also presents minor but important problems. Assume we have a window that has a high score, and has passed through non-maximum suppression. The procedure that generated the window does not do a detailed assessment of all pixels in the window (otherwise we wouldn't have needed the classifier), so this window likely does not represent the best localization of the object. A better estimate can be obtained by predicting a new bounding box using a feature representation for the pixels in the current box. It's natural to use the feature representation computed by the classifier for this **bounding box regression** step.

> **Remember This:**   *Object detectors work by passing image boxes that are likely to contain objects into a classifier. The classifier gives scores for each possible object in the box. Multiple detections of the same object by overlapping boxes can be dealt with by non-maximum suppression, where higher-scoring boxes eliminate lower-scoring but overlapping boxes. Boxes are then adjusted with a bounding box regression step.*

18.2.2 Selective Search

The simplest procedure for building boxes is to slide a window over the image. This is simple, but works rather badly. It produces a large number of boxes, and the boxes themselves ignore important image evidence. Objects tend to have quite clear boundaries in images. For example, if you are looking at a picture of a horse in a field, there's usually no uncertainty about where the horse ends and where the field begins. At these boundaries, a variety of image properties change quite sharply. At the boundary of the horse, color changes (say, brown to green); texture changes (say, smooth skin to rough grass); intensity changes (say, dark brown horse to brighter green grass); and so on.

Making boxes by sliding windows ignores this information. Boxes that span a boundary probably contain only part of an object. Boxes that have no boundaries nearby likely don't contain anything interesting. It is still quite difficult to actually find the boundaries of objects, because not every boundary has a color change (think of a brown horse in a brown field), *and* some color changes occur away from boundaries (think about the stripes on a zebra). Nonetheless, it has been known for some time that one can use boundaries to score boxes for their "objectness." The best detectors are built by looking only at boxes that have a high enough objectness score.

The standard mechanism for computing such boxes is known as **selective search**. A quick description is straightforward, but the details matter (and you'll need to look them up). First, one breaks up the image into **regions**—groups of pixels that have coherent appearance—using an agglomerative clusterer. The agglomerative clusterer is quite important, because the representation it produces allows big regions to be made of smaller regions (so, for example, a horse might be made of a head, body, and legs). Second, one scores the regions produced by the clusterer for "objectness." This score is computed from computing a variety of region features, encoding color, texture, and so on. Finally, the regions are ranked by the score. It isn't safe to assume that regions with a score over some threshold are objects and the others aren't, but the process is very good at reducing the number of boxes to look at. One does not need to go very deep into the ranked list of regions to find all objects of interest in a picture (2000 is a standard).

> **Remember This:** *Image boxes that are likely to contain objects are closely related to regions. Selective search finds these boxes by building a region hierarchy, then scoring regions for objectness; regions with good objectness score produce bounding boxes. This gives an effective way of finding the boxes that are likely to contain objects.*

18.2.3 R-CNN, Fast R-CNN and Faster R-CNN

There is a natural way to build a detector using selective search and an image classifier. Use selective search to build a ranked list of regions. For each region in the ranked list, build a bounding box. Now warp this box to a standard size, and pass the resulting image to an image classifier. Rank the resulting boxes by the predicted score for the best object, and keep boxes whose score is over a threshold. Now apply non-maximum suppression and bounding box regression to that list. Figure 18.6 shows this architecture, known as **R-CNN**; it produces a very successful detector, but a speedup is available.

The problem with R-CNN is that one must pass each box independently through an image classifier. There tends to be a high degree of overlap between the boxes. This means the image classifier has to compute the neural network features at a given pixel for every box that overlaps the pixel, so doing unnecessary redundant work. The cure produces a detector known as **Fast R-CNN**. Pass the whole image through a convolutional neural network classifier (but ignore the fully connected layers). Now take the boxes that come from selective search, and use them to identify regions of interest (ROIs) in the feature maps. Compute class probabilities from these regions of interest using image classification machinery.

The ROIs will have different sizes, depending on the scale of the object. These need to be reduced to a standard size; otherwise, we cannot pass them into the usual machinery. The trick is a **ROI pooling layer**, which produces a standard size summary of each ROI that is effective for classification. Decide on a standard size to which the ROIs will be reduced (say $r_x \times r_y$). Make a stack of grids this size, one per ROI. For each ROI, break the ROI into an $r_x \times r_y$ grid of evenly sized blocks. Now compute the maximum value in each block, and place that value in the corresponding location in the grid representing the ROI. This stack of grids can then be passed to a classifier.

The culmination of this line of reasoning (so far!) is **Faster R-CNN**. It turns out that selective search slows down Fast R-CNN. At least part of this slowdown is computing features, etc., for selective search. But selective search is a process that predicts boxes from image data. There is no particular reason to use special features for this purpose, and it is natural to try and use the same set of features to predict boxes and to classify them. Faster R-CNN uses image features to identify important boxes (Fig. 18.7).

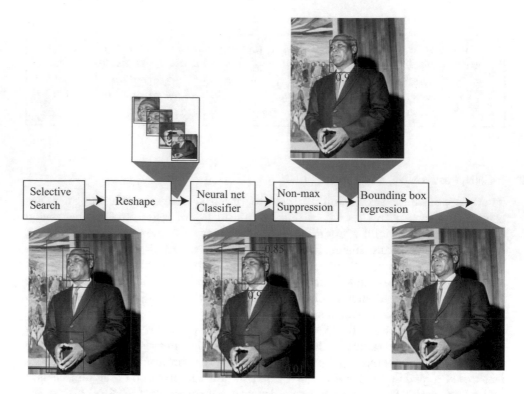

Figure 18.6: A schematic picture of how R-CNN works. A picture of Inkosi Albert Luthuli is fed in to selective search, which proposes possible boxes; these are cut out of the image, and reshaped to fixed size; the boxes are classified (scores next to each box); non-maximum suppression finds high scoring boxes and suppresses nearby high scoring boxes (so his face isn't found twice); and finally bounding box regression adjusts the corners of the box to get the best fit using the features inside the box

Convolutional neural networks aren't particularly good at making lists, but are very good at making spatial maps. The trick is to encode a large collection of image boxes in a representation of fixed size that can be thought of as a map. The set of boxes can be represented like this. Construct a 3D block where each spatial location in the block represents a point on a grid in the image (a stride of 16 between the grid points in the original). The third coordinate in the block represents an **anchor box**. These are boxes of different size and aspect ratio, centered at the grid location (Fig. 18.8; 9 in the original). You might be concerned that looking at a relatively small number of sizes, locations, and aspect ratios creates problems; but bounding box regression is capable of dealing with any issues that arise. We want the entries in this map to be large when a box is likely to contain an object (you can think of this as an "objectness" score) and small otherwise. Thresholding the boxes and using non-maximum suppression yields a list of possible boxes, which can be handled as above.

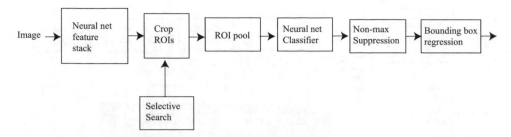

Figure 18.7: Fast R-CNN is much more efficient than R-CNN, because it computes a single feature map from the image, then uses the boxes proposed by selective search to cut regions of interest (ROIs) from it. These are mapped to a standard size by a ROI pooling layer, then presented to a classifier. The rest should be familiar

A significant attraction of this approach is that the process that makes boxes can be trained at the same time as the classifier—box proposals can take classifier eccentricities in mind, and vice versa. At training time, one needs two losses. One loss measures the effectiveness of the box proposal process and the other measures the accuracy of the detector. The main difference is that the box proposal process needs to give a high score to any box with a good IoU against any ground truth bounding box (whatever the object in the box). The detector needs to name the object in the box.

> **Remember This:** *R-CNN, Fast R-CNN, and Faster R-CNN are strongly performing object detection systems that differ by how boxes are proposed. R-CNN and Fast R-CNN use selective search; Faster R-CNN scores anchor boxes. As of writing, Faster R-CNN is the reference object detector.*

18.2.4 YOLO

All the detectors we have seen so far come up with a list of boxes that are likely to be useful. **YOLO** (*You Only Look Once*) is a family of detectors (variants pay off accuracy against speed) that uses an entirely different approach to boxes. The image is divided into an $S \times S$ grid of tiles. Each tile is responsible for predicting the box of any object whose center lies inside the tile. Each tile is required to report B boxes, where each box is represented by the location of its center in the tile together with its width and its height. For each of these boxes (write b), each tile must also report a box confidence score $c(b(\text{tile}))$. The method is trained to produce a confidence score of zero if no object has its center in the tile, and the IoU for the box with ground truth if there is such an object (of course, at run time it might not report this score correctly).

Each tile also reports a class posterior, $p(\text{class}|\text{tile})$ for that tile. The score

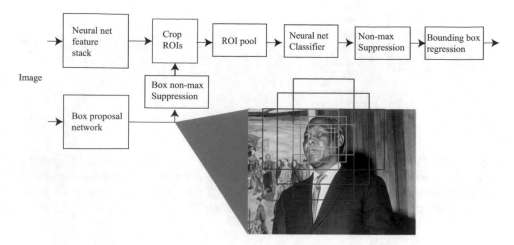

Figure 18.8: Faster RCNN uses two networks. One uses the image to compute "objectness" scores for a sampling of possible image boxes. The samples (called "anchor boxes") are each centered at a grid point. At each grid point, there are nine boxes (three scales, three aspect ratios). The second is a feature stack that computes a representation of the image suitable for classification. The boxes with highest objectness score are then cut from the feature map, standardized with ROI pooling, then passed to a classifier. Bounding box regression means that the relatively coarse sampling of locations, scales, and aspect ratios does not weaken accuracy

linking each of the boxes b in a tile to a class is then computed as

$$c(b(\text{tile})) \times p(\text{class}|\text{tile}).$$

Notice how the box scoring process has been decoupled from the object class process. Each tile is scoring *what* object overlaps the tile and also scoring which boxes linked to the tile are important. But these scores are computed separately—the method does not know which box is being used when it computes the object scores. This means the method can be extremely fast, and YOLO offers relatively easy trade-offs between speed and accuracy, which are often helpful (for example, one can use more or fewer network layers to make features; more or fewer boxes per tile; and so on).

Decoupling boxes from classes comes with problems. YOLO tends to handle small objects poorly. There is a limited number of boxes, and so the method has difficulties with large numbers of small objects. The decision as to whether an object is present or not is based on the whole tile, so if the object is small compared to the tile, the decision might be quite inaccurate. YOLO tends not to do well with new aspects or new configurations of familiar objects. This is caused by the box prediction process. If the method is trained on (say) all vertical views of trees (tall thin boxes), it can have trouble with a tree lying on its side (short wide box).

> **Remember This:** *The YOLO family of detectors works very differently from the R-CNN family. In Yolo, image tiles produce objectness scores for boxes and a classification score for objects independently; these are then multiplied. The advantage is speed, and tunable payoffs between speed and accuracy. The disadvantages are that many small objects are hard to detect, and new configurations of familiar objects are often missed.*

18.2.5 Evaluating Detectors

Evaluating object detectors takes care. An object detector takes an image, and, for each object class it knows about, produces a list of boxes each of which has a score. Evaluating the detector involves comparing these boxes with ground truth boxes that have been marked on the image by people. The evaluation should favor detectors that get the right number of the right object in the right place. It should discourage detectors that just propose an awful lot of boxes. Getting this right takes a fair amount of careful work, which won't appeal to (or be useful to) all. The rest of the section is skippable if you're not that interested in object detection.

To start, assume the detector responds to only one kind of object. You now have two lists: one ($\mathcal{G}$) is the list of ground truth boxes, the other ($\mathcal{D}$) is the list of boxes the detector produces, which has already been subject to non-maximum suppression, bounding box regression, and anything else the team that created the detector can think of. You should think of the detector as a search process. The detector has searched a huge collection of boxes, and produced some boxes that it asserts are relevant, in order of relevance (this is the list $\mathcal{D}$). This list needs to be scored. The evaluation must mark boxes in $\mathcal{D}$ with `relevant` if they match ground truth boxes and `irrelevant` otherwise, and then summarize the lists.

The boxes that the detector predicts are unlikely to match ground truth exactly, and we need some way of telling whether the boxes are good enough. The standard method for doing this is to test the **IoU** (intersection over union). Write B_g for the ground truth box and B_p for the predicted box. The IoU is

$$\text{IoU}(B_p, B_g) = \frac{\text{Area}(B_g \cap B_p)}{\text{Area}(B_g \cup B_p)}.$$

Choose some threshold t. If $\text{IoU}(B_p, B_g) > t$, then B_p could match the ground truth box B_g.

The detector should be credited for producing a box that has a high score and matches a ground truth box. But the detector should not be able to improve its score by predicting many boxes on top of a ground truth box. The standard way to handle the problem is to mark the overlapping box with highest score `relevant`. The procedure is:

- Choose a threshold t.
- Order $\mathcal{D}$ by the score of each box, and mark every element of $\mathcal{D}$ with `irrelevant`. Choose a threshold t.

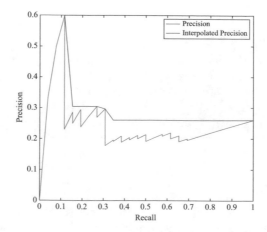

Figure 18.9: Two plots for an imaginary search process. The precision plotted against recall shows a characteristic sawtooth shape. Interpolated precision measures the best precision you can get by increasing the recall, and so smoothes the plot. Interpolated precision is also a more natural representation of what one wants from search results—most people would be willing to add items to get higher precision. Interpolated precision is used to evaluate detectors

- For each element of $\mathcal{D}$ in order of score, compare that box against all ground truth boxes. If any ground truth box has IoU > t, mark the detector box **relevant** and remove that ground truth box from $\mathcal{G}$. Proceed until there are no more ground truth boxes.

Now every box in $\mathcal{D}$ is tagged either **relevant** or **irrelevant**.

There are standard evaluations for search results like those produced by our detector. The first step is to merge the lists for each evaluation image into a single list of results. The **precision** of a set of search results $\mathcal{S}$ is given by

$$\mathbf{P}(\mathcal{S}) = \frac{\text{number of relevant search results}}{\text{total number of search results}}.$$

The **recall** is given by

$$\mathbf{R}(\mathcal{S}) = \frac{\text{number of relevant search results}}{\text{total number of relevant items in collection}}.$$

As you move down the list $\mathcal{D}$ in order of score, you get a new set of search results. The recall never decreases as the set gets larger, and so you could plot the precision as a function of recall (write $\mathbf{P}(\mathbf{R})$). These plots have a characteristic sawtooth structure (Fig. 18.9). If you add a single irrelevant item to the set of results, the precision will fall; if you then add a relevant item, it jumps up. The sawtooth doesn't really reflect how useful the set of results is—people are usually willing to add several items to a set of search results to improve the precision—and so it is better to use **interpolated precision**. The interpolated precision at some recall value R_0 is given by

$$\hat{\mathcal{P}}(R_0) = \max_{R \geq R_0} \mathbf{P}(R)$$

(Fig. 18.9). By convention, the **average precision** is computed as

$$\frac{1}{11} \sum_{i=0}^{10} \hat{\mathcal{P}}\left(\frac{i}{10}\right).$$

This value summarizes the recall–precision curve. Notice this averages in interpolated precision at high recall. Doing so means a detector cannot get a high score by producing only very few, very accurate boxes—to do well, a detector should have high precision even when it is forced to predict every box.

Average precision evaluates detection for one category of object. The **mean average precision** (mAP) is the mean of the average precision for each category. The value depends on the IoU threshold chosen. One convention is to report mAP at $IoU = 0.5$. Another is to compute mAP at a set of 10 IoU values ($0.45 + i \times 0.05$ for $i \in 1 \dots 10$), then average the mAPs. These evaluations produce numbers that tend to be bigger for better detectors, but it takes some practice to have a clear sense of what an improvement in mAP actually means.

Remember This: *Evaluating object detectors should favor detectors that get the right number of the right objects in the right places, and should discourage detectors that just produce a lot of boxes. Evaluation scores boxes produced by the detector for relevance (is this the right box in the right place?) using IoU scores to evaluate how well boxes overlap with ground truth. The average precision is computed from an interpolated precision curve for each type of object. This is then averaged over object types to yield mAP.*

18.3 Further Reading

To proceed further, you really should be reading original papers, which is how this subject is communicated. Here's a reading list to get started with.

- **Origins of CNNs:** *Gradient-based learning applied to document recognition*, by Yann LeCun, Léon Bottou, Yoshua Bengio, and Patrick Haffner, Proceedings of the IEEE 86 (11), 2278–2324.
- **Batch normalization:** *Batch Normalization: Accelerating Deep Network Training by Reducing Internal Covariate Shift*, by Sergey Ioffe, Christian Szegedy, Proc Int. Conf. Machine Learning, 2015. You can find a version at https://arxiv.org/abs/1502.03167.
- **ImageNet:** *ImageNet Large Scale Visual Recognition Challenge*, by Olga Russakovsky, Jia Deng, Hao Su, Jonathan Krause, Sanjeev Satheesh, Sean Ma, Zhiheng Huang, Andrej Karpathy, Aditya Khosla, Michael Bernstein, Alexander C. Berg, and Li Fei-Fei in International Journal of Computer Vision December 2015, Volume 115, Issue 3, pp. 211–252.

- **Pascal:** *The Pascal Visual Object Classes (VOC) Challenge*, by Mark Everingham, Luc Van Gool, Christopher K. I. Williams, John Winn, and Andrew Zisserman, International Journal of Computer Vision, June 2010, Volume 88, Issue 2, pp. 303–338.

- **VGGNet:** *Very Deep Convolutional Networks for Large-Scale Image Recognition* by Karen Simonyan and Andrew Zisserman, Proc. Int. Conf. Learned Representations, 2015. You can find a version of this at https://arxiv.org/pdf/1409.1556.pdf.

- **Inception:** *Going Deeper with Convolutions*, by Christian Szegedy, Wei Liu, Yangqing Jia, Pierre Sermanet, Scott Reed, Dragomir Anguelov, Dumitru Erhan, Vincent Vanhoucke, and Andrew Rabinovich, Proc Computer Vision and Pattern Recognition, 2015. You can find a version of this at https://arxiv.org/abs/1409.4842.

- **ResNets:** *Deep Residual Learning for Image Recognition* by Kaiming He, Xiangyu Zhang, Shaoqing Ren, and Jian Sun, Proc Computer Vision and Pattern Recognition, 2015. You can find a version of this at https://arxiv.org/abs/1512.03385.

- **Selective search:** *Selective Search for Object Recognition* by J. R. R. Uijlings, K. E. A. van de Sande, T. Gevers, and A. W. M. Smeulders, International Journal of Computer Vision September 2013, Volume 104, Issue 2, pp. 154–171.

- **R-CNN:** *Rich feature hierarchies for accurate object detection and semantic segmentation*, by R. Girshick, J. Donahue, T. Darrell, and J. Malik, IEEE Conf. on Computer Vision and Pattern Recognition, 2014. You can find a version of this at https://arxiv.org/abs/1311.2524.

- **Fast R-CNN:** *Fast R-CNN*, by Ross Girshick, IEEE Int. Conf. on Computer Vision (ICCV), 2015, pp. 1440–1448. You can find a version of this at https://www.cv-foundation.org/openaccess/content_iccv_2015/html /Girshick_Fast_R- CNN_ICCV_2015_paper.html.

- **Faster R-CNN:** *Faster R-CNN: Towards Real-Time Object Detection with Region Proposal Networks*, by Shaoqing Ren, Kaiming He, Ross Girshick, and Jian Sun, Advances in Neural Information Processing Systems 28 (NIPS 2015). You can find a version of this at http://papers.nips.cc/paper/5638-faster-r-cnn-towards-real-time-object-detection-with-region-proposal-networks.pdf.

- **YOLO:** *You Only Look Once: Unified, Real-Time Object Detection*, by Joseph Redmon, Santosh Divvala, Ross Girshick, and Ali Farhadi, Proc Computer Vision and Pattern Recognition, 2016. You can find a version of this at https://www.cv-foundation.org/openaccess/content_cvpr_2016/papers/Redmon_You_Only_Look_CVPR_2016_paper.pdf. There's a home page at https://pjreddie.com/darknet/yolo/.

18.4 You Should

18.4.1 Remember These Terms

18.4.2 Remember These Facts

18.4.3 Be Able to

- Run an image classifier in your chosen environment.
- Explain how current object detectors work.
- Run an object detector in your chosen environment.

Problems

18.1. Modify the backpropagation algorithm to deal with directed acyclic graphs like that of Fig. 18.3. Note the layers are numbered there, and I will denote the parameters of the i'th layer as θ_i.

    (a) The first step is to deal with layers that have one output, but two inputs. If we can deal with two inputs, we can deal with any number. Write the inputs of layer 6 as $\mathbf{x}_1^{(6)}$ and $\mathbf{x}_2^{(6)}$. Write $J_{\mathbf{o}^{(6)};\mathbf{x}_i^{(6)}}$ for the Jacobian of the output with respect to the i'th input. Explain how to compute $\nabla_{\theta_4}\mathcal{L}$ using this Jacobian.

    (b) Layer 2 has two outputs. Write the outputs $\mathbf{o}_1^{(2)}$ and $\mathbf{o}_2^{(2)}$. Write $J_{\mathbf{o}_i^{(2)};\mathbf{x}^{(2)}}$ for the Jacobian of the i'th output with respect to its input. Explain how to compute $\nabla_{\theta_2}\mathcal{L}$ using this Jacobian (and others!).

    (c) Can you backpropagate through a layer that has two inputs *and* two outputs?

    (d) What goes wrong with backpropagation when the computation graph has a cycle?

Programming Exercises

    General Remark: *These exercises are suggested activities, and are rather open ended. It will be difficult to do them without a GPU. You may have to deal with some fun installing software environments, etc. It's worthwhile being able to do this, though.*

    Minor Nuisance: *At least in my instance of the ILSVRC-2012 validation dataset, some images are grey level images rather than RGB. Ensure the code you use turns them into RGB images by making the R, G, and B channel the same as the original intensity channel, or funny things can happen.*

18.2. Download a pretrained VGGNet-19 image classifier for your chosen programming framework.

    (a) Run this classifier on ILSVRC-2012 validation dataset. Each image needs to be reduced to 224×224 block. Do this by first resizing the image uniformly so that the smallest dimension is 224, then cropping the right half of the image. Ensure that you do whatever preprocessing your instance of VGGNet-19 requires on this crop (this should be subtracting the mean RGB at each pixel from each pixel; i.e., follow the procedure on page 416, but don't divide by the standard deviation). In this case, what is the top-1 error rate? What is the top-5 error rate?

    (b) Now investigate the effect of multiple crops. For each image in the validation dataset, crop to 224×224 for five different crop windows. One of these is centered in the image; the other four are obtained by placing a corner of the crop window at each corner of the image, respectively. Ensure that you do whatever preprocessing your instance of VGGNet-19 requires on each crop (this should be subtracting the mean RGB at each pixel from each pixel; i.e., follow the procedure on page 416, but don't divide by the standard deviation). Pass each crop through the network, then average the predicted class posteriors, and use that score. In this case, what is the top-1 error rate? What is the top-5 error rate?

18.3. Download a pretrained ResNet image classifier for your chosen programming framework.

    (a) Run this classifier on ILSVRC-2012 validation dataset. Each image needs to be reduced to 224×224 block. Do this by first resizing the image uniformly so that the smallest dimension is 224, then cropping the right half of the image. Ensure that you do whatever preprocessing your instance of ResNet requires on this crop (this should be subtracting the mean RGB at each pixel from each pixel; i.e., follow the procedure on page 416, but don't divide by the standard deviation). In this case, what is the top-1 error rate? What is the top-5 error rate?

(b) Now investigate the effect of multiple crops. For each image in the validation dataset, crop to 224×224 for five different crop windows. One of these is centered in the image; the other four are obtained by placing a corner of the crop window at each corner of the image, respectively. Ensure that you do whatever preprocessing your instance of ResNet requires on each crop (this should be subtracting the mean RGB at each pixel from each pixel; i.e., follow the procedure on page 416, but don't divide by the standard deviation). Pass each crop through the network, then average the predicted class posteriors, and use that score. In this case, what is the top-1 error rate? What is the top-5 error rate?

18.4. Download both a pretrained ResNet image classifier and a pretrained VGG-19 for your chosen programming framework. For each image in the validation dataset, use a center crop to 224×224. Ensure that you do whatever preprocessing your instances require. Record for every image the true class, the class predicted by ResNet, and the class predicted by VGG-19.

(a) On average, if you know VGG-19 predicted the label correctly or not, how accurately can you predict whether ResNet gets the label right? Answer this by computing $P(\text{ResNet right}|\text{VGG right})$ and $P(\text{ResNet right}|\text{VGG wrong})$ using your data.

(b) Both networks are quite accurate, even for top-1 error. This means that their errors must be correlated, because each gets most examples right. We would like to know whether the result of the previous subexercise is due to this effect, or something else. Write the VGG-19 error rate as v, and the ResNet error rate as r. Write $\mathbf{v}$ for a 50,000 dimensional binary vector, with v 1's, where the entries are IID samples from a Bernoulli distribution with mean v. This is a model of randomly distributed errors with the same error rate as VGG-19. A similar $\mathbf{r}$ models random errors for ResNet. Draw 1000 samples of $(\mathbf{v}, \mathbf{r})$ pairs, and compute the mean and standard error of $P(r_i = 0|v_i = 1)$ and $P(r_i = 1|v_i = 0)$. Use this information to determine whether ResNet "knows" something about VGG-19 errors.

(c) What could cause the effect you see?

(d) How are errors distributed across categories?

(e) **Hard!** *(but interesting).* Obtain instances of several different ImageNet classification networks. Investigate the pattern of errors. In particular, for images that one instance mislabels, do other instances mislabel it as well? If so, how many different labels are used in total? I have found surprisingly strong agreement between instances that mislabel an image (i.e., if network A thinks an image of a dog is a cat, *and* network B gets the same image wrong as well, then network B will likely think it's a cat, too).

18.5. Choose 10 ImageNet classes. For each class, download 50 example images of items that belong to those classes from an internet image source (images.

google.com or images.bing.com are good places to start; query using the name of the category).

(a) Classify these images into the ImageNet classes using a pretrained network. What is the top-1 error rate? What is the top-5 error rate?

(b) Compare the results of this experiment with the accuracy on the validation set for these classes. What is going on?

CHAPTER 19

Small Codes for Big Signals

This chapter explores a different kind of use of neural networks. Rather than classifying or detecting patterns directly, we try to build low dimensional representations of high dimensional signals. The simplest reason to do so is to build a map of a dataset. We've already seen one procedure for doing so. It turns out that procedure has problems; this chapter starts with two alternative procedures. These are useful in their own right for mapping datasets.

The next step is to try to learn a function that will take new data items and predict a low dimensional representation for them. One could see this as a regression problem (i.e., learn this function from example pairs of high-d/low-d representations). But it's more productive to see it as an encoding problem. Learn an encoding of the input data that preserves what is important. To evaluate the encoding, we also learn to decode the code to produce the original data item. This leads us to an autoencoder—a pair of encoder and decoder, which are trained together, to produce a good mapping to a low dimensional code. We have already seen how to build networks that make data blocks smaller using stride or pooling. A decoder needs to make a data block bigger, which requires some new tricks. Ensuring that an autoencoder produces good looking images requires some work on the training loss and the training process, but can be done.

If you have a working decoder, it is natural to feed it random codes to see if it makes new images. This doesn't work as stated. The difficulties are that encoders tend to produce codes with odd distributions, and that decoders tend to panic when fed with unfamiliar codes. Each effect can be mitigated with tricks. Producing images out of random numbers using decoder-like strategies requires clever management of the loss function used for training, but works moderately well for special cases as of writing.

19.1 Better Low Dimensional Maps

One really important use for small representations of big signals is building maps. We've already seen one algorithm (Sect. 6.2). We start with N d dimensional points $\mathbf{x}$, where the i'th point is $\mathbf{x}_i$. We would like to build a map of this dataset, to visualize its major features. We would like to know, for example, whether it contains many or few blobs; whether there are many scattered points; and so on. We might also want to plot this map using different plotting symbols for different kinds of data points. For example, if the data consists of images, we might be interested in whether images of cats form blobs that are distinct from images of dogs, and so on. I will write $\mathbf{y}_i$ for the point in the map corresponding to the $\mathbf{x}_i$. The map is an M dimensional space. If one is trying to build a map, M is almost always two or three in applications. Sometimes one just wants a lower dimensional representation that

© Springer Nature Switzerland AG 2019
D. Forsyth, *Applied Machine Learning*,
https://doi.org/10.1007/978-3-030-18114-7_19

preserves important information. In this case, M can be large and the procedure is often referred to as **embedding**.

Principal coordinate analysis used eigenvectors to identify a mapping of the data that made low dimensional distances similar to high dimensional distances. I argued that the choice of map should minimize

$$\sum_{i,j} \left(\|\mathbf{y}_i - \mathbf{y}_j\|^2 - \|\mathbf{x}_i - \mathbf{x}_j\|^2 \right)^2,$$

then rearranged terms to produce a solution that minimized

$$\sum_{i,j} \left(\mathbf{y}_i^T \mathbf{y}_j - \mathbf{x}_i^T \mathbf{x}_j \right)^2.$$

But the choice of cost function is not a particularly good idea. The map will be almost entirely determined by points that are very far apart. This happens because squared differences between big numbers tend to be a lot bigger than squared differences between small numbers, and so distances between points that are far apart will be the most important terms in the cost function. In turn, this could mean our map does not really show the structure of the data—for example, a small number of scattered points in the original data could break up clusters in the map (the points in clusters are pushed apart to get a map that places the scattered points in about the right place with respect to each other).

19.1.1 Sammon Mapping

Sammon mapping is a method to fix these problems by modifying the cost function. We attempt to make the small distances more significant in the solution by minimizing

$$C(\mathbf{y}_1, \ldots, \mathbf{y}_N) = \left(\frac{1}{\sum_{i<j} \|\mathbf{x}_i - \mathbf{x}_j\|} \right) \sum_{i<j} \left[\frac{(\|\mathbf{y}_i - \mathbf{y}_j\| - \|\mathbf{x}_i - \mathbf{x}_j\|)^2}{\|\mathbf{x}_i - \mathbf{x}_j\|} \right].$$

The first term is a constant that makes the gradient cleaner, but has no other effect. What is important is we are biasing the cost function to make the error in small distances much more significant. Unlike straightforward multidimensional scaling, the range of the sum matters here—if i equals j in the sum, then there will be a divide by zero.

No closed form solution is known for this cost function. Instead, choosing the $\mathbf{y}$ for each $\mathbf{x}$ is by gradient descent on the cost function. You should notice there is no unique solution here, because rotating, translating, or reflecting all the $\mathbf{y}_i$ will not change the value of the cost function. Furthermore, there is no reason to believe that gradient descent necessarily produces the best value of the cost function. Experience has shown that Sammon mapping works rather well, but has one annoying feature. If one pair of high dimensional points is very much closer together than any other, then getting the mapping right for that pair of points is extremely important to obtain a low value of the cost function. This should seem like a problem to you, because a distortion in a very tiny distance should not be much more important than a distortion in a small distance.

Worked Example 19.1   *Sammon Mapping MNIST Data*

Prepare a Sammon mapping of 1000 examples, drawn at random, from the MNIST dataset to two dimensions.

Solution: The problem has 2000 variables (the unknown $\mathbf{y}$'s). The cost function depends on $\mathbf{x}$ only through the distances, so it isn't clear if reducing the dimension of the $\mathbf{x}$ helps or not. I tried two cases: in one, I just used the 784 dimensional vectors for the MNIST digits, and in the second, I used principal coordinate analysis to map the digits to 30 dimensions. I then computed the Sammon mapping for each case. I used MATLAB's `fmincon` optimizer (which can do an approximate version of Newton's method called LBFGS that speeds things up). Figure 19.1 shows results. Note that the dimension reduction doesn't seem to have changed anything important, and it didn't make the method notably faster, either.

Remember This: *Sammon mapping produces an embedding of high dimensional data into a lower dimensional space that reduces the emphasis that principal coordinate analysis places on large distances. It does so by solving an optimization problem to choose coordinates in a low dimensional space for each data point. Sammon mappings are often biased by very small distances, however.*

19.1.2   T-SNE

We will now build a model by reasoning about probability rather than about distance (although this story could likely be told as a metric story, too). We will build a model of the probability that two points in the high dimensional space are neighbors, and another model of the probability that two points in the low dimensional space are neighbors. We will then adjust the locations of the points in the low dimensional space so that the KL divergence between these two models is small.

We reason first about the probability that two points in the high dimensional space are neighbors. Write the conditional probability that $\mathbf{x}_j$ is a neighbor of $\mathbf{x}_i$ as $p_{j|i}$. Write

$$w_{j|i} = \exp\left(\frac{\|\mathbf{x}_j - \mathbf{x}_i\|^2}{2\sigma_i^2}\right).$$

We use the model

$$p_{j|i} = \frac{w_{j|i}}{\sum_k w_{k|i}}.$$

Notice this depends on the scale at point i, written σ_i. For the moment, we assume this is known. Now we define p_{ij} the joint probability that $\mathbf{x}_i$ and $\mathbf{x}_j$ are neighbors

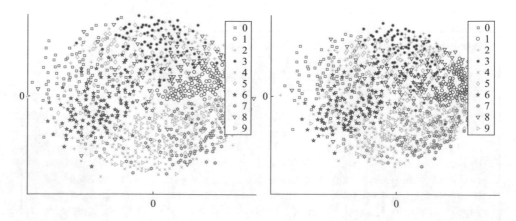

Figure 19.1: Sammon mappings of 1000 samples of a 784 dimensional MNIST digits. On the **left**, the mapping used the whole digit vector, and on the **right**, the data was reduced to 30 dimensions using PCA, then subjected to a Sammon mapping. The class labels were not used in training, but the plot shows class labels. This helps to determine whether the visualization is any good—you could reasonably expect a visualization to put items in the same class close together and items in very different classes far apart. As the legend on the side shows, the classes are moderately well separated. Reducing dimension doesn't appear to make much difference

by assuming $p_{ii} = 0$, and for all other pairs

$$p_{ij} = \frac{p_{j|i} + p_{i|j}}{2N}.$$

This is an $N \times N$ table of probabilities; you should check that this table represents a joint probability distribution (i.e., it's non-negative, and sums to one).

We use a slightly different probability model in the low dimensional space. We know that, in a high dimensional space, there is "more room" near a given point (think of this as a base point) than there is in a low dimensional space. This means that mapping a set of points from a high dimensional space to a low dimensional space is almost certain to move some points further away from the base point than we would like. In turn, this means there is a higher probability that a distant point in the low dimensional space is still a neighbor of the base point. Our probability model needs to have "long tails"—the probability that two points are neighbors should not fall off too quickly with distance. Write q_{ij} for the probability that $\mathbf{y}_i$ and $\mathbf{y}_j$ are neighbors. We assume that $q_{ii} = 0$ for all i. For other pairs, we use the model

$$q_{ij}(\mathbf{y}_1, \ldots, \mathbf{y}_N) = \frac{\frac{1}{1 + \|\mathbf{y}_i - \mathbf{y}_j\|^2}}{\sum_{k,l,k \neq l} \frac{1}{1 + \|\mathbf{y}_l - \mathbf{y}_k\|^2}}$$

(where you might recognize the form of Student's t-distribution if you have seen that before). You should think about the situation like this. We have a table representing the probabilities that two points in the high dimensional space are neighbors, from our model of p_{ij}. The values of the $\mathbf{y}$ can be used to fill in an

$N \times N$ joint probability table representing the probabilities that two points are neighbors. We would like this tables to be like one another. A natural metric of similarity is the KL divergence, of Sect. 15.2.1. So we will choose $\mathbf{y}$ to minimize

$$C_{tsne}(\mathbf{y}_1, \ldots, \mathbf{y}_N) = \sum_{ij} p_{ij} \log \frac{p_{ij}}{q_{ij}(\mathbf{y}_1, \ldots, \mathbf{y}_N)}.$$

Remember that $p_{ii} = q_{ii} = 0$, so adopt the convention that $0 \log 0/0 = 0$ to avoid embarrassment (or, if you don't like that, omit the diagonal terms from the sum). Gradient descent with a fixed steplength and momentum was sufficient to minimize this in the original papers, though likely the other tricks of Sect. 16.4.4 might help.

There are two missing details. First, the gradient has a quite simple form (which I shall not derive). We have

$$\nabla_{\mathbf{y}_i} C_{tsne} = 4 \sum_j \left[(p_{ij} - q_{ij}) \frac{(\mathbf{y}_i - \mathbf{y}_j)}{1 + \|\mathbf{y}_i - \mathbf{y}_j\|^2} \right].$$

Second, we need to choose σ_i. There is one such parameter per data point, and we need them to compute the model of p_{ij}. This is usually done by search, but to understand the search, we need a new term. The **perplexity** of a probability distribution with entropy $H(P)$ is defined by

$$\mathrm{Perp}(P) = 2^{H(P)}.$$

The search works as follows: the user chooses a value of perplexity; then, for each i, a binary search is used to choose σ_i such that $p_{j|i}$ has that perplexity. Experiments currently suggest that the results are quite robust to wide changes in the users choice. In practical examples, it is quite usual to use PCA to get a somewhat reduced dimensional version of the $\mathbf{x}$.

Worked Example 19.2   *T-SNE on MNIST Data*

Prepare a T-SNE mapping of 1000 examples, drawn at random, from the MNIST dataset to two dimensions.

Solution: The problem has 2000 variables (the unknown $\mathbf{y}$'s). I used the very nice MATLAB code provided by Laurens van der Maaten at https://lvdmaaten. github.io/tsne/. There are codes for a variety of other environments on that page, too. I tried two cases: in one, I used principal coordinate analysis to map the digits to 30 dimensions and in the other, I mapped to 200 dimensions. I then computed the T-SNE mapping for each case. Figure 19.2 shows results. Note that the dimension reduction doesn't seem to have changed anything important, and it didn't make the method notably faster, either.

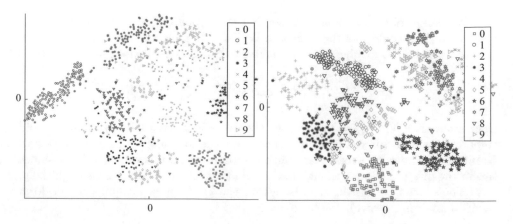

Figure 19.2: A T-SNE mapping of 1000 samples of a 784 dimensional dataset. On the **left**, the data was reduced to 30 dimensions using PCA, then subjected to a T-SNE mapping. On the **right**, the data was reduced to 200 dimensions using PCA, then mapped. The class labels were not used in training, but the plot shows class labels. This helps to determine whether the visualization is any good—you could reasonably expect a visualization to put items in the same class close together and items in very different classes far apart. As the legend on the side shows, T-SNE separates the classes much more effectively than Sammon mapping (Fig. 19.1)

> **Remember This:** *T-SNE produces an embedding of high dimensional data into a lower dimensional space. It does so by solving an optimization problem to choose coordinates in a low dimensional space for each data point. The optimization problem tries to make the probability a pair of points are neighbors in the low dimensional space similar to that probability in the high dimensional space. T-SNE appears less inclined to distort datasets than either principal coordinate analysis or Sammon mapping.*

19.2 Maps That Make Low-D Representations

T-SNE and Sammon mapping establish that we can produce low dimensional representations from high dimensional data. These representations are helpful, but there are problems. First, while we have low-d representations, we don't have a map to low-d—for example, we have no process to construct the **y** corresponding to a new **x** (other than applying the whole procedure for a new set of points. Second, we don't have any way of talking about whether the representation is "right"; in fact, we don't have any way of telling a good representation from a bad one.

The natural way to fix the first problem is to try and build a map that accepts the **x** and produces a **y**. This could be a network. We could (say) train the network to produce the $\mathbf{y}_i$ made by a set of $\mathbf{x}_i$ and smooth for other $\mathbf{x}$. This isn't really appealing, because we don't know whether the $\mathbf{y}_i$ are right. One way to fix this is to insist that each $\mathbf{y}_i$ can be used to reconstruct its original $\mathbf{x}_i$. We train two networks together: One produces an $\mathbf{y}$ from an $\mathbf{x}$; and another reconstructs the $\mathbf{x}$ from the $\mathbf{y}$ alone.

19.2.1 Encoders, Decoders, and Autoencoders

An **encoder** is a network that can take a signal and produce a code. Typically, this code is a description of the signal. For us, signals have been images and I will continue to use images as examples, but you should be aware that all I will say can be applied to sound and other signals. The code might be "smaller" than the original signal—in the sense it contains fewer numbers—or it might even be "bigger"—it will have more numbers, a case referred to as an **overcomplete** representation. You should see our image classification networks as encoders. They take images and produce short representations. A **decoder** is a network that can take a code and produce a signal. We have not seen decoders to date.

An **autoencoder** is a coupled pair of encoder and decoder. The encoder maps signals into codes, and the decoder reconstructs the original signal from those codes. The pair is trained so that the reconstruction is accurate—if you feed a signal $\mathbf{x}$ into an encoder $\mathcal{E}$ to get $\mathbf{y} = \mathcal{E}(\mathbf{x})$, then the decoder $\mathcal{D}$ should ensure that $\mathcal{D}(\mathbf{y})$ is close to $\mathbf{x}$. Autoencoders have great potential to be useful, which we will explore in the following sections. One application is in unsupervised feature learning, where we try to construct a useful feature set from a set of unlabelled images. We could use the code produced by the autoencoder as a source of features. Another possible use for an autoencoder is to produce a clustering method—we use the autoencoder codes to cluster the data. Yet another possible use for an autoencoder is to generate images. Imagine we can train an autoencoder so that (a) you can reconstruct the image from the codes and (b) the codes have a specific distribution. Then we could try to produce new images by feeding random samples from the code distribution into the decoder.

We will describe one procedure to produce an autoencoder. The encoder is a set of layers that produces a block of data we shall call a code. For concreteness, we will discuss grey level images, and assume the encoder consists of convolutional layers. Write $\mathcal{I}_i$ for the i'th input image. All images will have dimension $m \times m \times 1$. We will assume that the encoder produces a block of data that is $s \times s \times r$. It is usual to have these layers produce a block of data where the spatial dimensions are smaller than the input. This is the result of stride and pooling—s is likely a lot smaller than m. Write $\mathcal{E}(\mathcal{I}, \theta_e)$ for the encoder applied to image $\mathcal{I}$; here θ_e are the weights and biases of the units in the encoder. Write $Z_i = \mathcal{E}(\mathcal{I}_i, \theta_e)$ for the code produced by the encoder for the i'th image. Decoder architectures typically mimic encoder architectures (but with the data flow in the opposite direction!). This leads to a characteristic appearance in pictures of the network, and encoder–decoder architectures are often called **hourglass networks**. Figure 19.3 shows a simple autoencoder architecture derived from our MNIST classifier of Sect. 17.2.1.

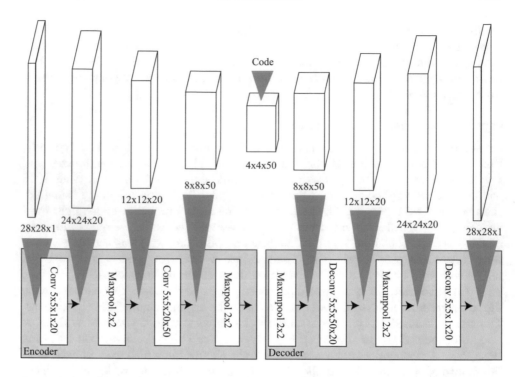

Figure 19.3: A simple autoencoder architecture following the network architecture of Sect. 17.2.1. Note the hourglass shape of the picture of the data blocks—the encoder makes them smaller, and the decoder then makes them bigger again

In the left half of this figure, big data blocks get smaller; in the right, small blocks get bigger.

> **Remember This:** *An autoencoder consists of an encoder, which is trained to produce low dimensional codes from high dimensional data, and a decoder, which is trained to recover that data from the codes. The two are trained together to produce codes that allow reconstruction.*

19.2.2 Making Data Blocks Bigger

We know how to make a big block of data smaller. But building a decoder requires making a small block of data bigger. This requires new tricks (or, rather, modified versions of old tricks). A **transposed convolution layer** or **deconvolution layer** increases the spatial dimensions of a block of data. In the simplest case (stride 1, as in the example of Fig. 19.3) padding will do the trick. So padding an 8×8 feature map by four on top, bottom, left, and right, then applying a 5×5 kernel will lead to a 12×12 feature map.

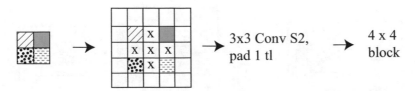

Figure 19.4: A deconvolutional layer or transposed convolutional layer takes a spatially small block of data and enlarges it. The most common case involves stride 2 and a 3×3 kernel, illustrated here. An $m \times m$ feature map arrives. A $(2m+1) \times (2m+1)$ intermediate feature map is created, and populated with values from the input map. The values can be placed at every second location (a larger stride would place them further apart), as illustrated. Various methods use the locations marked with an "x" differently (one could leave them at zero; copy input pixels; or interpolate input pixels). The intermediate feature map is padded with one row and one column, then a 3×3 kernel is applied. The result is $2m \times 2m$

Dealing with stride in the layers requires more care. Figure 19.4 illustrates the process for a 2×2 feature map. There are several options for enlarging the feature map. For the simplest, take an $m \times m$ feature map, and create an intermediate map of zeros that is $(2m+1) \times (2m+1)$. Place an input feature map value at every second location of the intermediate map (Fig. 19.4). Alternatively, some locations in the intermediate map can be reconstructed with bilinear interpolation, a standard upsampling procedure (marked with "x"s in Fig. 19.4). This is very slightly slower, but can suppress some artifacts in reconstruction. Either procedure produces a $(2m + 1) \times (2m + 1)$ feature map. Now pad top and left with zeros, and apply a 3×3 convolution to obtain a $2m \times 2m$ map.

Encoders often have pooling layers. A pooling layer loses information by choosing the largest (resp. average) of a set of values. When the encoder and decoder have mirrored layers, it is straightforward to build an **unpooling** layer. We will assume that pooling windows do not overlap. For average pooling, the pooling layer takes a window, computes an average, and reports that at a location. Unpooling average pooling is straightforward *if* the pooling windows don't overlap— take the average value at an input location and copy it to each of the elements of the corresponding window in the output location (Fig. 19.5). Unpooling max pooling is slightly more elaborate, and works only if the pooling and unpooling mirror one another (Fig. 19.5). Adjust the pooling layer so that it records which of the pooling window elements was the max. To unpool, create a zero intermediate feature map of the appropriate size. Now use the pointers created when the feature map was pooled to place the pooled value in the corresponding window location on the unpooled image.

One last trick is important. Think about the final layer of the decoder, which will be a convolutional layer followed by a non-linearity. Using a ReLU for the non-linearity should seem odd to you, because we know that image pixel values are bounded above as well as below. It is usual to use a **sigmoid layer** here. The sigmoid function is

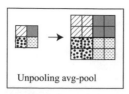

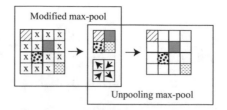

Figure 19.5: When the pooling windows do not overlap, unpooling an avg-pool layer is straightforward (**left**). The layer makes an intermediate map the size of the pooling layers input, with one copy of the average for each location in the pooling window. Unpooling a max-pooling layer is more intricate (**right**). The max-pooling layer is modified to record *which* of the elements of the window is the max; this is passed forward with the pooling results. At unpooling time, this information is used to place pixel values in a map of zeros. The "x"s in the figure are pixels whose values are ignored by the pooling process

$$\sigma(x) = \frac{1}{1 + e^{-x}}.$$

For very small values of x, the sigmoid is close to zero, and for large values it is close to one. There is a smooth transition around zero. A sigmoid layer accepts a block of data and applies the sigmoid non-linearity to each element of the block. Using a sigmoid layer means that the output is in the range 0 to 1. If your images are scaled 0 to 255, you will need to rescale them to compute a sensible loss like this.

We have $Z_i = \mathcal{E}(\mathcal{I}_i, \theta_e)$, and would like to have $\mathcal{D}(Z_i, \theta_d)$ close to $\mathcal{I}_i$. We could enforce this by training the system, by stochastic gradient descent on θ_e, θ_d, to minimize $\|\mathcal{D}(Z_i, \theta_d) - \mathcal{I}_i\|^2$. One thing should worry you. If $s \times s \times r$ is larger than $m \times m$, then there is the possibility that the code is redundant in uninteresting ways. For example, if $s = m$, the encoder could consist of units that just pass on the input, and the decoder would pass on the input too—in this case, the code is the original image, and nothing of interest has happened. This redundancy might be quite hard to spot and could occur even if $s \times s \times r$ was smaller than $m \times m$ if the network "discovered" some clever representational trick. There is a good chance that an autoencoder trained with this loss alone will behave badly.

Remember This: *A decoder makes small data blocks—the codes— larger using a transposed convolution layer or deconvolution layer. The input block is increased in size using one of the several methods, then subjected to a convolution.*

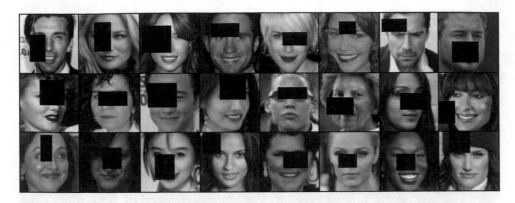

Figure 19.6: Three batches of face images from the widely used Celeb-A dataset, which you can find at http://mmlab.ie.cuhk.edu.hk/projects/CelebA.html, with black boxes over the faces where the inpainting autoencoder is required to reconstruct the image without seeing it. Results in Fig. 19.7. Figure courtesy of Anand Bhattad, of UIUC

19.2.3 The Denoising Autoencoder

There is a clever trick to avoid this problem. We can require the codes to be robust, in the sense that if we feed a noisy image to the encoder, it will produce a code that recovers *the original image*. This means that we are requiring a code that not only describes the image, but is not disrupted by noise. Training an autoencoder like this results in a **denoising autoencoder**. Now the encoder and decoder can't just pass on the image, because the result would be the noisy image. Instead, the encoder has to try and produce a code that isn't affected (much) by noise, and the decoder has to take the possibility of noise into account while decoding.

Depending on the application, we could use one (or more) of a variety of different noise models. These impose slightly different requirements on the behavior of the encoder and decoder. There are three natural noise models: add independent samples of a normal random variable at each pixel (this is sometimes known as additive Gaussian noise); take randomly selected pixels, and replace their values with 0 (masking noise); and take randomly selected pixels and replace their values with a random choice of brightest or darkest value (salt and pepper noise).

Some extreme training tricks are possible, and sometimes justified. Figures 19.6 and 19.7 illustrates an autoencoder trained to fill in large blocks of an image (an **inpainting autoencoder**). This can work for, say, faces, because the missing piece of face can be predicted moderately well from the remaining face.

Remember This: *Autoencoder training can be tricky, because training can discover trivial encodings, particularly if codes are large. Training can be improved by requiring the autoencoder denoise its inputs.*

Figure 19.7: **Top:** shows three batches of face images from the widely used Celeb-A dataset, which you can find at http://mmlab.ie.cuhk.edu.hk/projects/CelebA.html. **Bottom:** shows the output of an inpainting autoencoder on these images. You should notice that the autoencoder does not preserve high spatial frequency details (the faces have been slightly blurred), but that it mostly reproduces the faces rather well. The inputs are in Fig. 19.6; notice there are large blocks of face missing, which the autoencoder is perfectly capable of supplying. Figure courtesy of Anand Bhattad, of UIUC

It remains tricky to get really nice images from the decoder. We will discuss some of the tricks later, but using sum of squared errors as a reconstruction loss tends to produce somewhat blurry images (e.g., Fig. 19.8). This is because the square of a small number is very small. As a result, the sum of squared error loss tends to prefer errors that are small, but somewhat widely distributed. This is a quirk that is quite easy to spot with a little practice (Fig. 19.9 sketches why small widely distributed errors tend to result in blur). The loss is not a particularly good representation of what is important in the appearance of an image. For example, shifting an image left by one pixel leads to a huge sum of squares error, without really changing the image in any important way.

A **perceptual loss** between two images $\mathcal{I}_1$ and $\mathcal{I}_2$ is computed as follows. Obtain an image classification network (VGGNet is popular for this purpose). Write $D_i(\mathcal{I}_1)$ for the block of data that leaves the i'th layer of this network when $\mathcal{I}_1$ is

Figure 19.8: **Top:** shows three batches of face images from the widely used Celeb-A dataset, which you can find at http://mmlab.ie.cuhk.edu.hk/projects/CelebA.html. **Bottom:** shows the output of a simple autoencoder on these images. You should notice that the autoencoder does not preserve high spatial frequency details (the faces have been slightly blurred), but that it mostly reproduces the faces rather well. It is traditional to evaluate image autoencoders on face datasets, because mild blurring often makes a face look more attractive. Figure courtesy of Anand Bhattad, of UIUC

passed in. This block has size $W_i \times H_i \times F_i$. Reshape the block into a vector $\mathbf{d}_i(\mathcal{I}_1)$. Then the feature reconstruction loss for the i'th layer is

$$\mathcal{L}_{\mathrm{fr},i}(\mathcal{I}_1, \mathcal{I}_2) = \left(\frac{1}{W_i H_i F_i} \right) \left(\| \mathbf{d}_i(\mathcal{I}_1), \mathbf{d}_i(\mathcal{I}_2) \|^2 \right).$$

Choose a set of weights w_i for each layer. The perceptual loss between $\mathcal{I}_1$ and $\mathcal{I}_2$ is then

$$\mathcal{L}_{\mathrm{per}}(\mathcal{I}_1, \mathcal{I}_2) = \sum_i w_i \mathcal{L}_{\mathrm{fr},i}(\mathcal{I}_1, \mathcal{I}_2).$$

It is usual to use only the first few layers in computing this loss. The loss works because the early layers of a classification network look for local patterns in the image, and forcing the layer outputs to be similar means that local patterns are

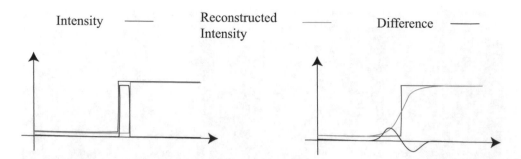

Figure 19.9: Autoencoders trained with a sum of squared error loss tend to produce rather blurry images. This is because placing a sharp edge in slightly the wrong place is expensive with this loss (even though it doesn't change the image much), but blurring an edge is cheap. On the **left**, a stylized image edge with one possible reconstruction error. The top graph shows the image intensity along some line (x is distance along the line); the middle graph shows one possible reconstruction, with the edge correctly reproduced but in the wrong place; and the lower graph shows the resulting error. This error will have a large sum of squares, because it consists of large values. On the **right**, a stylized image edge with a different reconstruction error, which makes the edge blurry. Notice how the error is small and spread out; as a result, the sum of squared errors is small. We can safely assume that an autoencoder will make some kind of error. Using the sum of squared errors means that blurring edges is relatively cheap, and so likely to occur

similar. This preserves edges, for example, because a smoothed edge doesn't look the same as a sharp edge to a pattern detector. Strong autoencoders can be trained with a loss

$$\mathcal{L}_{\text{gen}}(\mathcal{I}_1, \mathcal{I}_2) = \lambda_1 \mathcal{L}_{\text{per}}(\mathcal{I}_1, \mathcal{I}_2) + \lambda_2 \|\mathcal{I}_1 - \mathcal{I}_2\|^2.$$

Now write $\mathtt{noise}(\mathcal{I}_i)$ to mean the result of applying noise to image I_i. The training loss, for example, i, encoder parameters θ_e, and decoder parameters θ_d is

$$\mathcal{L}_{\text{gen}}(\mathcal{D}(Z_i, \theta_d), \mathcal{I}_i), \text{ where } Z_i = \mathcal{E}(\mathtt{noise}(\mathcal{I}_i), \theta_e).$$

You should notice that masking noise and salt and pepper noise are different to additive Gaussian noise, because for masking noise and salt and pepper noise *only some pixels* are affected by noise. It is natural to weight the least-square error at these pixels *higher* in the reconstruction loss—when we do so, we are insisting that the encoder learn a representation that is really quite good at predicting missing pixels. Training is by stochastic gradient descent, using one of the gradient tricks of Sect. 16.4.4. Note that each time we draw a training example, we construct a new instance of noise for that version of the training example, so the encoding and decoding layer may see the same example with different sets of pixels removed, etc.

Autoencoders have a variety of interesting uses. Because the codes are small, but represent the signals that are encoded, they can be used to index the signals or to cluster them. They are often used to learn features that might be useful for some purpose other than reconstruction. One important case occurs when we have little labelled image data. There aren't enough labels to learn a full convolutional neural network, but we could hope that using an autoencoder would produce usable features. The process involves: fit an autoencoder to a large set of likely relevant image data; now discard the decoders, and regard the encoder stack as something that produces features; pass the code produced by the last layer of the stack into a fully connected layer; and fine-tune the whole system using labelled training data. The features aren't necessarily well adapted to classification—with a lot of labelled data, it's better to use the methods of Chap. 18—but they tend to be quite good.

> **Remember This:** *Using only sum of square errors to train image autoencoders tends to result in blurry reconstructed images. The perceptual loss helps to improve this.*

19.3 Generating Images from Examples

Assume Alice has a large dataset of images which need to remain private. For example, these could be medical images, and Alice might not have permission to show pictures of other peoples' insides to the world. Bob wants to build systems for classifying medical images. If Alice could train a method to make images (an image generator) using her images, she could then generate pictures. The method would need to produce images that were like Alice's images, but not the same. If Alice could prove that releasing the generated pictures doesn't violate the privacy rights of her patients, and that they're "right," then she could release them to Bob. Proving it's acceptable to release the images requires some work—for example, if the generator just reorganizes or copies the training data, there is a problem. Proving that the images are "right" may also require some work. But Alice may have a way to show Bob something like her data that helps Bob *and* doesn't hurt Alice's patients.

It's worth understanding why it's hard to generate images. The natural strategy is to build a model of the probability distribution of images, $P(X)$, then draw samples from that model. Such a model is hard to build directly, because there is a lot of structure in an image. For most pixels, the colors nearby are about the same as the colors at that pixel. At some pixels, there are sharp changes in color. But these **edge points** are very highly organized spatially, too—they (largely) demarcate shapes. There is coherence at quite long spatial scales in images, too. For example, in an image of a donut sitting on a table, the color of the table inside the hole is about the same as the color outside. All this means that the overwhelming majority of arrays of numbers are not images. If you're suspicious, and not easily bored, draw samples from a multivariate normal distribution with unit covariance and see how long it will take before one of them even roughly looks like an image (hint: it won't happen in your lifetime, but looking at a few million samples is a fairly harmless way to spend time). This section will be a very broad survey, and will focus on image generation. But procedures for generating images from examples will apply to other kinds of signal, too.

19.3.1 Variational Autoencoders

Here is one natural strategy for generating images. Build an autoencoder. Now generate random codes, and feed them into the decoder. It's worth trying this to reassure yourself that it really doesn't work. It doesn't work for two reasons. First, the codes that come out of a decoder have a complicated distribution, and generating codes from that distribution is difficult because we don't know it. Notice that choosing one code from the codes produced by a training dataset isn't good enough—the decoder will produce something very close to a training image, which isn't what we're trying to achieve. Second, the decoder has been trained to decode the training codes *only*. The training procedure doesn't force it to produce sensible outputs for codes that are *near* training codes, and most decoders in fact don't do so.

It is possible to train an autoencoder to overcome both difficulties. These are known as **variational autoencoders**. Rather than producing codes, the encoder produces a representation of a probability distribution. Typically, the encoder takes an input image and produces a mean value for the code that would be produced by that distribution and a set of standard deviations. It does so to represent the distribution of codes for images "near" the input image. The set of all codes produced by a decoder like this has a probability distribution that is a very large mixture of normal distributions (one per training example).

The next step is to write a training loss that scores the KL divergence between the code distribution and a standard distribution. This is typically a standard normal distribution. This loss would drive the encoder–decoder pair so that the code distribution is a standard normal distribution, meaning it is easy to draw a sample from the code distribution.

We must now ensure that the decoder can decode codes that are not those seen in training. Here is how to do this. When an image is passed into the encoder, it produces a mean and standard deviation. Rather than showing the decoder this mean, draw a sample from the normal distribution represented by that mean and standard deviation. This is a code that (a) comes from the code distribution and (b) is "close" to the code that results from the original training image. Now write a loss that requires the decoder takes that code, and produces an image "close" to the original training image.

I have omitted the details, because they are somewhat delicate and are not for everyone. However, assume you have trained an encoder/decoder pair like this (there is good evidence that this can be done). Then you can generate images by drawing samples of a standard normal distribution, and passing them through the decoder.

Variational autoencoders tend to produce blurry images. Adjusting the loss to improve the images is unattractive (you might spend all of your time adding terms to the loss to suppress new effects). An alternative is to try and show that some competent classifier can't tell the difference between a real image and a generated image.

> **Remember This:** *Variational autoencoders are trained to produce codes of a known distribution. The training procedure tries to ensure that the decoder produces sensible results for codes that are close to codes observed in training. As a result, the decoder can produce sensible images from random numbers, though the results are often blurry.*

19.3.2 Adversarial Losses: Fooling a Classifier

Here is a strategy for generating specialized images, for example, images of faces. Construct a decoder. Feed it with a stream of random codes, drawn as IID samples from some convenient distribution. Now train the decoder by requiring that a competent adversary can't tell the difference between the generated images and real images of faces. In this scheme, the decoder is usually called a **generator** and the adversary is a classifier usually called a **discriminator**. The discriminator will need training, too. The natural procedure is to construct a discriminator using the best generator you have; then adjust the generator to fool that; then readjust the discriminator; then the generator; and so on. In this scheme, the decoder is usually called a **generator**, and networks trained like this are usually called **generative adversarial networks** or **GANs**. Actually imposing these requirements involves important technical difficulties.

Write $G(\mathbf{z})$ for an image generated from a code $\mathbf{z}$; write $D(\mathbf{x})$ for the discriminator applied to some image $\mathbf{x}$. We assume the discriminator produces a number between 0 and 1, and we would like it to produce a 1 for any real image, and a 0 for any synthetic image. Now consider the cost function

$$\mathcal{C}(D, G) = \frac{1}{N_r} \sum_{\mathbf{x}_i \in \text{real images}} \log\left(D(\mathbf{x}_i)\right) + \frac{1}{N_s} \sum_{\mathbf{z}_j \in \text{codes}} \log\left(1 - D(G(\mathbf{z}_j))\right).$$

If the discriminator works very well (i.e., can tell the difference between real and synthetic images) this will be large. If the generator works very well (i.e., can fool the discriminator), the cost will be small. So we could try and find $\hat{D}$ and $\hat{G}$ that are obtained as

$$\operatorname*{argmin}_{G} \operatorname*{argmax}_{D} \mathcal{C}(D, G).$$

Here G would be some form of decoder, and D would be some form of classifier. It seems natural to try using stochastic gradient descent/ascent on this problem. One scheme is to repeatedly fix the G, and take some uphill steps in D; now fix D, and take some downhill steps in G. This is where the term *adversarial* comes from: the generator and the discriminator are adversaries, trying to beat one another in a game.

This apparently simple scheme is fraught with practical and technical difficulties. Here is one important difficulty. Imagine G isn't quite right, but D is perfect, and so reports 1 for every possible true image and 0 for every possible synthetic image. Then there is no gradient to train G, because any small update of G will

Figure 19.10:   Three batches of face images generated by a variant of the GAN strategy described in Sect. 19.3.2, using the Celeb-A dataset (which you can find at http://mmlab.ie.cuhk.edu.hk/projects/CelebA.html) as training data. You should notice that these images really look like faces "at a glance," but if you attend you'll see various slightly creepy eyes, small global distortions of the face shape, odd mouth shapes, and the like. Figure courtesy of Anand Bhattad, of UIUC

still produce images that aren't quite right. This means that we may want a D that isn't very good. In fact, we are requiring that D has an important property: if you make an image "more real," then D will produce a larger value, and if you make it "less real," D will produce a smaller value. This is a much more demanding requirement than requiring D as a classifier.

Here is a second difficulty. Imagine there are two clusters of faces that are quite different. I will use "glasses" and "no glasses" as an example. In principle, if the generator does not produce "glasses" faces, then the discriminator has an easier job (any face with "glasses" can be classified as real). But this is a very weak signal—it may be hard to use this information to force the generator to produce faces with "glasses," particularly if they're uncommon in the training data. This leads to a quite common phenomenon, sometimes called **mode collapse**, where a generator will produce some kinds of image but not others. This is particularly difficult to identify experimentally, because it is hard to know what is missing from the generated images.

Despite these caveats, it has been possible to train networks like this. There is good evidence that they are capable of producing rather good images (Fig. 19.10), if the contents are specialized (i.e., one can produce images of faces, of rooms, or of lungs as below, but not some generic image of anything). There is also good evidence that the general idea of an adversarial loss can be used to tune other generators rather well. For example, efforts to improve VAE-like networks or autoencoders by imposing an adversarial loss are often successful. The discriminator can easily spot that real images aren't fuzzy; and the caveats above are mitigated by the use of other losses to ensure the generator starts in about the right place.

> **Remember This:** *An adversarial loss balances a generator—which makes images—with a discriminator—which tries to tell a real image from a generated image. The two compete: the generator tries to fool the discriminator, and the discriminator tries to spot the generator. Training with adversarial losses can be tricky.*

19.3.3 Matching Distributions with Test Functions

Here is an alternative view of our training requirements. You can see the images made by a generator as samples of a probability distribution $P(R)$ (R for reconstruct). It is hard to write out the form of this distribution, but easy to sample it—just draw a random code and pass it through the decoder. The true images are samples of another probability distribution, $P(X)$. We would like to adjust the generator so that $P(R)$ is "the same" as $P(X)$.

One way to test whether two distributions are "the same" is to look at expectations. For example, think about two probability distributions $P(x)$ and $Q(x)$ on the closed interval from 0 to 1. Choose a sufficiently large set of functions ϕ_k, indexed by k. As a concrete example, you could think of the monomials, where $\phi_0 = 1$, $\phi_1 = x$, $\phi_2 = x^2$, and so on. Now assume $\mathbb{E}_{\phi_k}[P(x)] = \mathbb{E}_{\phi_k}[Q(x)]$ for all of these functions. This implies that, for any other function $f(x)$, $\mathbb{E}_f[P(x)]$ must be arbitrarily close to $\mathbb{E}_f[Q(x)]$. This is because you can represent $f(x)$ with a series to arbitrary precision, so that

$$f(x) = a_o\phi_o(x) + a_1\phi_1(x) + \cdots + \text{arbitrarily small error.}$$

In turn, $P(x)$ and $Q(x)$ are "the same" for all practical purposes. If you've seen some formal analysis and probability, you'll notice that I've fudged on a variety of details here, but you'll be able to fill them in.

This all suggests the following strategy. Come up with a collection of test functions ϕ_k. Choose the generator to force

$$\sum_k \left[\frac{1}{N_r} \sum_{\mathbf{x}_i \in \text{real images}} \phi_k(\mathbf{x}_i) - \frac{1}{N_s} \sum_{\mathbf{z}_j \in \text{codes}} \phi_k(G(\mathbf{z}_j)) \right]^2$$

to be small. There are difficulties here, too. First, the collection of test functions might need to be very large, creating problems with gradients and the like. Second, these test functions will need to be "useful" in some reasonable way. So, for example, a test function that extracts the value of a single pixel is unlikely to be much help. The right way to proceed is to search for a **witness function**, which is a test function that emphasizes the difference between the distributions. Notice this is quite like the adversarial interaction in the previous section: we adjust the generator so that $P(R)$ is close to $P(X)$ using our current test functions; then we search for a witness function that emphasizes the difference, and add it to the set of test functions; then adjust using the new set of tests; and so on.

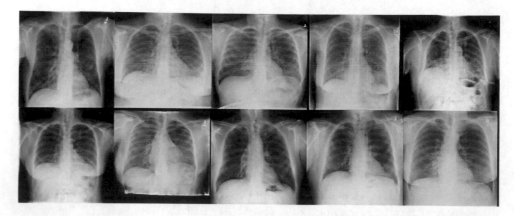

Figure 19.11: Images of chest X-Rays generated by a GAN using the one dimensional method sketched in the text. Figure courtesy of Ishan Deshpande and Alex Schwing, of UIUC. I can show these images *because they're not medical images of real humans*—they were made by a computer program!

> **Remember This:** *The distribution of images made by a generator should match the distribution of training images. One way to evaluate this is to ensure that there are not test functions that have a different expectation on the training and generated images. This criterion has an adversarial component, because the training process needs to identify test functions that highlight the difference between training and generated images.*

19.3.4 Matching Distributions by Looking at Distances

An alternative method for comparing $P(R)$ and $P(X)$ reasons about nearby points. We will think about two sets of samples $\{R_i\}$ and $\{X_j\}$ *in one dimension*. For simplicity, we will reason about sample sets that have the same size. If these samples come from the same distribution, there should be an X close to any R, and an R close to any X. In particular, a reasonable measure of similarity is to pair X's with R's, then sum the distance between pairs. We choose pairs so that: each X (resp. R) has exactly one R (resp. X); and the sum of the distances is minimized. It turns out that one can evaluate this particular distance in an easy way. Sort the R_i in descending order; sort the X_j in descending order; then obtain the pairs by pairing the first R_i with the first X_j, the second with the second, and so on. We now sum the squared distances between pairs.

This trick extends to multiple dimensions in a simple way. Assume we have high dimensional R_i (resp. X_j). Now choose some random direction in this high dimensional space, and project the R_i (resp. X_j) onto that direction. If the distributions are the same, the projected distributions are the same. So we should obtain a "small" value of the sum for that—and any—projection. In turn, this justifies averaging the distances over many random projections. Of course, what we need is a projection that emphasizes the difference between the generator and true images, and that involves an adversary again. The procedure looks like this: adjust the generator so that $P(R)$ and $P(X)$ are close using the current projections; now find a projection that makes them look different; now adjust the generator using that projection as well; and proceed. This line of reasoning leads to rather good generative models, as Fig. 19.11 suggests.

> **Remember This:** *A strong measure of similarity between two distributions in one dimension can be obtained by reasoning about the distances between points. This can be extended to handle high dimensional distributions by projecting down to one dimension. This criterion has an adversarial component, because the training process needs to identify projections that highlight the difference between training and generated images.*

19.4 You Should

19.4.1 Remember These Terms

19.4.2 Remember These Facts

19.4.3 Be Able to

- Construct a Sammon mapping of a set of data points.
- Construct a T-SNE mapping of a set of data points.
- Train and evaluate a simple autoencoder, using downloaded software.
- Understand what makes a variational autoencoder useful.
- Give a brief account of why images can be generated from random numbers.

Programming Exercises

General Remark: *These exercises are suggested activities, and are rather open ended. It will be difficult to do them without a GPU. You may have to deal with some fun installing software environments, etc. It's worthwhile being able to do this, though.*

19.1. Download an autoencoder for MNIST data for your preferred programming environment. For MNIST, there are two kinds of autoencoder. One treats MNIST data as images (so reports each pixel in the range 0–1 or 0–255) and the other treats them as binary images (so reports either 0 or 1 at each pixel). You want the first kind. Start with a pretrained model.

(a) For each image in the MNIST test dataset, compute the residual error of the autoencoder. This is the difference between the true image and the reconstruction of that image by the autoencoder. It is an image itself. Prepare a figure showing the mean residual error, and the first five principal components. Each is an image. You should preserve signs (i.e., the mean residual error may have negative as well as positive entries). The way to show these images most informatively is to use a mid grey value for zero, then darker values for more negative image values and lighter values for more positive values. The scale you choose matters. You should show mean and five principal components on the same grey scale for all six images, chosen so the largest absolute value over all six images is full dark or full light, respectively, and mean and five principal components on a scale where the grey scale is chosen for each image separately.

(b) Determine the mean and covariance of the codes on training data. Now draw random samples from a normal distribution with that mean and covariance, and feed these samples into the decoder. Do the results look like images?

(c) Model the codes with a mixture of normals in the following way. For each class, determine the mean and covariance of the codes on training data. Your mixture distribution is then an evenly weighted sum of the ten class distributions. Now draw random samples from this mixture distribution, and feed these samples into the decoder. Do the results look like images?

(d) Classify the MNIST images using a random decision forest applied to the autoencoder codes. Compare the results of this approach with using a straightforward neural network classifier. In particular, how much labelled training data do you need to classify the MNIST images with reasonable accuracy like this?

19.2. Download an autoencoder for MNIST data for your preferred programming environment. For MNIST, there are two kinds of autoencoder. One treats MNIST data as images (so reports each pixel in the range 0–1 or 0–255) and the other treats them as binary images (so reports either 0 or 1 at each pixel). You want the first kind.

(a) Now train the autoencoder. Can you improve the error on the test set by data augmentation? You should investigate small scales and small rotations of the images.

(b) Modify the autoencoder by adding a layer to the encoder and a layer to the decoder, and retrain it. Does it get better?

19.3. Download an autoencoder for CIFAR-10 data for your preferred programming environment. For CIFAR-10, there are two kinds of autoencoder. One treats CIFAR-10 data as images (so reports each pixel in the range 0–1 or 0–255) and the other treats them as binary images (so reports either 0 or 1 at each pixel). You want the first kind. Start with a pretrained model.

(a) For each image in the CIFAR-10 test dataset, compute the residual error of the autoencoder. This is the difference between the true image and the reconstruction of that image by the autoencoder. It is an image itself. Prepare a figure showing the mean residual error, and the first five principal components. Each is an image. You should preserve signs (i.e., the mean residual error may have negative as well as positive entries). The way to show these images most informatively is to use a mid grey value for zero, then darker values for more negative image values and lighter values for more positive values. The scale you choose matters. You should show mean and five principal components on the same grey scale for all six images, chosen so the largest absolute value over all six images is full dark or full light, respectively, and mean and five principal components on a scale where the grey scale is chosen for each image separately.

(b) Determine the mean and covariance of the codes on training data. Now draw random samples from a normal distribution with that mean and covariance, and feed these samples into the decoder. Do the results look like images?

(c) Model the codes with a mixture of normals in the following way. For each class, determine the mean and covariance of the codes on training data. Your mixture distribution is then an evenly weighted sum of the ten class distributions. Now draw random samples from this mixture distribution, and feed these samples into the decoder. Do the results look like images?

(d) Classify the CIFAR-10 images using a random decision forest applied to the autoencoder codes. Compare the results of this approach with using a straightforward neural network classifier. In particular, how much labelled training data do you need to classify the CIFAR-10 images with reasonable accuracy like this?

19.4. Download an autoencoder for CIFAR-10 data for your preferred programming environment.

(a) Now train the autoencoder. Can you improve the error on the test set by data augmentation? You should investigate small scales and small rotations of the images.

(b) Modify the autoencoder by adding a layer to the encoder and a layer to the decoder, and retrain it. Does it get better?

19.5. We will evaluate a variational autoencoder applied to the MNIST dataset. Obtain (or write! but this isn't required) a for a variational autoencoder. Train this autoencoder on the MNIST dataset. Use only the MNIST training set.

(a) We now need to determine how well the codes produced by this autoencoder can be interpolated. For ten pairs of MNIST test images of the same digit, selected at random, compute the code for each image of the pair. Now compute seven evenly spaced linear interpolates between these codes, and decode the result into images. Prepare a figure showing this interpolate. Lay out the figure so each interpolate is a row. On the left of the row is the first test image; then the interpolate closest to it; etc.; to the last test image. You should have a ten rows and nine columns of images.

(b) For ten pairs of MNIST test images of different digits, selected at random, compute the code for each image of the pair. Now compute seven evenly spaced linear interpolates between these codes, and decode the result into images. Prepare a figure showing this interpolate. Lay out the figure so each interpolate is a row. On the left of the row is the first test image; then the interpolate closest to it; etc.; to the last test image. You should have a ten rows and nine columns of images.

(c) Determine the mean and covariance of the codes on training data. Now draw random samples from a normal distribution with that mean and covariance, and feed these samples into the decoder. Do the results look like images?

(d) Model the codes with a mixture of normals in the following way. For each class, determine the mean and covariance of the codes on training data. Your mixture distribution is then an evenly weighted sum of the ten class distributions. Now draw random samples from this mixture distribution, and feed these samples into the decoder. Do the results look like images?

(e) Classify the MNIST images using a random decision forest applied to the variational autoencoder codes. Compare the results of this approach with using a straightforward neural network classifier. In particular, how much labelled training data do you need to classify the MNIST images with reasonable accuracy like this?

Index

© Springer Nature Switzerland AG 2019
D. Forsyth, *Applied Machine Learning*,
https://doi.org/10.1007/978-3-030-18114-7

Index: Useful Facts

© Springer Nature Switzerland AG 2019
D. Forsyth, *Applied Machine Learning*,
https://doi.org/10.1007/978-3-030-18114-7

485

Index: Procedures

© Springer Nature Switzerland AG 2019
D. Forsyth, *Applied Machine Learning*,
https://doi.org/10.1007/978-3-030-18114-7

Index: Worked Examples

© Springer Nature Switzerland AG 2019
D. Forsyth, *Applied Machine Learning*,
https://doi.org/10.1007/978-3-030-18114-7

Index: Remember This

© Springer Nature Switzerland AG 2019
D. Forsyth, *Applied Machine Learning*,
https://doi.org/10.1007/978-3-030-18114-7

Printed in the United States
By Bookmasters